Marine Pollution

Sources, Fate and Effects of Pollutants in Coastal Ecosystems

Ricardo Beiras
University of Vigo

D0206961

ELSEVIER

Elsevier
Radarweg 29, PO Box 211, 1000 AE Amsterdam, Netherlands
The Boulevard, Langford Lane, Kidlington, Oxford OX5 1GB, United Kingdom
50 Hampshire Street, 5th Floor, Cambridge, MA 02139, United States

Notices
Knowledge and best practice in this field are constantly changing. As new research and experience broaden our
understanding, changes in research methods, professional practices, or medical treatment may become necessary.

Practitioners and researchers must always rely on their own experience and knowledge in evaluating and using any
information, methods, compounds, or experiments described herein. In using such information or methods
they should be mindful of their own safety and the safety of others, including parties for whom they have a
professional responsibility.

To the fullest extent of the law, neither the Publisher nor the authors, contributors, or editors, assume any liability
for any injury and/or damage to persons or property as a matter of products liability, negligence or otherwise, or
from any use or operation of any methods, products, instructions, or ideas contained in the material herein.

Library of Congress Cataloging-in-Publication Data
A catalog record for this book is available from the Library of Congress

British Library Cataloguing-in-Publication Data
A catalogue record for this book is available from the British Library

ISBN: 978-0-12-813736-9

For information on all Elsevier publications visit our website at
https://www.elsevier.com/books-and-journals

Working together
to grow libraries in
developing countries

www.elsevier.com • www.bookaid.org

Publisher: Candice Janco
Acquisition Editor: Louisa Hutchins
Editorial Project Manager: Michelle W. Fisher
Production Project Manager: Omer Mukhthar
Designer: Christian J. Bilbow

Typeset by TNQ Technologies

Marine Pollution

Dedication

To Domingo

Contents

Foreword

Marine pollution is at present part of the media circus. Who has not been shocked by the images of moribund seabirds spreading their coal-black wings, or dolphins strained on a plastic-invaded beach? The advantage of this media focus is the public support gained for pollution remediation and prevention initiatives, but the disadvantage is the lack of scientific rigor in the debates concerning pollution. According to H.L. Windom, "the focus of attention on coastal [pollution] problems has been based more upon public perceptions than on sound scientific evaluations of sources, fates and environmental effects."[1] J. S. Gray illustrated the same problem with the case of the planned dumping of the disused oil rig *Brent Spar* in deep water off the Scottish coast. Eventually the dumping was stopped after a Greenpeace campaign against it, but the decision did not include any rational elements since neither Greenpeace nor Shell gave any data on the environmental risk of the options—sinking or disposal on shore—and the decision-makers did not consider any scientific study, though they were available.[2]

In 2009, at the onset of the economic crisis in Spain, President Zapatero declared on TV, "I want to make a call to the citizens […] they must keep on consuming." Consumption is considered by conventional wisdom as the engine of the economy. In fact, this ignores the most basic principles of thermodynamics. Since the 1980s, E. Odum called our attention to the need to change focus from maximizing production (and thus consumption of resources and generation of wastes) to maximizing efficiency, the ratio between production and consumption. Slowly—perhaps too slowly?—this true wisdom permeates societies, but the effects on the decision-makers are so far more cosmetic than real. The part of this question that is scientists' responsibility is to conduct hard science to study environmental issues. Scientists replaced priests as advisors of the empowered leaders only because their predictions were more reliable. The higher the certitude of the scientific predictions the more influential they will be for decision-makers. Ecotoxicology must be just as rigorous as medicine, and nobody conceives discussing in the media the diagnostic of an ill patient or the most suitable drug and correct dose to be prescribed.

In fact, this book has a practical and applied vocation. I am an empirical scientist fascinated by the elegant simplicity of the scientific method based on contrasting hypotheses at the light of observation and experimentation. Excess of theoretical apparatus has been identified as one of the limitations of ecological sciences, and the debates on the effects of environmental factors, including the nonconcept of "global change," on the stability of ecosystems seem to me a good example of this. As R.H. Peters complained, logic, i.e., the set of possible alternatives, replaced theory, the set of probable alternatives, and this eventually constrained some ecological theories to tautological formulations whose implications are included in the premises, and thus not suitable to experimental contrast.[3]

This book intends to be useful to a wide range of readers: academic audiences seeking a basic theoretical background on marine pollution, but also professionals involved in the daily routine of managing the marine environment and seeking applied knowledge related to specific issues on pollution prevention, monitoring, effects, and abatement. As a result, the book admits two levels of reading. The advanced reader is offered with a broad selection of specialized scientific references that back the statements made throughout the text, listed at the end of each chapter. For didactic purposes, the learning reader can ignore those references and look for more basic information in the Suggested Further Reading section, and review the essential contents in the Key Ideas section at the end of each chapter.

In short, the hopefully not-too-ambitious aim of this book is to provide a rigorous tool to train marine ecotoxicologists and contribute to make them familiar with the contrasted theories and quality-controlled methods that may provide solid scientific foundations to their current or future work.

Ricardo Beiras

References

1. Windom HL. Contamination of the Marine Environment from Land-based Sources. Marine Pollution Bulletin 1992;25(1—4):32—6.

2. Gray JS. Chapter 17. Risk assessment and management in the exploitation of the seas. In: Calow P, editor. Handbook of environmental risk assessment and management. Oxford: Blackwell Science; 1998. p. 453—74.

3. Peters RH. A critique for ecology. Cambridge University Press; 1991.

Acknowledgments

I thank my colleagues Marion Nipper, Paula Sánchez Marín, Juan Bellas, Inés Viana, Filipe M.G. Laranjeiro, Miren B. Urrutia, Enrique Navarro, Silvia Messinetti, Leo Mantilla, Iria Durán, and Leticia Vidal Liñán for their useful comments and discussion on several parts of this book. Many ancient and hardly available bibliographic references were readily obtained thanks to the efficient work of the librarians at the University of Vigo. I apologize to Leticia, Xulia, Roi, and Valentina for the time taken for this project.

Abbreviations and Symbols

4-MBC	4-Methylbenzylidene camphor
ABS	Alkyl-benzene sulfonate
AChE	Acetylcholinesterase
AE	Absorption efficiency/assimilation efficiency
AF	Assessment factor
AhR	Aryl hydrocarbon receptor
ALA-D	δ-Aminolevulinic acid dehydratase
Ant	Anthracene
ANZECC	Australian and New Zealand Environment Conservation Council
ASP	Amnesic shellfish poison
ASTM	American Society for Testing and Materials
ATP	Adenosine triphosphate
AVS	Acid-volatile sulfide
BaA	Benzo-a-Anthracene
BAC	Background Assessment Concentration
BAF	Bioamplification Factor
BaP	Benzo-a-pyrene
BbF	Benzo-b-Fluoranthene
BC	Background concentration
BCF	Bioconcentration factor
BDE	Brominated diphenylether
BeP	Benzo-e-Pyrene
BEWS	Biological Early Warning System
BghiP	Benzo-g,h,i-Perylene
BkF	Benzo-k-Fluoranthene
BMF	Biomagnification factor
BMF_{TW}	Trophic web biomagnification factor
BOD	Biological oxygen demand
BOD_5	5-days biological oxygen demand
BOD_L	Ultimate biological oxygen demand
BP	Benzophenone
BPA	Bisphenol A

(Continued)

BTEX	Benzene, toluene, ethylbenzene, xylene
CAT	Catalase
CB	Chlorinated biphenyl
CBB	Critical body burden
CCME	Canadian Council of Ministers of the Environment
cDNA	Complementary deoxyribonucleic acid
CEMP	Coordinated Environmental Monitoring Programme
CEP	Caribbean Environment Programme
CF	Contamination factor
CFU	Colony-forming units
Chry	Chrysene
CLC	Civil liability convention
COD	Chemical oxygen demand
CPI	Chemical pollution index
Cpn60	Chaperon 60
CYP	Cytochrome P450
Cys	Cysteine
DDD	Dichlorodiphenyldichloroethane
DDE	Dichlorodiphenyldichloroethylene
DDT	Dichlorodiphenyltrichloroethane
DEHP	Diethylhexyl phthalate
DIN	Dissolved inorganic nitrogen
DNA	Deoxyribonucleic acid
DO	Dissolved oxygen
DOC	Dissolved organic carbon
DOM	Dissolved organic matter
DW	Dry weight
EC	European Commission
EC_{50}	Median effective concentration
ECHA	European Chemicals Agency
EDC	Endocrine disrupting compound
EEA	European Environment Agency
EF	Enrichment factor
EHMC	Ethylhexyl methoxycinnamate
ELISA	Enzyme-linked immunosorbent assay
ELS	Early life stages
EMSA	European Maritime Safety Agency
EPA	Environmental Protection Agency
EQC	Environmental quality criteria
EQC/S	Environmental quality criteria and standards
EQR	Ecological quality ratio
EQS	Environmental quality standards

ER	Estrogen receptor
ERA	Ecological risk assessment
ERL	Effects range low
ERM	Effects range median
EROD	Ethoxyresorufin-O-deethylase
EU	European Union
FC	Fecal coliforms
FIAM	Free ion activity model
f_L	Weight proportion of lipids
Flu	Fluoranthene
f_{OC}	Weight proportion of organic carbon
FR	Filtering rate
GPx	Glutathione peroxydase
GSH	Glutathione
GST	Glutathione transferase
HBCD	Hexabromocyclododecane
HC_5	Hazard concentration for 5% of species
HELCOM	Helsinki Commission
HRA	Health risk assessment
H_S	Shannon diversity index
ICES	International Council for the Exploration of the Sea
IF	Interaction factor
IFREMER	Institut Français de Recherche pour l'Exploitation de la Mer
IMO	International Maritime Organization
IOPC	International oil pollution compensation
IPy	Indenepyrene
IR	Ingestion rate
ISO	International Organization for Standardization
K_{OC}	Organic carbon-water partition coefficient
K_{OW}	Octanol-water partition coefficient
LAS	Linear alkylbenzene sulfonate
LC_{50}	Median lethal concentration
LMS	Lysosomal membrane stability
LW	Lipid weight
MDS	Multidimensional scaling
Me−Hg	Methylmercury
MFO	Mixed function oxidase or monooxygenase
MLVSS	Mixed liquor volatile suspended solids
MPN	Most probable number
mRNA	Messenger ribonucleic acid
MSFD	Marine strategy framework directive
MSW	Municipal solid waste

(Continued)

MT	Metallothionein
NADH	Nicotinamide adenine dinucleotide
NADPH	Nicotinamide adenine dinucleotide phosphate
NOAA	National oceanic and atmospheric administration
NP	Nonylphenol
NPE	Nonylphenol ethoxylate
NRRT	Neutral red retention time
NSP	Neurotoxic shellfish poison
OC	Organochlorine
OD-PABA	Octyl dimethyl-paraaminobenzoic acid
OPA	Oil pollution act
OSPAR	Oslo—Paris commission
PA	Polyamide
PAH	Polyaromatic hydrocarbon
PBDE	Polybrominateddiphenylethers
PBT	Persistent bioaccumulable toxic
PC	Polycarbonate
PCA	Principal components analysis
PCB	Polychlorinatedbiphenyls
PCDD	Polychlorinateddibenzo-p-dioxins
PCDF	Polychlorinateddibenzofurans
PCR	Polymerase chain reaction
PE	Polyethylene
PEC	Predicted environmental concentration
PEL	Probable effect level
PET	Polyethylene terephthalate
PFOA	Perfluorooctanoic acid
PFOS	Perfluorooctane sulfonate
PFOSA	Perfluorooctane sulfonamide
PFU	Plaque-forming units
Phe	Phenanthrene
PLA	Polylactic acid
PNEC	Predicted no-effect concentration
PNR	Proportion net response
POM	Particulate organic matter
POP	Persistent organic pollutant
PS	Polystyrene
PSP	Paralytic shellfish poison
PP	Polypropylene
PUR	Polyurethane
PVC	Polyvinyl chloride
Pyr	Pyrene

QA	Quality assurance
QC	Quality control
QSAR	Quantitative structure-activity relationship
R	Risk quotient
RBC	Rotating biological contactor
RDA	Redundancy analysis
REACH	Registration, evaluation, authorization, and restriction of chemicals
RNA	Ribonucleic acid
RNO	Réseau National d'Observation
ROCCH	Réseau d'Observation de la Contamination Chimique
ROS	Reactive oxygen species
RPLI	Relative penis length index
RT-PCR	Reverse transcription-polymerase chain reaction
RTR	Ratio to reference
S	Species richness
SARA	Saturated, aromatics, resins, asphaltenes
SDS	Sodium dodecylsulfate
SEM	Simultaneously extracted metals
SER	Smooth endoplasmic reticulum
SET	Sea-urchin embryo test
SOD	Superoxide dismutase
SQC	Sediment quality criteria
SS	Suspended solids
SSD	Species sensitivity distribution
SW	Seawater
$T_{1/2}$	Environmental half-life
T_{90}	90% die-off time
TBT	Tributyl-tin
TC	Total coliforms
TCDD	2,3,7,8-tetrachlorodibenzo-p-dioxin
TCEP	Tris(2-chloroethyl) phosphate
TCPP	Tris(chloropropyl) phosphate
TCS	Triclosan
TDCPP	Tris(1,3-dichloro-2-propyl)phosphate
TEL	Threshold effect level
TL	Trophic level
TOC	Total organic carbon
TPM	Total particulate matter
TSCA	Toxic substances control act
TT	Toxicity threshold
TTF	Trophic transfer factor
TU	Toxic units

(Continued)

UDP	Uridine diphosphate
UDPGT	Uridine diphosphate glucuronosyltransferase
UK	United Kingdom
UN	United Nations
UNCLOS	United Nations Convention on the Law of the Sea
UNEP	United Nations Environment Program
US	United States
UV	Ultraviolet radiation
VDSI	Vas deferens sequence index
VTG	Vitellogenin
WFD	Water framework directive
WHO	World Health Organization
WOE	Weight of evidence
WQC	Water quality criteria
WW	Wet weight
WWTP	Wastewater treatment plant

Part I: Pollutants in Marine Ecosystems

CHAPTER 1

Basic Concepts

1.1 POLLUTION, AN ANTHROPOGENIC PROCESS

We normally understand as pollution the unwanted presence in the environment of diverse classes of toxic substances generated by human activities. As we will soon discuss, because of the main circulation pathways of matter in the environment, those inputs frequently end up in the sea. In the context of marine science, a more formal definition provided by a United Nations advisory board, though strongly anthropocentric, was very successful and quoted in the scientific literature. Marine pollution, according to that group of experts, is "**the introduction by man of substances into the marine environment resulting in such deleterious effects as harm to living resources, hazards to human health, hindrance to marine activities including fishing, impairment of quality for use of seawater and reduction of amenities**" (GESAMP 1969).[1] Latter developments of this definition added the introduction of **energy** to make clear that heat and radioactivity, already contemplated in the original definition, could also be considered pollutants, and specified that the introduction into the sea might also be indirect via riverine or atmospheric pathways.

A formal definition of marine pollution

In the context of maritime transportation, the same board[2] produced a list of 166 substances of major concern (Category 1), and their escape into the marine environment should universally be prevented because they may cause long-term or permanent damage, and 231 additional substances (Category 2) that because of their short-term effects represented a hazard only in certain scenarios. From this seminal report stems the many lists of so-called **priority pollutants** subsequently identified by agencies and institutions committed to environmental protection worldwide.

The first aspect inherent to pollution thus is its human origin, i.e., pollution is an **anthropogenic** process derived from human activities. Climatic, geological, or oceanographic natural events (floods, earthquakes, red tides, etc.), even when they can be extremely harmful for the environment, are specifically excluded from the definition of pollution. Therefore, it is not surprising that the most polluted places were those supporting the highest human population

Pollution is quantitatively related to population density and energy consumption

3

densities. But not all human societies pollute the same. Since many physical and chemical pollutants are originated by industrial activities, industrialization is also quantitatively related to pollution. A good quantitative subrogate for the degree of industrialization is **energy consumption**. As illustrated in Fig. 1.1, the per capita energy consumption may be up to two orders of magnitude higher in industrialized societies compared to rural ones. According to this source, the average American consumes approximately twice the energy than a person from Europe, 10 times that of a person from India, and 100 times that of a person from South Sudan.

Environmental regulations are more strict in developed countries

Another societal factor affecting the environmental impact of its inhabitants is **environmental awareness**, which is directly related to the cultural level. This issue has been much less explored and quantified but can be illustrated by a

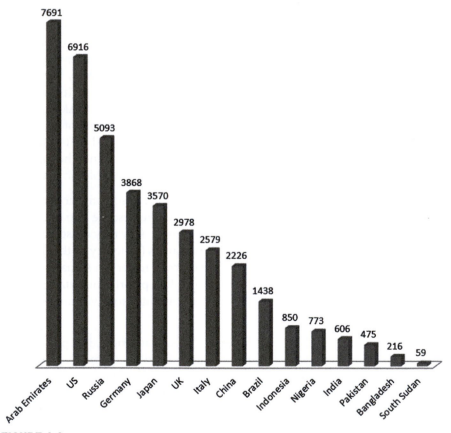

FIGURE 1.1

Per capita energy consumption in the world in 2013. Units are Kg of oil equivalents per person and year. *Data source: World Bank.*

few examples. Environmentalism was borne in the most developed countries, and under its influence environmental protection regulations are far stricter in those countries. This translates into the fact that many chemicals that cause environmental concern and were banned in the most developed countries, such as persistent organochlorine pesticides, are still used in other parts of the world with laxer environmental standards.

In 2002 the world's largest mercury mine (Almadén, Spain) was closed as a result of the different restrictions imposed by the European Commission to the use of this metal in thermometers and many other applications. Currently the global mercury production is largely dominated by China, and 79% of global Hg emissions are located in Asia, Africa, and South America.[3] Another illustration of that is the export of waste from electronic equipment originated in industrialized European countries to West Africa and other underdeveloped countries, giving rise there to the uncontrolled and unhealthy "e-waste" graveyards.

PHOTOGRAPH 1.1
Landfill of electronic equipments and other discarded appliances from all over the world in Accra (West Africa). *Photograph: Daily Mail (http://www.dailymail.co.uk/news/article-3049457/Where-computer-goes-die-Shocking-pictures-toxic-electronic-graveyards-Africa-West-dumps-old-PCs-laptops-microwaves-fridges-phones.html).*

In short, for a given geographical region, a conceptual model can express pollution (P) as a function of human population density (N), directly related to the degree of industrialization (i), and inversely related to cultural level (c):

$$P = N * \frac{i}{c}$$

(Eq. 1.1)

An interesting implication follows; pollution effects cannot be quantitatively reduced in a scenario of continuous industrial development and population growth.

Pollution implies deleterious effects

The second aspect inherent to the formal definition of pollution is the **deleterious** effects caused, either to the environment or directly to people. If pollution became an issue in developed societies it is because it deteriorates the natural environment, impairs its utility for humankind, and even poses a risk to human health. We are now used to car traffic restrictions in the large cities caused by excessive levels of atmospheric particles, or beaches closed to swimming due to detectable inputs of fecal waters. Accidental spillages or chronic chemical pollution render some fisheries commercially valueless because of accumulation of toxic substances in the animals. However, in most instances the deleterious effects of contaminants are more subtle. Due to factors such as environmental persistence, continuous input, chronic effects, or trophic transfer, the manifestation of the harmful effects can be largely delayed in time and space, and a direct link between input of a substance in the environment and measurable deleterious effects on living organisms is normally difficult to establish.

Some authors advocated a formal distinction between pollutant (or pollution) and contaminant (or contamination), the latter not necessarily harmful. Walker et al.[4] identify at least three difficulties with this distinction. First, harm is related to dose and not an inherent property of a substance. Second, there is no general agreement about what constitutes environmental harm or damage. For example, concerning ecological effects, the changes in ecosystem structure and functioning caused by pollutants may be perceived as deleterious or not depending on the perspective. A large anthropogenic input of organic matter in an estuary may cause hypereutrophication and replacement of sensitive by tolerant benthic species. Biodiversity may be reduced but some commercially exploitable species may proliferate and the ecosystem may become more productive. Third, pollution has been traditionally assessed by measuring chemical concentrations, and corresponding biological effects are seldom known. Therefore, the distinction between pollution and contamination is not useful in theoretical nor in applied science.

1.2   CLASSIFICATIONS OF POLLUTANTS ACCORDING TO ORIGIN AND PERSISTENCE

Until the late 19th or early 20th century, waste disposal practices in the Western world, in contrast with the progress made by Eastern civilizations in ancient times, were fully careless, and turned the Thames or Seine rivers into open sewers transmitting cholera and many other infectious diseases. Even the inhabitants of the largest European cities got rid of their urine and feces at best through pipes draining on the street or more frequently straight through the window. The Spanish expression "aguava!," warning away the window because of that, is still remembered by older generations. The generalization of sewerage systems simply applied the dilution principle to solve the problem and transferred the contaminants to larger and more distant water bodies.

Ancient civilizations already used water to get rid of waste

As reflected in the motto: "dilution is the solution to pollution," physical dispersal was generally accepted as waste management strategy until a particular kind of pollutants entered the scene. In 1962 Rachel Carson published *Silent spring*, an influential book that put the focus on the **persistent chemical pollutants** that rather than diluting and disappearing in the environment accumulated into the birds and caused them unexpected side-effects such as reproductive failure. Advanced societies grew aware of the dangers posed by chemicals freely disposed in the environment, including risks to human health. Only when environmental concerns rose worldwide from the 1960s on did waste treatment and pollution abatement become priority issues in the political agendas.

Persistence and accumulation challenge dilution as an effective strategy against pollution

Many persistent pollutants have been found to accumulate in organisms at concentrations orders of magnitude above those found in their physical environment, a phenomena termed bioconcentration or bioaccumulation, dealt with in Chapter 11. Bioconcentration factor (BCF) is the ratio between the concentration in the organism and the concentration in its environment. According to most regulations, a substance is considered as bioaccumulative if its BCF exceeds a limit between 2000 and 5000. Environmental persistence and accumulation in the organisms challenge dilution as an effective strategy against pollution and stress the need for reduction of industrial use and environmental disposal of the so-called **PBT substances**, from persistent, bioaccumulative, and toxic.

Considering the broad definition of pollution explained earlier in the chapter, no universal method of pollution control is available, and the different characteristics and sources will require different management approaches. Pollution is immediately associated to chemicals, but according to their nature, pollutants can be classified into **physical** (heat and radioactivity, light, noise, particles,

Pollutants can be of physical, chemical, or biological nature …

plastic objects), **chemical**, and **biological** (pathogen microorganisms transmitted from fecal waters; see Chapter 4) pollutants.

According to the source, we can first make the difference between urban, agriculture, and industrial waste, and each of them can be classified into solid or liquid waste. Thus, solid urban waste, for instance, is specifically managed separately from other types of more dangerous solid wastes such as those of industrial origin not suitable for recycling or reutilization and that may require more careful disposal practices. Similarly, urban effluents are currently treated in wastewater treatment plants (WWTP), normally not designed to receive industrial wastewaters containing toxicants that would spoil the biological treatment of the WWTP. Generally speaking, the more we could separate different wastes the more efficient will be their treatment.

... or dissipating, degradable, or persistent according to their environmental life

Another useful classification attends at the environmental persistence of the contaminants. **Dissipating pollutants** are those that rapidly lose their damaging properties once released into the aquatic environment. Any potential effects are thus only local, and physical dispersion, instantaneous chemical reaction, or rapid biological uptake solves the problem. Heat (cooling water from power and nuclear stations), acids and alkalis, or cyanide are examples of this type of contaminants. Nitrates and phosphates are also quickly depleted from the environment by plant or microbial activity, and thus can also be classified within this category.

Biodegradable pollutants are those susceptible of biological oxidation and eventual mineralization to CO_2, reduced nitrogenous and phosphorus, and water under environmental conditions. Natural organic compounds, including oil, are all biodegradable because bacteria and fungi have evolved to obtain energy from virtually every natural molecule, but the degradation rates and hence environmental persistence may be very different, and in practice some types of natural organic matter such as lignin or humic substances can be considered as persistent. Environmental conditions such as temperature, oxygen availability, and microbial flora may greatly affect degradation rates.

Persistent or conservative pollutants are those not very chemically reactive and not readily subject to microbial attack either. Halogenated hydrocarbons, synthetic polymers, radioactive isotopes, and trace metals fall within this class. Conventionally, a contaminant is considered as environmentally persistent when its half-life is in the order of months or even years. Environmental persistence, as already stated, depends on phisic-chemical and microbiological properties of the environmental compartment, and hence half-life values must always be indicated for a particular set of environmental conditions, a requirement not always fulfilled in regulatory studies. According to several American and European regulations,[5] a substance is considered as persistent if its half-life in marine water is higher than 60 days, or higher than 180 days in marine sediment. Examples of persistent substances according to these criteria are diuron, lindane, polybrominated diphenyl-ethers, or polyfluorinated plastic additives.

1.3 IS AN ECOSYSTEM POLLUTED? A FIRST SCIENTIFIC ANSWER: BACKGROUND CONCENTRATION AND ENRICHMENT FACTOR

Moving on to practical issues, one is immediately interested in assessing pollution and answering the question whether a particular ecosystem is polluted or not. Apart from the two common traits of anthropogenic origin and potential harmful effects, little more is common to the wide range of processes labeled as "environmental pollution." In fact the question of whether or not an ecosystem is polluted is very difficult to answer if we are not more specific, since different kinds of pollutants according to their physical nature, persistence, or type of source may require very different techniques of assessment and pose very different levels of risk depending on the kind of organisms, including humans, potentially affected. For example, sewage waters may pose a high risk to human health because of microbial pathogens but be innocuous or even beneficial for the ecosystem production because of organic enrichment, whereas a potent chemical herbicide may pose imminent risk to primary producers while being innocuous to humans or perhaps beneficial for biodiversity in a hypereutrophic ecosystem. We thus should always pose questions such as, Is this ecosystem polluted by ...? to define the scientific tools capable to give an answer to the question and, if needed, implement the correct management tools for its abatement.

Different types of pollution require different methods of assessment

Once the type of pollution is defined, and provided the pollutant was a natural substance, the degree of pollution may be assessed by comparison of the current levels of the substance in the environment with those corresponding to pristine areas not subjected to strong human pressure. Within the field of aquatic sediments this approach was pioneered by Hakanson,[6] who proposed an index of contamination for each site based on the summation for all the substances measured of the rates between the mean measured concentration (C) and the **background concentration (BC)**, or reference value, defined as "the standard preindustrial reference level." The BC can be measured in samples from pristine areas of similar mineralogical composition or in deeper layers of the sediment corresponding to preindustrial times. The ratio C/BC has been later termed **enrichment factor**.

Preindustrial levels of chemicals are called background

This approach is not applicable for synthetic substances, for which the background level should be zero. Besides, the search for pristine areas is nowadays difficult, and requires resorting to very remote sites or historical data. Faced with these problems, the OSPAR Commission has established for common chemical pollutants **background assessment concentrations** (BAC), which are statistical tools derived from BC data that enable testing of whether observed concentrations can be considered to be near BCs.[7] The BAC, though derived from purely chemical data bases with no ecotoxicological information included, are an example of a useful concept in environmental management that will be discussed in Section 1.5: the environmental quality criteria.

1.4   POLLUTION IMPLIES A DELETERIOUS EFFECT: TOXICITY TESTS AND ECOTOXICOLOGICAL BIOASSAYS

Pollution implies a deleterious effect

Once we have set a suitable background level for our study area, an issue far from simple, and calculated an enrichment factor for the chemical pollutant of concern, we face a problem that chemistry is no longer capable to solve. Are those enrichment factors posing a risk to the native organisms or human populations? How can we know whether a given concentration of a chemical in an environmental compartment may be harmful? This question can only be answered by means of biological tools allowing the establishment of quantitative relationships between the levels of pollutants and their harmful **biological effects**.

Biological effects can be measured in the laboratory or by field studies

This issue can be addressed *a priori*, in the laboratory, with **toxicity tests** dosing known amounts of individual substances or combinations of substances on biological models, and recording biological responses at molecular, cellular, organismic, or micro- and mesocosm levels, or *a posteriori*, either in the laboratory exposing our biological models to known dilutions of environmental samples in **ecotoxicological bioassays**, or in the field, studying **biomarkers** and ecological **indices of communities** in the sites affected by the pollutants (see also Fig. 17.1). Therefore, the potential harm that pollution may cause on native organisms from affected sites can be the subject of prospective studies, aimed at prevention, and based on laboratory toxicity tests with the chemicals of concern (see Chapter 13), or the subject of retrospective studies conducted once the pollutants have been already discharged into the environment, aiming at the diagnosis of the ecological status of the affected sites. The latter use ecotoxicological bioassays with laboratory species exposed to environmental samples, biomarkers measured in native populations at different levels of organization (see Chapter 16), or community indices in affected sites (see Chapter 15).

Within the context of ecotoxicology, **toxicity tests** are laboratory assays consisting of the exposure of a population of test organisms to known amounts of the substance of concern under controlled conditions, including, exposure time, amount of the substance available to the test organisms, and all environmental variables known to affect the biological response recorded. The test organisms should belong to a standardized population of homogeneous biological traits (size, age, sex, reproductive condition, etc.), or even clones. The recorded biological response can be mortality or any sublethal response relevant for the Darwinian fitness of the test species (growth, reproduction, motility, etc.).

Seeking for high throughput biological models, miniaturization of techniques and an understanding of the biological mechanisms of toxicity, and also concerns about animal welfare, have encouraged toxicity testing with cell cultures, at the expense of ecological relevance of the results. On the other hand, criticism about lack of consideration of ecological interactions of single species assays prompted toxicity testing with more complex (and less controlled) micro- and mesocosms.

The quantitative relationship between the dose of a chemical and the deleterious biological response in the exposed organism is described by **dose: response curves** (see Section 13.2). Traditionally in medical toxicology, those curves are obtained from short-term (i.e., acute) exposures of groups of test organisms to increasing oral or injected doses of the chemical of concern. In aquatic toxicology, the exposure is more frequently waterborne and the testing organisms are exposed to the different concentrations of the dissolved chemical by bathing. Irrespective of the via of exposure and the biological model, a universal empirical finding is that the dose:response curves all present an S-shape (called sigmoid shape), such as that depicted for very different pollutants, test species, and experimental designs in Fig. 1.2.[8–10] As explained in detail in Chapter 13, the reason underlying this empirical finding is that the susceptibility to any pollutant in any population, quantified by the **individual lethal dose** or dose above which an individual dies, follows a log-normal distribution. When we record the mortality in a group of individuals exposed to a certain level of the pollutant we are actually recording the percentage of individuals whose individual lethal dose is lower or equal to that level, i.e., the cumulative frequency of the variable: "individual lethal dose." It is well known that the cumulative frequency curve for a normally distributed variable typically presents a sigmoid shape.

The information included in a toxicity curve is synthesized in two fundamental parameters that describe the effect of a substance on a particular organism under standardized exposure conditions: the **toxicity threshold (TT)** and the

Toxicity is described by two parameters: threshold and median effect, obtained from dose:response curves

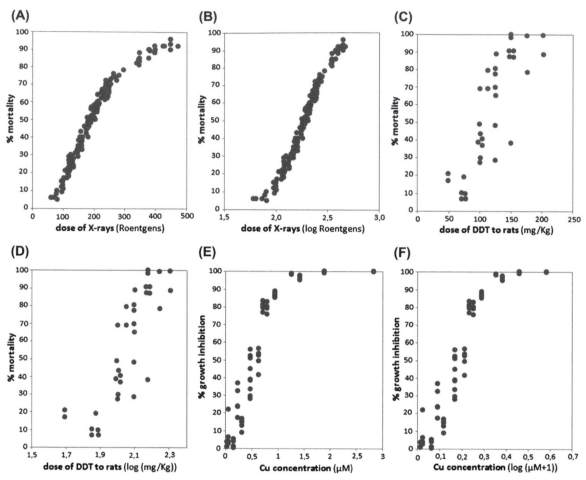

FIGURE 1.2

Dose:response toxicity curves obtained from three very different biological assays; (A—B): mortality of Drosophila eggs under X-ray irradiation (means of n > 100); (C—D): mortality of rats treated with oral doses of DDT (means of n = 10); E—F growth inhibition of sea-urchin larvae incubated at different copper concentrations (means of n = 125). Doses are presented in linear (A, C, E) and logarithmic (B, D, F) scale. Despite large differences in variability reflected in the degree of scattering of the points, notice the sigmoidal shape of all the curves that become more symmetrical using log scale. *Sources: Bliss Cl, Packard C. Stability of the standard dosage-effect curve for radiation. Am J Roentgenol Radium Ther 1941;46(3):400—404; Rozman KK, Doull J, Hayes Jr WJ. Dose and time determining, and other factors influencing, toxicity. Chapter 1, pp 3—101. In: Krieger R, editor. Hayes' handbook of pesticide toxicology. 3ʳᵈ ed. Elsevier; 2001; Lorenzo Jl. Metal speciation and bioavailability in the presence of organic matter studied using two marine biological models: the sea-urchin embryo (P. lividus) and the mussel (M. edulis) [PhD Thesis]. University of Vigo; 2003. 164 pp.*

median effective concentration. TT is the minimum concentration of the contaminant above which harmful effects begin to manifest in the test organisms. **Median effective concentration (EC_{50})** is the theoretical concentration of the contaminant causing a 50% reduction in the value of the biological response (survival, growth, reproduction, or any surrogate for any of them). It is interesting to note that both TT and EC_{50} are theoretical parameters with a single ideal value for each substance-organism pair under fixed exposure conditions, but experimental dose:response curves can only provide with estimates of that value. Thus, it is desirable that disregarding the statistical methods applied, EC_{50} values should always be reported with the corresponding statistical error measure to make possible comparisons among different data sets.

While traditional toxicology focuses on the estimation of LC_{50} values—the particular case of EC_{50} where the endpoint is mortality—environmental toxicology should be more interested in the TT since it will be more useful for environmental management. Section 13.2 deals with the theoretical aspects of these toxicity parameters, and Section 19.4 explains their use for derivation of environmental quality criteria.

1.5 ENVIRONMENTAL QUALITY CRITERIA AND STANDARDS

From the broad corpus of information obtained for a given substance using the biological and chemical tools described earlier (laboratory toxicity tests and field studies of chemical concentrations and corresponding biological effects), we can extract a series of figures representing the maximum concentrations that should not be exceeded to avoid deleterious effects on the organisms under a limited number of environmental scenarios (including different habitats, seasonal conditions, species to be protected, uses of water, etc.). Those figures are the **EQC**, derived from databases encompassing all the available ecotoxicological research about the chemical of concern, and recommended by scientific consensus. Water and sediment EQC are independently derived, resulting in **water quality criteria (WQC)** and **sediment quality criteria (SQC)**, also known sometimes as guidelines. For environmental assessment purposes, and not intended to the protection of human health, the OSPAR Commission issued assessment tools known as environmental assessment concentrations **(EAC)** for metals, organochlorines, and PAHs that include not only criteria for sediments but also for native biota, derived from ecotoxicological information.[11]

Environmental quality criteria (EQC) are scientific recommendations

The administrations with competences in environmental regulations (frequently national authorities) are entitled to go one step further and on the basis of the available EQC and the environmental policies favored they can implement **EQS** that, unlike EQC, are reflected in laws enforced by the competent authorities, and their compliance is thus compulsory. For example,

Environmental quality standards (EQS) are compulsory maximum levels endorsed by national authorities

in Europe member states endorsed the maximum levels of microbial pollution in bathing water set by the Directive 2006/7/EC, and if levels exceed these EQS a bathing prohibition must be introduced, which becomes permanent if standards are exceeded for five consecutive years.

In short, to come up with a solid EQS value for a given substance much prenormative research must be previously conducted on that substance, as illustrated in Fig. 1.3. Unfortunately, while many authorities set EQS, not always those values are originated from the bottom-up flow of information depicted in Fig. 1.3, which warrants ecological relevance to the figures, and hence effective protection of the environment.

Enrichment above environmental quality criteria measures pollution with ecological relevance

In Section 1.3 we made a first attempt to measure pollution on the basis of purely chemical information, by calculating the enrichment above background levels. Now we are in a position to assess pollution taking into account also the potential deleterious effects, as stated in the formal definition of pollution. With this aim we may use as reference value the ecotoxicologically based criterion instead of the background level, and calculate the ratio C/EQC, where C is the concentration of the pollutant measured. These ratios are more suitable for further analysis when expressed as logarithms. First, concentrations of pollutants in ecosystems frequently follow a log-normal distribution, allowing standard statistical testing. Second, the logarithm takes a positive value when the

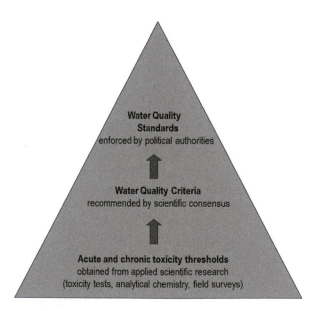

FIGURE 1.3

Bottom-up flow of information for the implementation of environmental quality standards on scientific grounds.

measured concentration exceeds the criterion and negative otherwise, which is very convenient for visualization of the results.

Thus, for a given pollutant, i, the contamination factor (CF)[12] is:

$$CF_i = \log(C_i/EQC_i) \qquad \text{(Eq. 1.2)}$$

The resulting CF values may be summed up to integrate the information obtained for all pollutants studied. When the pollutants are chemical substances the resulting parameter has been termed **Chemical Pollution Index** (CPI), that for a series of n pollutants is:

$$CPI = \sum_{i=1}^{n} CF_i = \sum_{i=1}^{n} \log(C_i/EQC_i) \qquad \text{(Eq. 1.3)}$$

The choice of EQC is a key aspect in the computation of this index, and internationally accepted criteria, such as the ERL and TEL for sediments, are recommended.[13]

1.6 PATHWAYS AND DISTRIBUTION OF POLLUTANTS IN COASTAL ECOSYSTEMS

We will turn our attention now to the study of pollution in a particular environment, the coastal ecosystems. Although oceanic and atmospheric circulation spread pollutants along the entire oceanic ecosystems, there are at least two reasons why pollution affects coastal zones more than open-ocean environments. The first simply derives from the fact that pollution, an anthropogenic process, is a direct consequence of human activities, and about 44% of human population, including 8 of the 10 world's largest cities, concentrates in the coast line.[14] The second stems from the patterns of physical transport of matter in the aquatic environments. Contaminants are largely conveyed to the sea via freshwater inputs at the near shore. Ocean margins are efficient filters of dissolved and particulate substances. The mixing zones of estuaries, in particular, act in many cases as traps for riverine-transported materials that end up in the estuarine sediments.[15] According to some estimations, nearly all of the nutrients from land-based sources delivered by rivers and 90% transported by the atmosphere are trapped in ocean margins,[16] and 90%–95% of the inputs of trace elements to the sea accumulate in coastal environments.[17]

Marine pollution mainly affects coastal zones

Moreover, compared to the rivers, the residence time of pollutants in tidal estuaries is highly elevated, and thus the ability to dilute point discharges decreased, as a consequence of the influence of tides. This can be illustrated with figures from the Thames estuary, where the overall seaward flow speed decreases from more than 20 km per day to barely 2 km per day in the mouth, as a result of the upstream flow during flood tide, with the corresponding increase in residence time.[18]

In addition, the relevance of sediment pollution in coastal zones is higher than in deeper waters, since the benthic-pelagic coupling and resuspension characteristic of shallow environments may favor recycling of contaminants from sediments, which does not happen in oceanic waters.[19] Coastal ecosystems are also especially productive, and thus the ecological and even the economic impacts of pollution are also especially harmful in these environments. The neritic zone, from low-tide level to a depth of 200 m approximately, is the area most densely populated by benthic organisms owing to the penetration of sunlight.

The main routes of pollutants into the sea are riverine inputs, atmospheric deposition, and direct dumping

Provided pollution is an anthropogenic process and most human activities take place on land, the terrestrial habitats are the origin of most environmental contaminants. However, these contaminants are transported to the sea by means of three main routes: via atmosphere, by riverine input, or from direct spillage into the sea (see Fig. 10.1).

The **riverine input**, that includes run-off water from polluted urban and rural areas and effluents of liquid wastes from sewage and industry, is the most important route of transport of both dissolved and particulate contaminants from land to sea. As explained earlier, the estuaries, the interface between fluvial and marine ecosystems, act as filters that retain a high percentage of the materials transported by the rivers, including the contaminants. Coastal marine sediment is one of the major sinks of contaminants at global scale. Relevant mechanisms accounting for this are **precipitation** of dissolved elements associated to the salinity increase, flocculation of colloids, and **sedimentation** of suspended matter due to the decrease in current speed and tidal dynamics at the river mouth (see also Sections 9.3 and 10.1).

The dissolved inorganic substances can precipitate as a result of the sharp increase in salinity of the medium, due to the common ion effect. For example, the solubility of $PbCl_2$ in pure water, considering complete dissociation into Pb^{2+} and Cl^-, is 0.016M[a]. However, seawater has approximately 19 g Cl^- per L, i.e., 0.54 M. As a result, the equilibrium between the solid $PbCl_2$ and the aqueous ions Pb^{2+} and Cl^- is strongly displaced toward the former species, and solubility is reduced by a factor of 300 approximately.

Suspended particulate matter can sediment as a consequence of the decrease in current velocity and processes related to estuarine circulation and tidal dynamics. The finest particles (clay and colloids) can form aggregates with increased sinking velocity due to flocculation. This process is facilitated by

---

[a] This is calculated from a solubility product constant $K_{sp} = 1.7 \times 10^{-5}$. Due to partial dissociation into other species such as $PbCl^+$ experimental solubility is c. twice higher.

the sea water positive ions that neutralize the negative charge of the clay particle surface.

The **atmospheric input** is relevant for contaminants with low vapor pressure that evaporate from terrestrial sources, such as light hydrocarbons, for particles and gases resulting from combustion of organic matter, including fossil fuels and their trace metal contents, and for pesticides and other products used as aerosols. Far-reaching transportation processes driven by atmospheric dynamics are responsible for the occurrence of chemical pollutants even in uninhabited regions of Earth. All these contaminants can be washed by rainfall on the sea surface. In polluted places an oil slick on the water surface creates a hydrophobic phase where neutral organics accumulate at concentrations orders of magnitude above those in the underlying water.

Direct input of contaminants into the sea can be locally very intense, for example as a result of intentional **dumping of wastes**, oil platform activities, or due to navigation accidents involving oil tankers or other ships transporting hazardous substances. Nuclear testing in the Pacific Ocean was conducted by the United States, the United Kingdom, and France until 1996.[20] The oceanic depths were the final destination for industrial and radioactive wastes until the London Convention on the Prevention of Marine Pollution, in force since 1975. From 1949 until 1982 eight European countries dumped 140,000 T of radioactive wastes at c. 700 km from the northwest Iberian coast. Marine dumping is still a common procedure for dredged materials. The discarding of fishing nets made from plastic materials not degraded in the sea is another unresolved environmental issue. Other direct input of pollutants as a result of bad practices and navigation routines are more diffuse, though they may globally account for a large part of the hydrocarbon or plastic input into the oceans.

According to their degree of dispersion, sources of contaminants can also be classified into **point sources** and **diffuse sources**. This classification is operationally useful since the first are easier to abate than the latter. Point-source pollution can be reduced by the introduction of new technologies and the investment in wastewater treatment, but discharges from diffuse sources are much more difficult to control. For example, fight against hyper-eutrophication of water bodies caused by excess of nutrients frequently target phosphates, with point sources (urban sewage carrying detergents), in contrast to nitrates whose main source is much more diffuse, run-off waters contaminated by the in-land application of agriculture fertilizers. In Europe, discharges of both nitrogen and phosphorus from point sources have decreased significantly over the past 30 years, mainly due to improved treatment of urban wastewater, the lowered phosphate content of detergents, and reduced industrial discharges, whereas measures to reduce the nitrogen surplus from agricultural land have only had limited success so far.[21]

Point-source contaminants are easier to control

KEY IDEAS

- Environmental pollution is anthropogenic (i.e., caused by human activities). Natural phenomena (harmful algal blooms, storms, floods, etc.) are never considered as pollution.

- The impact of pollution in a particular area is directly related to human population density and degree of industrialization, and inversely related to societal awareness of environmental issues.

- Pollution implies potential harmful effects on human health, biological communities or ecosystem properties useful for humans (exploitation of marine resources, leisure, etc.). However, the direct link between presence of a particular pollutant and deleterious effects caused by that pollutant is rarely demonstrated.

- Quantitative tools are available to assess and control pollution. When the pollutant is a natural substance the enrichment above natural BC can be recorded. However, the best way to assess pollution is to study the maximum admissible concentrations above which harmful effects are triggered: EQC.

- EQC should be based on the TTs of pollutants, i.e., the levels above which the pollutant begin to harm the growth, reproduction, or survival ability of the organisms. These thresholds can be estimated from laboratory toxicity tests with organisms representative of marine ecosystems.

- Some pollutants are not readily degraded by chemical nor biological processes and thus they persist for years in the environment. Many of them in addition tend to accumulate in the organisms and may cause toxicity. Environmental persistence and bioaccumulation challenge the notion of dilution as effective strategy against pollution.

- Coastal ecosystems are normally more polluted than oceanic waters due to population density and proximity to sources. Consequences of pollution are especially relevant in these productive ecosystems.

- Estuarine sediments trap and concentrate materials transported by rivers, including chemical pollutants. Precipitation due to increased salinity and sedimentation due to reduced current speed and tidal dynamics explain this accumulation.

Endnotes

1. GESAMP. Report of the first session (London, 17—21 march 1969). Joint IMCO/FAO/ UNESCO/WMO Group of Experts on the Scientific Aspects of Marina Pollution. GESAMP I/11 17; July 1969. 28 pp. VI Annexes.

2. GESAMP. Report of the third session held at FAO headquarters, Rome, 22—27 February 1971. IMCO/FAO/UNESCO/WMO Joint Group of Experts on the Scientific Aspects of Marina Pollution. GESAMP III/19; 13, May 1971. 9 pp. IX Annexes.

3. UNEP Global mercury assessment (2013). Available online: http://wedocs.unep.org/bitstream/ handle/20.500.11822/7984/-Global%20Mercury%20Assessment-201367.pdf?sequence=3&is Allowed=y.

4. Walker CH, Sibly RM, Hopkin SP, Peakall DB. Principles of ecotoxicology. 4th ed. CRC Press; 2012.

5. Matthies M, Solomon K, Vighi M, et al. The origin and evolution of assessment criteria for persistent, bioaccumulative and toxic (PBT) chemicals and persistent organic pollutants (POPs). Environmental Science: Processes & Impacts 2016;18:1114—1128

6. Hakanson L. An ecological risk index for aquatic pollution control. A sedimentological approach. Wat Res 1980;14:975—1001.

7. OSPAR Comission. Agreement on background concentrations for contaminants in seawater, biota and sediment. OSPAR agreement 2005—2006; 2005.

8. Bliss CI, Packard C. Stability of the standard dosage-effect curve for radiation. Am J Roentgenol Radium Ther 1941;46(3):400—404.

9. Rozman KK, Doull J, Hayes Jr WJ. Dose and time determining, and other factors influencing, toxicity. Chapter 1, pp 3—101. In: Krieger R, editor. Hayes' handbook of pesticide toxicology. 3rd ed. Elsevier; 2001.

10. Lorenzo JI. Metal speciation and bioavailability in the presence of organic matter studied using two marine biological models: the sea-urchin embryo (*P. lividus*) and the mussel (*M. edulis*) [PhD Thesis]. University of Vigo; 2003. 164 pp.

11. OSPAR Commission. Agreement on CEMP assessment criteria for the QSR 2010. OSPAR agreement 2009—2012; 2009.

12. Bellas J, Nieto O, Beiras R. Integrative assessment of coastal pollution: development and evaluation of sediment quality criteria from chemical contamination and ecotoxicological data. Cont Shelf Res 2011;31:448—456.

13. Beiras R, Durán I, Parra S, et al. Linking chemical contamination to biological effects in coastal pollution monitoring. Ecotoxicology 2012;21:9—17.

14. UN Atlas of the Oceans. http://www.oceansatlas.org/servlet/CDSServlet?status=ND0xODc3 JjY9ZW4mMzM9KiYzNz1rb3M~.

15. See Table 9.5, p. 194, in: Chester R, Jickells T. Marine Geochemistry 3rd ed. Wiley-Blackwell; 2012.

16. Windom HL. Contamination of the marine environment from land-based sources. Mar Pollut Bull 1992;25(1-4):32—36.

17. Martin JM, Windom HL. Present and future roles of ocean margins in regulating marine biogeochemical cycles of trace elements. pp 45—67. In: Mantoura RFC, Martin J-M, Wollast R, editors. Ocean Margin Processes in Global Change. Wiley Interscience; 1991.

18. See p. 39 in: Clark RB. Marine Pollution 5th ed. Oxford University Press; 2001.

19. Valiela I. Marine ecological processes. 2nd ed. Springer; 1995.

20. http://www.sea.edu/spice_atlas/nuclear_testing_atlas/french_nuclear_testing_in_polynesia.

21. See p. 107 in: EEA 2007. Europe's environment. The fourth assessment. European Environment Agency, Copenhagen.

Suggested Further Reading

- Islam Md S, Tanaka M. Impacts of pollution on coastal and marine ecosystems including coastal and marine fisheries and approach for management: a review and synthesis. Mar Pollut Bull 2004;48:624—649.
- Weiss JS. Marine pollution. What everyone needs to know. Oxford University Press; 2015. 273 pp.

CHAPTER 2

Nonpersistent Organic Pollution

2.1 THE CARBON CYCLE AND THE BALANCE OF OXYGEN IN AQUATIC ENVIRONMENTS; HYPOXIA

Carbon is the key element for life and the structural basis of all organic molecules. The biogeochemical cycle of this element is mainly driven by the opposite biological processes of organic matter production and consumption, synthetically schemed in the following equation:

Biological production and consumption drive the carbon cycle

$$CO_2 + H_2O \leftrightarrow CH_2O + O_2 \qquad \text{Eq. 2.1}$$

Organic matter production, mainly by **photosynthesis**, traps inorganic carbon into the organic molecules that build up the living organisms, while metabolic **respiration** and mineralization of organic detritus by heterotrophic organisms work in the opposite direction. Notice that the balance between both processes will determine also the net production or deficit of oxygen, an element essential for animal life.

Indeed, the equilibrium in Eq. (2.1) is easily disturbed in small or poorly renewed water bodies, for example by the input of domestic wastewater rich in organic matter from toilets and sinks (Fig. 2.1).

Excess of nonliving (either dissolved or particulate) organic matter unbalances Eq. (2.1) equilibrium toward the left, due to the increased heterotrophic activity of aerobic microorganisms, and causes deficit of dissolved oxygen (DO). For the mineralization of 1 mol of glucose ($C_6H_{12}O_6$) 6 mol of oxygen are consumed, at a rate thus of 2.7 g oxygen per g of carbon. Since solubility of oxygen in water is limited to a few mg/L (Table 2.1), the diffusion of oxygen in water is a slow process, limited by the repletion from the atmosphere which is negligible in stratified deep waters.

Excess of organic matter in water causes oxygen depletion

The effect of temperature is remarkable, with about a 2% reduction in O_2 solubility per 1°C increase, due to the increase of the Henry's constant value that accounts for increased volatilization. Salinity also reduces O_2 solubility, and seawater shows saturation values c. 20% lower than freshwater.

21

Marine Pollution. https://doi.org/10.1016/B978-0-12-813736-9.00002-7

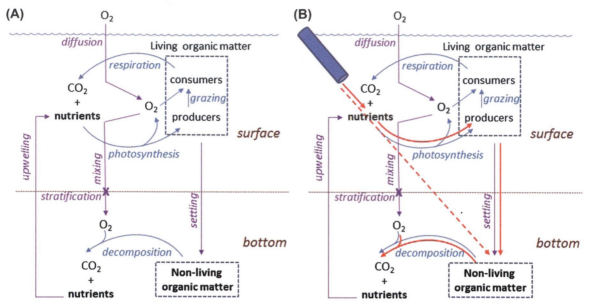

FIGURE 2.1

Biomass production, consumption and mineralization, nutrient cycling, and oxygen balance in an aquatic ecosystem. Physical processes are marked in purple and biological processes are marked in blue. (A) shows a natural system and (B) the same system submitted to anthropogenic discharges of nutrients and residual organic matter. Both nutrients and organic matter contribute through different mechanisms to the same overall depletion of oxygen in the bottom. Notice that, since photosynthesis is not present in the bottom, stratification can lead there to seasonal hypoxia even in the natural system scenario.

Table 2.1 Solubility of Oxygen in Water (mg/L)

Salinity (ppt)	0°C	5°C	10°C	15°C	20°C	25°C	30°C
0	14.6	12.8	11.3	10.1	9.1	8.3	7.6
5	14.2	12.4	11.0	9.8	8.9	8.0	7.4
10	13.7	12.0	10.7	9.5	8.6	7.8	7.2
15	13.3	11.7	10.3	9.3	8.4	7.6	7.0
20	12.9	11.3	10.0	9.0	8.1	7.4	6.8
25	12.4	10.9	9.7	8.7	7.9	7.2	6.6
30	12.0	10.5	9.4	8.4	7.6	7.0	6.4
35	11.5	10.2	9.0	8.2	7.4	6.8	6.2

Adams and Bealing (1994).

According to the oxygen concentration, aquatic habitats may experience the following conditions: **normoxia**, when aerobic metabolism prevails, **hypoxia**, when the anaerobic metabolism plays a remarkable role and the oxygen deficit is likely to produce a deleterious impact on benthic macrofauna (arbitrarily <2 mg/L), and **anoxia**, when DO values are near zero (<0.1 mg/L). Hypoxia in the benthic habitats promotes denitrification and production of H_2S, which can be toxic, as it reacts strongly with iron-containing enzymes inhibiting enzymatic processes in the cells; and a concomitant reduction in pH. Anoxic environments are azoic from the macrofauna standpoint, and microbial flora is dominated by obligate anaerobic microbiota such as the methanogen bacteria.

Anoxia can be traced using an isotopic approach, as sulfate fractionated during sulfate reduction and sulfide accumulated in the sediment pore waters have a more depleted $\delta^{34}S$ signature compared to the more enriched (+21‰) of sulfate from the seawater. This sulfate is taken up and assimilated by marine rooted plants or benthic fauna and incorporated in their tissues.

The deleterious effects of oxygen depletion on the biological communities, particularly for streams or enclosed water bodies, are very well known. DO levels <2 mg/L cause acute lethal toxicity in sensitive freshwater fish such as trout, while sustained values below 1 mg/L cause mass mortalities in benthic communities. Early life stages of salmonids may show even higher oxygen requirements, in the order of 4−5 mg/L.[1] For marine organisms, the 96h LC_{50} in juvenile invertebrates and fish ranged from 0.5 to 1.7 mg/L, while longer exposures retarded growth of larvae and juveniles at concentrations from 2.3 to 7.7 mg/L, with crustacean larvae requiring particularly high DO levels.[2] Based on this information some countries developed environmental quality guidelines for DO specific for marine waters that range from 4.8 to 8.0 mg/L.[3]

And early life stages require high dissolved oxygen levels

With the onset of sewerage facilities that injected virtually untreated domestic effluents with high loads of organic matter, many instances of drastic oxygen depletion in receiving waters were reported. Early attempts to simulate this process at small laboratory scale resulted in the development of the **Biological Oxygen Demand (BOD)** test, whose basic principle dates back to 1912,[4] but it is still used as a simple and inexpensive method for the estimation of organic matter in the water. Thus, a sample of problem water is incubated in a closed recipient in the dark at a standard temperature (20°C), and the oxygen consumption after 5 days (BOD_5) is recorded, originally by the Winkler method. With several methodological allowances, including previous aeration and appropriate dilution with a mineral salt medium, to the amount of biodegradable organic matter in the water sample may be estimated from the measured BOD_5. Indeed, organic substances occurring in natural or residual waters are extremely heterogeneous, but the overall organic content of an effluent can be easily estimated by taking advantage of the equimolar oxygen consumption associated to its mineralization, illustrated in Eq. (2.1). the advantages and limitations of the BOD test and some alternative methods will be discussed in Section 2.3.

Biodegradable organic matter is estimated by the Biological Oxygen Demand (BOD) test

2.2   NATURAL AND ANTHROPOGENIC SOURCES OF ORGANIC MATTER IN THE WATER

Natural organic matter may derive from local primary production or allochthonous

Natural organic matter in aquatic environments is classified into dissolved (DOM) and particulate (POM), the latter being operationally defined as that retained in 0.45 μM filters. A high proportion of the natural DOM is of refractory character, meaning that it is resistant to microbial biodegradation, and hence both ecologically inert and undetectable to the BOD test.

In open sea DOM exceeds by several orders of magnitude POM,[5] and the latter is mostly restricted to surface waters. Deep water DOM is mostly refractory, with an average "age" of thousands of years, while in the photic zone (c. 100 m) DOM has an important additional semilabile component derived from cellular lysis and plankton exudates and a minor labile fraction rapidly recycled by bacteria.

In coastal zones POM is relatively more abundant. In addition, to the authochtonous primary production, especially high in upwelling areas it must be added the allochthonous organic matter of continental origin -both dissolved and particulate-transported by river run-off. Due to purely natural reasons related to estuarine circulation (see Section 1.6), nearshore sediments are the main sink for that organic matter, with global estimations reaching $9 \cdot 10^{12}$ mol C y^{-1}.

Anthropogenic organic matter comes from urban and industrial sewage

In densely populated coastal areas, anthropogenic organic matter must be added up. Every person contributes with a load of c. 70 g BOD/day to the organic matter disposed as liquid wastes through the domestic sewage. Industrial wastewaters, particularly effluents from food processing (farms, canned food, slaughterhouses, etc.) and paper mills typically show even higher contents of organic matter. The organic residues generated by one cow are equivalent to those of 16.4 humans, and thus the demographic equivalent of the organic liquid waste produced by US farms, in terms of BOD, was 216 million habitant equivalents, and for the paper mills 1900 million.[6]

2.3   ESTIMATES OF ORGANIC MATTER IN EFFLUENTS AND RECEIVING WATERS: BIOLOGICAL OXYGEN DEMAND (BOD), CHEMICAL OXYGEN DEMAND (COD), AND TOTAL ORGANIC CARBON (TOC)

BOD$_5$ estimates the amount of organic matter in water

The BOD basic principle is such a useful and practical parameter that despite its limitations, characterization of natural and residual waters, methods for testing the design of sewage treatment, or current environmental regulations determining whether discharges are permissible, all make use of it. The BOD$_5$ for some common effluents are shown in Table 2.2.

One of the few updates suggested for the BOD method consists of the continuous measurement of the oxygen concentration by means of an oxygen probe

Table 2.2 Typical Ranges of BOD_5 Values (mg/L) for Some Natural and Residual Waters

Water	BOD_5 (mg/L)
Clean natural waters	<3
Run-off water	10–15
Untreated municipal wastewater	200–400
Paper mill wastewaters	300–700
Meat processing raw effluent	500–1500
Dairy processing raw effluent	2500
Bakery processing raw effluent	2000–4000
Treated urban effluents	<25*

Directive 91/271/EEC

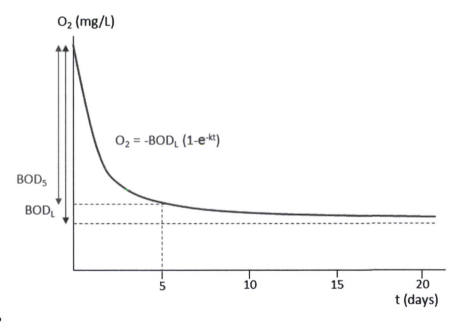

FIGURE 2.2

Oxygen depletion curve of a typical water sample affected by untreated urban sewage incubated under the conditions of the BOD_5 test. The ultimate carbonaceous BOD value (BOD_L) can be calculated by adjusting the data to the equation shown.

or a manometer, in order to obtain a curve like the one depicted in Fig. 2.2. The oxygen demand increases as the duration of the incubation increases. The five-day standard period provides an approximation of the actual oxygen demand that would be obtained at much longer incubation times, called limiting or ultimate (BOD_L).

Assuming the appropriate heterotrophic microbial flora in exponential growth phase is present in the water sample, the microorganisms will grow at a decelerating rate caused by the consumption of the organic matter. The resulting time-course depletion of DO can thus be fitted to an asymptotic equation of negative slope, according to the expression: $O_2 = -BOD_L (1-e^{-kt})$.[7] This expression allows analytical calculation of the BOD_L. As shown in Fig. 2.2, the higher the absolute value of the slope, k, the closer BOD_5 will estimate BOD_L. High k values will be expected when the constituents of the organic matter were readily biodegradable, no other nutrients than carbon would limit oxygen consumption, and the appropriate heterotrophic flora was present. However, one or several of these conditions can fail. Carbohydrates and proteins, which account for 90% of the organic load of domestic wastewaters, are rapidly mineralized, but fats and oils may require longer incubations. The k value for natural surface waters is 2.5 times lower than that for a glucose solution.[8] Certain hydrocarbons and celluloses need special microbial flora that may require lag times for blooming or even be absent from the sample. In the latter case the sample must be inoculated with a previously acclimatized flora.

Limitations of the BOD₅ method

Typical time courses for microbially driven oxygen depletion in enclosed water samples are depicted in Fig 2.3. The shape of the curve depends on the specific composition of the organic matter present. Curves A and C represent samples with the same amount of organic matter, but while in sample A the detrital substances are easily biodegradable, in sample C they require growth of specific microbiota after a lag time. In that case BOD_5 is a very poor estimate of BOD_L.

Also, even when previously aerated to saturation levels, samples with high organic loads must be diluted with mineral water before incubation, otherwise

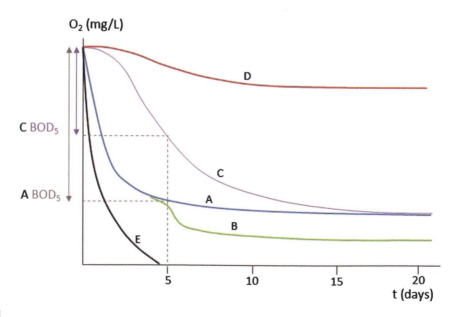

FIGURE 2.3

Different patterns of oxygen depletion curves of water samples incubated under the conditions of the BOD₅ test. BOD₅ from samples C, D, and E provide incorrect estimations of the dissolved organic matter. See text for details.

they will run anoxic preventing BOD_5 calculation (curve E). This requires educated guesswork to decide a range of dilution factors appropriate for each particular sample type. The ideal oxygen depletion ranges between 30% and 70% of the starting O_2 concentration.[9]

Another methodological problem is illustrated by curve B. The oxygen consumption has an initial component derived from organic matter mineralization, the carbonaceous oxygen demand, and a subsequent component derived from the nitrification activity, the nitrogenous oxygen demand that causes the additional oxygen demand between curves B and A. Oxygen depletion by nitrification activity is proportional to inorganic ammonia, and if it is considered in the BOD calculation the organic matter will be overestimated. Fortunately, ammonia-oxidizing microorganisms are slow-growing and not abundant in natural waters, so their levels only increase after a lag time that normally comprises the five-day standard incubation period. However, for certain samples such as treated effluents the initial load of nitrifiers may be already high and these interference relevant, so standard methods may include the addition of a nitrification inhibitor such as allylthiourea.[10]

Industrial effluents may include toxic chemicals that inhibit microbial degradation of the organic matter, rendering the BOD test unfeasible (curve D). If the toxicity is mild this problem may be overcome by serial dilutions.

Some kinds of organic matter such as celluloses and lignin are not easily biodegradable, and hence they are not quantifiable by the BOD_5 test. An alternative method is the **Chemical Oxygen Demand (COD)** that relies on the chemical oxidation of the organic matter by a strong oxidant agent such as potassium dichromate. The consumption of the chemical agent is in stoichiometric relationship to the original amount of organic matter in the sample.

Another alternative method to estimate the organic load in liquid samples is the **Total Organic Carbon (TOC)** analyzer. The instrument measures the CO_2 produced after strong chemical oxidation or after catalytic combustion at temperatures of $900-1000°C$. The latter has the advantage of rapidity but the limitation of underestimation due to incomplete combustion.

2.4 ECOLOGICAL EFFECTS OF HYPOXIA AND ANOXIA

The distribution of oxygen in natural waters is determined by the balance between the biological processes of synthesis and consumption of organic matter, the input across the surface from the atmosphere, and the physical transport and mixing of water masses caused by hydrodynamics. In the sea surface, oxygen levels reach saturation—in fact slight supersaturation[11]—due to both atmospheric input and photosynthesis. Below the photic layer, as a result of microbial respiration fueled by the settling detrital organic matter, DO markedly falls down with depth.

Oxygen is saturated in surface and sharply decreases with depth

In oceanic environments an oxygen minima is reached at around 1000 m depth. Underneath the colder water masses at the bottom that originally sank at high latitudes are again richer in oxygen. However, due to limited vertical mixing and advection during very prolonged periods of time, naturally hypoxic deep-sea regions, termed oxygen minimum zones, with DO values below 20 µM (0.64 mg/L) or even reaching anoxia e.g., in the East Pacific have been described.

Hypoxia is a reversible but widespread deleterious anthropogenic influence that currently ranks among the main environmental issues in coastal environments. The extent of ecosystem degradation is determined by the duration of the hypoxic period and the DO concentration reached.

In shallow waters fluctuations in the degree of stratification combined with seasonal blooms may origin transient periods of hypoxia with drastic effects on community structure but quick recovery after vertical mixing. The recovery path from severe anoxia is different than the declining path because of the hysteresis-like progression of successional dynamics.[12]

Normoxic, hypoxic, and anoxic aquatic environments strongly differ in ecological richness

Transient hypoxia favors opportunistic benthic species with shorter life spans and smaller body sizes.[13] Megafauna leave from hypoxic zones as soon as DO reaches concentrations of around 40%–25% saturation for cod and whiting, and 15% for dab and flounder.[14] The density and biomass of macrofauna decreases at DO levels below 1.4 mg/L, while meiofauna is still unaffected. Functional diversity and species richness are reduced, and communities are dominated by very few but highly abundant species.[15] Echinoderms and most crustaceans are particularly affected by oxygen depletion, while annelids, molluscs, and cnidarians are typically more tolerant. Usually surface-deposit feeders are favored over subsurface feeders,[16] although some carnivorous, epifaunal species such as polynoid polychaetes may persist.

Decrease in diversity is exacerbated by the drastic reduction of spatial heterogeneity caused by the extinction of large sediment-dwelling megafauna and loss of structures formed by foundation species such as seagrass beds,[17] oyster reefs,[18] or mussel beds.[19]

Mass mortalities of benthic fauna eventually take place at DO levels below 1–0.7 mg/L.

Low DO levels also affect energy flows

In normoxic environments typically 25%–75% of the macrobenthic carbon will be transferred to higher-level predators. Hypoxia enhances the diversion of energy flows into microbial pathways to the detriment of higher trophic levels. Demersal fish predators will avoid hypoxic environments since they are generally less tolerant to low oxygen levels than benthic fauna. If oxygen deficit becomes persistent the proportion of benthic energy production transferred to microbes rapidly increases and eventually, under anoxia, microbes will process all benthic energy as hydrogen sulfide.[20]

KEY IDEAS

- Direct input of residual organic matter or excessive primary production caused by large inputs of anthropogenic nutrients both may lead to the same ecological unbalance between photosynthesis and respiration that cause hypoxia or even anoxia in bottom waters.

- Since residual organic matter has a complex composition it is difficult to quantify directly by chemical analysis, and the level in water can be indirectly estimated by measuring the consumption of oxygen needed for its biological or chemical mineralization.

- In deep marine water hypoxic zones have been described, characterized by low DO levels, loss of ecological diversity, and reduced spatial heterogeneity in the benthic communities.

Endnotes

1. See p. 735 in: Adams N, Bealing D. 11: Organic pollution: biochemical oxygen demand and ammonia. In: Calow P, editor. Handbook of ecotoxicology. Blackwell Science Ltd; 1998.

2. US EPA. Ambient aquatic life water quality criteria for dissolved oxygen (saltwater): Cape Cod to Cape Hatteras. EPA-822-R00-012; 2000.

3. CCME 1999. Canadian Water Quality Guidelines for the Protection of the Aquatic Life. Canadian Council of Ministers of the Environment; US EPA (2000) op. cit.

4. See p. 742 in: Adams & Bealing (1998) op. cit.

5. See p. 185 in: Chester R, Jickells T. Marine Geochemistry 3rded. Wiley-Blackwell; 2012.

6. Wadleish CH. Wastes in relation to agriculture and forestry. Miscellaneous Publication N. 1065. United States Department of Agriculture. Washington DC; 1968.

7. See p. 304 in: Pepper IL, Gerba CP, Brusseau ML. editors.Environmental and Pollution Science. 2nd ed. Elsevier; 2006.

8. Sawyer and McCarty. Cited by Henry JG, Heinke GW, editors. Environmental Science and Engineering. 2nded. Prentice Hall; 1996.

9. See p. 277 in: Adams & Bealing (1994) op. cit.

10. ISO 5815−1:2003. Water quality − Determination of biochemical oxygen demand after n days (BODn) − Part 1: Dilution and seeding method with allylthiourea addition.

11. See p. 181 in: Chester & Jickells (2012) op. cit.

12. Diaz RJ, Rosenberg R. Spreading dead zones and consequences for marine ecosystems. Science 2008;321:926−929.

13. Diaz RJ, Rosenberg R. 1995. Marine benthic hypoxia: a review of its ecological effects and the behavioural responses of benthic macrofauna. Oceanogr Mar Biol Annu Rev 1995;33: 245−303.

14. Gray JS. 1992. Eutrophication in the sea. In: G Colombo, I Ferrari, VU Ceccherelli, R Rossi, editors. Marine eutrophication and population dynamics. p. 3−15.

15. Middelburg JJ, Levin LA. Coastal hypoxia and sediment biogeochemistry. Biogeosciences 2009;6:1273−1293.

16. Levin LA, Gage JD, Martin C, Lamont PA. Macrobenthic community structure within and beneath the oxygen minimum zone, NW Arabian Sea. Deep-Sea Research II 2000;47: 189−226.

17. Burkholder JM, Tomasko DA, Touchette BW. Seagrasses and eutrophication. J Exp Mar Biol Ecol 2007;350:46—72.

18. Breitburg DL, Pihl L, Kolesar SE. Effects of low dissolved oxygen on the behavior, ecology and harvest of fishes: A comparison of the Chesapeake Bay and Baltic-Kattegat Systems. In: Rabalais NN, Turner RE, editors. Coastal and Estuarine Studies: Coastal Hypoxia Consequences for Living Resources and Ecosystems, American Geophysical Union, Washington, D. C., 293—310; 2001.

19. Mee D, Friedrich J, Gomoiu M. Restoring the Black Sea in times of uncertainty, Oceanography 2005;18:100—112.

20. Diaz & Rosenberg (2008) op. cit.

Suggested Further Reading

- Adams N, Bealing D. 11: organic pollution: biochemical oxygen demand and ammonia. In: Calow P, editor. Handbook of ecotoxicology. Blackwell Science Ltd; 1998. p. 728—49.

- Breitburg DL, Pihl L, Kolesar SE. Effects of low dissolved oxygen on the behavior, ecology and harvest of fishes: a comparison of the Chesapeake Bay and Baltic-Kattegat Systems. In: Rabalais NN, Turner RE, editors. Coastal and estuarine studies: Coastal hypoxia consequences for living resources and ecosystems. Washington, D. C.: American Geophysical Union; 2001. p. 293—310.

CHAPTER 3

Nonpersistent Inorganic Pollution

3.1 NATURAL AND ANTHROPOGENIC SOURCES OF NITROGEN AND PHOSPHORUS IN THE WATER

Eq. (2.1) describes primary production in a quite simplistic way, taking into account the cycling of carbon and oxygen balance only. In fact, carbon is very abundant in the water, either as dissolved CO_2 or as bicarbonate and carbonate ions, and this element does not limit algal growth. Other essential macronutrients such as K and S are also rather abundant in the seawater compared to the biological requirements. The nutrients essential for all forms of life more often limiting primary production are N and P. The C:N:P stoichiometric ratios of the living organisms are fairly constant, and termed **Redfield ratios**. For both phytoplankton and zooplankton organic matter 106:16:1 is applicable. This means that if the N:P atomic ratio of inorganic nutrients dissolved in the water exceed 16 the algal growth may be limited by phosphorus, and otherwise by nitrogen. In the ocean, limitation of phytoplankton biomass by iron has also been demonstrated, and the growth of certain taxa such as diatoms but not the overall production, may be limited by silicate. It is generally assumed that N is usually less harmful than P in inland waters where primary production is often limited by the latter, but it may cause problems when discharged to the seas.

Mineralization of organic matter by heterotrophic decomposers produces ammonium that in aerobic environments is oxidized to nitrites and nitrates by the chemotrophic nitrifying bacteria. Nitrate is the thermodynamically stable form of nitrogen in the well-oxygenated seawater. Except for a few species of cyanobacteria capable of N_2 fixation, nitrates are the most abundant source of nitrogen for primary producers in the sea. Other inorganic salts of N also available for phytoplankton are nitrites and ammonium, the latter the preferred source since cellular nitrogen is mostly in the same oxidation state (III) in the amine ($-NH_2$) groups of proteins.[1] Some forms of organic nitrogen such as urea and amino acids are also utilizable by phytoplankton to satisfy their nitrogen demands.

In anaerobic environments the denitrifying bacteria can mediate the reduction of nitrate to nitrogen gas, which in sediments displace the oxygen and contributes to further anoxia.

Primary production is limited by N and P

Marine Pollution. https://doi.org/10.1016/B978-0-12-813736-9.00003-9

Compared to nitrogen, the phosphorus cycle is much simpler, with a single inorganic form, the salts of the phosphoric acid, commonly orthophosphate ions, HPO_4^{2-}.

The vertical distribution of N and P shows depleted values in the photic zone and enrichment in deep waters

Mineral dissolved nutrients are depleted from the photic zone by primary producers and regenerated in deep waters by decomposers. As a result the vertical distribution of both nitrates and phosphates show a surface minimum, sharply increases with depth during the first 100–500 m, and it is approximately steady hence on. This typical nutrient-like behavior reflecting a dominant role for biological cycling is common to the distributions of several trace metals in the ocean (see Section 9.3). Table 3.1 shows typical nitrate and phosphate concentrations in natural rivers and deep seawater.

In marine surface waters there is a very high variability depending on geographical factors and oceanographic conditions. In coastal areas nutrient levels are enhanced by the supply from on land and from upwelling, while in central oceanic gyres concentrations of nutrients may be undetectable even for modern analytical methods. Therefore, there is not a constant background level of nutrients in surface marine waters, the range of natural scenarios overlaps with the water quality criteria established by different international agencies (Table 3.1), and the interpretation of measured concentrations must be conducted in light of complete environmental information to try to make the difference between anthropogenic impact and natural variability.

The main anthropogenic sources of N and P are fertilizers and detergents

The main source of anthropogenic nitrogen comes from the use of artificial **fertilizers** in agriculture, typically contributing 50–80 percent of the total load.[2] Nitrogen surplus from agricultural land, computed as fertilizer input minus harvest output, ranges in Europe from 10 to 200 Kg/Ha/year depending on the country.[3] Nitrate from fertilizers is very mobile in soil, and is easily leached to groundwater or surface water. Due to the implementation of national and international regulations such as the EU Nitrate Directive (1991), the levels of nitrate in inland waters have been reduced in some regions over the past

Table 3.1 Levels of Dissolved N and P Common in Different Kinds of Water Bodies

Nutrient	Inland Waters[a]	Atlantic Deep Seawater[b]	Threshold for Poor/ Bad Quality Status in Surface Seawater (EEA)	Raw Urban Wastewater[c]	Treated Urban Wastewater[d]
N	0.1–1 mg N/L	20 µM NO_3	>9/>16 µM NO_3	25–40 mg/L N	10–15 mg/L N
P	10–50 µg P/L; hypereutrophic: >96 µg/L	1.5 µM PO_4	>0.7/>1.1 µM PO_4	7–10 mg/L P	1–2 mg/L P

[a]Henry & Heinke (1996) op. cit.
[b]Chester & Jickells (2012) op. cit.
[c]EEA (2007) op. cit.; hypereutrophic: Carlson (2007) op. cit.
[d]Directive 98 /15 /EC.

15 years,[4] but this did not reflect in reduced N contents in coastal environments (see Fig. 3.2). Indeed, agriculture is a diffuse source, difficult to control and to quantify, and N levels in water courses are highly dependent on rainfall.

In contrast, the most important contributors to pollution by phosphorous are generally the point sources, such as wastewater treatment plants and industrial outlets. Detergents are the major source of P in municipal wastewater, although in many countries the phosphorus content of the detergents has been lowered by substitution with other substances (see Section 3.4). Industries producing fertilizers may emit quantities of P equal to the total emissions from small countries, though these emissions decreased significantly as a result of improved technology and wastewater treatment.[5]

The importance of different nitrogen sources and the identification of anthropogenic inputs in inert compartments (e.g., DIN, POM) or marine biota (e.g., phytoplankton, macroalgae) can be assessed from an isotopic standpoint. Nitrogen has two stable isotopes, the most common ^{14}N, and the heavier isotope with one additional neutron, the ^{15}N. The ratio of the N stable isotopes is termed $\delta^{15}N$ and this ratio changes as nitrogen circulates through certain metabolic routes since the light isotope may be mobilized faster; a process termed isotopic fractionation. Thus, nitrogen from sewage, groundwater, or fish farm discharges is often more enriched in ^{15}N than nitrogen from seawater. This is due to isotopic fractionation during nitrification and volatilization in the case of NH_4^+, or denitrification in the case of NO_3^-.[6] In contrast, nitrogen pools from most agricultural facilities are characterized by depleted $\delta^{15}N$ values, as they are synthesized from atmospheric N_2.[7] Besides nutrients from anthropogenic origin, different natural processes also affect inorganic nitrogen concentrations and their isotopic ratio. For instance, algae from mangrove habitats that were exposed to nitrogen derived from N_2 fixation were depleted in ^{15}N while those in habitats with frequent coastal upwelling were relatively enriched.[8]

Different N pools may be identified by their ^{15}N ratio

3.2 EUTROPHICATION AND HYPEREUTROPHICATION

The trophic status or **productivity** of aquatic ecosystems, in terms of organic carbon produced per unit of time and surface, shows very high natural variability. Waters with low productivity are termed oligotrophic and waters with high productivity eutrophic. Eutrophication can thus be defined as the process leading to eutrophic status, and it can be the result of an input of nutrients, but also to an increase in residence time, a decrease in turbidity or a decline in grazing pressure.

In lakes and other confined continental water bodies primary production is tightly controlled by the input of nutrients, and a scale of trophic status can be constructed as a function of either surface phosphorus or chlorophyll levels, since both variables correlate.[9] The load of nutrients, normally P, that the water body can admit is easily modelled as a function of physical characteristics such

Trophic status of aquatic ecosystems

as depth and flushing rate.[10] If this load is exceeded dystrophic scenarios leading to adverse ecological consequences take place. In this context, the states of **oligotrophic, mesotrophic, eutrophic, and hypereutrophic** correspond for P to <12, 12−24, 24−96, >96 µg/L, and for chlorophyll to the ranges <2.6, 2.6−20, 20−55, >55 µg/L.[11]

In marine systems the situation is more complicate and, although nitrogen is frequently considered as the limiting nutrient, simple quantitative relationships between nutrient input and primary production are not established. According to their levels of plankton chlorophyll, pelagic environments have been classified as oligotrophic (<0.1 µg/L), mesotrophic (0.1−1), and eutrophic (>1), approximately equivalent in terms of primary production to <50, 50−200, and > 200 gC/m^2 y.[12] The eutrophic condition in this case does not bear deleterious implications but results in increased productivity and ecological diversity. The issue will be further discussed later in the chapter.

Anthropogenic inputs of nutrients may lead to hypereutrophication

The biomass of primary producers is bottom-up controlled by nutrients availability and top-down controlled by consumption (grazing). Increased primary production due to nutrient supply may be channeled through consumers to increased production at higher trophic levels, with profound changes but no deleterious effects on the ecosystem. **Cultural eutrophication**[13] takes place when the nutrient supply is anthropogenic (agriculture fertilizers, detergents from domestic effluents, etc.). This is not inherently deleterious, and instances of increased marine productivity associated to anthropogenic enrichment of nutrients have been reported.[14] However, cultural eutrophication frequently takes place at such a large scale that the normal mechanisms of control of the plant standing crop by grazers are exceeded. In this case we talk of **hypereutrophication**.[15] Unfortunately, the terms cultural eutrophication and hypereutrophication are not consistently used, and many authors simply talk of eutrophication without any distinction of either natural or artificial causes, nor distinction between healthy increase in productivity and ecological unbalance leading to hypoxia.

Ecological effects of hypereutrophication

In hypereutrophic scenarios the organic biomass of decaying and not consumed primary producers causes oxygen depletion in bottom waters and hence **hypoxic** or even **anoxic environments**. Some of these areas where species richness is dramatically reduced were even called "dead zones" due to the absence of benthic macrofauna.[16] The ecological effects of hypoxia have been dealt with in depth in Section 2.4. Other potentially adverse ecological effects of hypereutrophication are: excessive biomass of phytoplankton (including harmful algal blooms) and macroalgae, shift of dominant phytoplankton species from diatoms to flagellates, increased dominance of green macroalgae (e.g., *Ulva, Cladophora, Enteromorpha*), blooms of gelatinous zooplankton, increased turbidity and reduction of light penetration, and reductions in harvestable fish and shellfish species.

Changes in phytoplankton species composition from dominance by diatoms to dominance by dinoflagellates has been linked to increased nutrient input.[17] Hypereutrophication-induced blooms of the dinoflagellate *Pfiesteria piscicida* caused massive fish mortalities in the coasts of North Carolina in the 1990s. Large algal blooms and excessive seaweed growth is also a nuisance for the recreational uses of the coasts, which is included within the standard definition of pollution (see Section 1.1). Several species of the unicellular alga *Phaeocystis* form colonies constituted by thousands of cells embedded in a matrix polysaccharides, and their blooms reach a very high biomass (up to $10 \text{ mg C } l^{-1}$)[18] of foam-like matter that yield affected beaches unpleasant for tourists.

Also, in continental waters, opportunistic species of algae include biotoxin producers that render the water not consumable, which is an important problem in reservoirs. However, it is important to remark that it is the excess of primary production, not the excess of nutrients, which causes the ecological unbalance, since nitrates and phosphates are innocuous for the organisms. Therefore, the oxygen depletion and further ecological effects of hypereutrophication are identical in nature to those derived from excessive organic matter, described in Chapter 2.

3.3 PREVENTION AND ABATEMENT OF HYPEREUTROPHICATION

Global spreading of cultural eutrophication-induced hypoxia stands along with overfishing and coastal zone management as one of the main marine environmental issues in political agendas. According to some sources, problems of hypoxia in marine systems lagged about 10 years behind the increased use of industrial N fertilizers that began in the 1940s and expanded during the 1960s and 1970s. In many regions of the North Adriatic, Baltic, and Black Sea, progressive hypoxia, either permanent or seasonal, has been recorded during those time periods. The issue becomes particularly relevant when commercial fisheries are affected, such as the decrease in catches of the Norway lobster in the Kattegat in 1983. Consequently, in Europe two legislative initiatives were promoted, the urban wastewater treatment Directive (91/271/EEC), and the nitrates Directive (91/676/EEC); the first promoting organic matter and when necessary, N and P removal from wastewaters, and the second targeting the pollution of waters caused by nitrates from agricultural sources. The nitrates Directive prompted the identification of vulnerable waters, good agricultural practices, and limitations of the land application of fertilizers. In 2007 The EEA reported that "The changes in point-source discharges are mainly due to improved treatment of urban wastewater […] the lowered phosphate content of detergents, and reduced industrial discharges. However, measures to reduce the nitrogen surplus from agricultural land have only had limited success so far."

P pollution was more effectively abated than N pollution

3.4   DOMESTIC DETERGENTS

Detergents form micelles that emulsify hydrophobic substances in water

Soap has been used for cleansing since ancient times, and it is produced from animal or plant fat through the reaction known as saponification, that consists of the hydrolysis of triglycerides with a strong base such as NaOH. The reaction produces the Na salts of the fatty acids and glycerol, a component that can remain in the soap as a softening agent. The fatty acid salts are amphiphilic molecules with a hydrophobic tail and a negatively charged polar head ($-COO^-$). These molecules are capable of washing out organic dirt by trapping it into micelles with the hydrophobic tails inside and the charged heads outside, thus diffusing in a polar medium such as water. Amphiphilic molecules with this property are also termed **surfactants**, from surface-active agents (Fig. 3.1).

Synthetic detergents replaced soap...

After World War II soap was progressively replaced by synthetic detergents that exhibited a higher cleaning performance.[19] This is because inorganic cations such as Ca^{2+} and Mg^{2+}, whose concentration is especially high in hard waters, bind to the polar heads and interfere with their cleaning action. To avoid this, commercial detergents, in addition to the surfactants, add to their composition anions called **builders** that sequester the calcium, magnesium, or iron

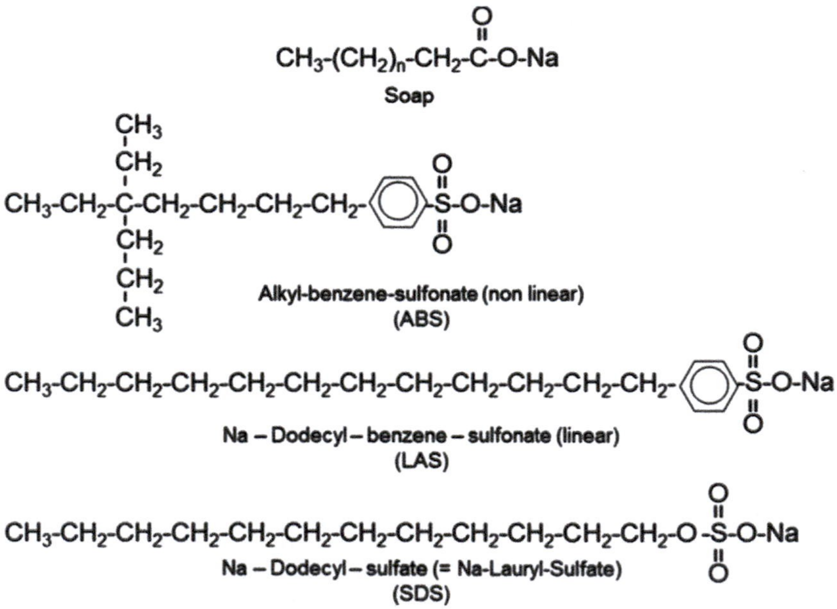

FIGURE 3.1

Molecular structures of common surfactants, all of them characterized by a polar group (on the right) and hydrophobic tails. The branched tails of the first synthetic surfactants, such as alkyl-benzene sulfonate (ABS), rendered these chemicals less biodegradable than current linear surfactants.

ions improving the overall washing performance. Polyphosphates such as tri-phosphoric acid ($H_5P_3O_{10}$) are the most common builders in commercial cleaning products. In water $H_5P_3O_{10}$ brakes down to $HPO_4{}^{2-}$ and $H_2P_3O_4{}^-$ ions, and this is why domestic detergents account for most of the inorganic phosphorus present in wastewaters.

Therefore, synthetic detergents may pose at least two completely different environmental problems in natural waters. One may stem from the environmental persistence and toxicity of synthetic surfactants. The first synthetic surfactants produced were ABS with branched apolar tails that caused **longer environmental half-lives**. The presence of surfactants in water bodies is very conspicuous because of the floating foam that causes public alert. Thus, ABS were progressively substituted by linear alkyl-benzene sulfonates (LAS) more easily biodegraded. Because of their interference with biological membranes, surfactants are toxic to aquatic organisms, particularly to naked early life stages. For example, the toxicity thresholds of surfactants for bivalve embryos and larvae generally range from 0.1 to 1 mg/L.

...but they are less biodegradable...

The second environmental problem is the hypereutrophication caused by the **excess of P** from the detergent polyphosphates that once in the water brake down to phosphate ions readily available to plants. In the 1960s laundry detergents contained about 10 percent phosphorus and continental waters suffered from excessive plant growth. The case of Lake Erie (North America) was well studied; the approximately 10,000 kg of phosphorus per day going into the lake resulted in about one-fourth the area of the lake with no oxygen within 3 m of the bottom.[20] As of 1967, mats of attached algae covered Lake Erie's shoreline, and desirable fish such as whitefish, blue pike, and walleye had either severely declined or disappeared altogether. This and other case studies prompted legislative initiatives aimed at, (1) substituting phosphates by alternative builders such as zeolites or citrate in commercial detergents, and (2) improving the depuration of urban wastewaters by introducing specific depuration steps for P removal. These initiatives have been fruitful according to temporal trend studies. For example in both the North Sea and Baltic Sea (Fig. 3.2) more than 30% of the studies for the 1985—2000 period reported decreasing trends of P, in contrast with the lack of progress in the combat of excess N.[21] As an unexpected result the N/P ratio is increasing in this coastal ecosystems.

...pollute water with P...

A third environmental problem related to synthetic detergents and other cleaning products is the environmental pollution with an estrogenic endocrine disrupter, the **nonylphenol** (see also Section 14.3). Alkylphenol ethoxylates are used as nonionic surfactants for several commercial applications, including detergents, and their degradation in aquatic environments may produce nonylphenol. This molecule is classified as a priority pollutant in Europe, and its commercial use has been restricted by Directive 2003/53/EC.

...and may contain toxic chemicals

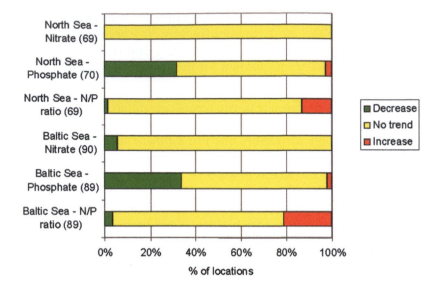

FIGURE 3.2

Temporal trends in the nitrate and phosphate levels in the North Sea and Baltic Sea. Notice the positive evolution of phosphates compared to nitrates and the resulting overall trend to increase the N/P ratios in both areas. *Source: European Environment Agency.*

KEY IDEAS

- When the input of mineral nutrients carried in residual waters blooms plant production above the levels that can be consumed by herbivores, the decomposition of the plant biomass excess causes lack of oxygen and presence of toxic reduced substances.

- This can be avoided by eliminating P from the wastewaters through specific tertiary treatments. The application of this treatment to waste water was successful in reducing P loads in some coastal environments.

- Synthetic detergents are composed by surfactants that mobilize organic particles in water by forming micelles, builders that facilitate the action of the surfactant by sequestering Ca^{2+} and Mg^{2+}, and different softeners and fragrances. Phosphates are the main detergent builders.

Endnotes

1. Chester R, Jickells T. Marine Geochemistry 3rd ed. Wiley-Blackwell; 2012. (p. 164).

2. EEA. Europe's environment. The fourth assessment. European Environment Agency, Copenhagen; 2007. (p. 107).

3. EEA (2007) op. cit. (p. 200).

4. EEA 2011. Europe's environment. An assessment of assessments. European Environment Agency, Copenhagen. (p. 83).

5. EEA 1998. Europe's environment: the second assessment. European Environment Agency.Copenhagen.

6. Montoya JP. Nitrogen stable isotopes in marine environments. In: Capone DG, Bronk DA, Mulholland MR, Carpenter EJ, editors. Nitrogen in the marine environment. San Diego, Academic Press; 2008. p 1277−1302.

7. Heaton THE. Isotopic studies of nitrogen pollution in the hydrosphere and atmosphere: a review. Chem Geol 1986;59:87−102.

8. Lamb K, Swart PK, Altabet MA. Nitrogen and carbon isotopic systematics of the Florida Reef Tract. Bull Mar Sci 2012;88:119−146.

9. Vollenweider. Quoted by Gray JS. Eutrophication in the sea. In: G Colombo, I Ferrari, VU Ceccherelli, R Rossi, editors. Marine eutrophication and population dynamics. 1992. p. 3−15.

10. Carlson RE. A trophic state index for lakes. Limnol Oceanogr 1977;22(2):361−369.

11. Carlson RE. Estimating trophic state. Lakeline. Spring 2007;2007:25−29.

12. Chester & Jickells (2012) op. cit. (p. 173 and 176).

13. Freedman, B. Environmental ecology, 2nd ed. Academic Press;1995. ((author's note: if needed, the city is San Diego)).

14. For example in the Seto Sea (Japan); see p. 40 in: Clark RB. Marine Pollution 5th ed. Oxford University Press, Oxford; 2001.

15. e.g. Benoit RJ. Self-purification in natural waters. In: Ciaccio LL, editor. Water and water pollution handbook. N.Y: Marcel Dekker Inc; 1971.

16. Diaz RJ, Rosenberg R. Spreading dead zones and consequences for marine ecosystems. Science 2008;321:926−929.

17. Bennekom et al. Quoted by Gray (1992) op. cit.

18. Schoemann V, Becquevort S, Jacqueline Stefels J, et al. *Phaeocystis* blooms in the global ocean and their controlling mechanisms: a review. Journal of Sea Research January 2005;53(1−2): 43−66.

19. Knud-Hansen C. 1994. http://www.colorado.edu/conflict/full_text_search/AllCRCDocs/94-54.htm.

20. Beeton AM. 1971. Eutrophication of the St. Lawrence Great Lakes. In: Detwyler TR, editor. Man's Impact on Environment. McGraw-Hill Book Co., New York, p. 233−245.

21. http://www.eea.europa.eu/data-and-maps/figures/trends-in-nitrate-and-phosphate-concentrations-and-n-p-ratio-in-the-north-sea-and-the-baltic-sea. See also: Schulz S, Ærtebjerg G, Behrends G et al. The present state of the Baltic Sea pelagic ecosystem − an assessment. p. 35−55. In: Colombo G, Ferrari I, Ceccherelli VU, Rossi R, editors. Marine eutrophication and population dynamics. Olsen & Olsen; 1992. p. 35−55. ((author's note: the city is Fredensborg))

Suggested Further Reading

• Gray JS. Eutrophication in the sea. In: Colombo G, Ferrari I, Ceccherelli VU, Rossi R, editors. Marine eutrophication and population dynamics. Olsen & Olsen; 1992. p. 3−15.

• Diaz RJ, Rosenberg R. Spreading dead zones and consequences for marine ecosystems. Science 2008;321:926−9.

CHAPTER 4

Microbial Pollution

4.1 PATHOGENIC MICROORGANISMS PRESENT IN MARINE WATERS

Contamination of natural waters by human sewage, including urine and feces of people suffering from infectious diseases, poses a risk to public health. The so-called **waterborne diseases** can be transmitted after contact with skin erosions or mucosa, deliberate or accidental ingestion of water, or consumption of contaminated seafood. Infectious organisms capable of causing human illness when present in water (Table 4.1) include viruses (0.01–0.1 µm), bacteria (0.1–10 µm), protozoa (1–100 µm), and helminths (from 1 µm to several cm). These organisms can produce respectively virions, spores, cysts, and eggs, which are environmentally resistant infectious stages largely more resistant to disinfectants than the reproductive stages.

> Viruses, bacteria, protozoa, and helminths cause waterborne infectious diseases

Several groups of viruses transmitted by the fecal–oral route are shed to natural waters via feces and may infect persons after ingestion of contaminated water or aquatic organisms. Of these, the norovirus and hepatitis A virus are currently recognized as the most important human food-borne pathogens regarding the number of outbreaks and people affected in the Western world.[1] Norovirus, a single-strand RNA virus member of the family *Caliciviridae* transmitted by fecally contaminated food or water, is the main cause of food-borne outbreaks of human gastroenteritis worldwide. The picornavirus are also single-strand RNA viruses. The genus Enterovirus are found in the intestine and can cause from mild infections to serious illness. The genus Hepatovirus cause infectious hepatitis A. The Reovirus (Reo from "Respiratory Enteric Orphan") family comprises double stranded RNA viruses that cause infections of the gastrointestinal system and the respiratory tract. When freely suspended in seawater they remain infectious for only a few days, but when attached to suspended solids they can retain the infectious quality for several weeks.[2] The adenovirus family comprises double stranded DNA viruses that normally cause infection of the eyes and respiratory tract, characterized by prolonged survival in the water.

Pathogenic species of *Salmonella*, *Shigella*, *Yersinia*, *Campylobacter*, and *Vibrio* bacteria are responsible for well-known waterborne diseases such as typhoid and cholera, or other diseases of fecal–oral transmission affecting the

Marine Pollution. https://doi.org/10.1016/B978-0-12-813736-9.00004-0

Table 4.1 Waterborne Human Pathogens

Group	Pathogen	Disease	Source
Viruses			
Fam. *Caliciviridae*	Gen. Norovirus (Norwalk virus)	gastroenteritis	Ingestion of polluted water
Fam. *Picornaviridae*	Gen. Enterovirus (polio, echo, coxsackie)	Meningitis, paralysis, rash, myocarditis, respiratory disease, gastroenteritis	Ingestion of polluted water
	Gen. Hepatovirus	Hepatitis A	Ingestion of polluted water or shellfish
Fam. *Hepeviridae*	Gen. Orthohepevirus	Hepatitis E	Ingestion of polluted water or food
Fam. *Reoviridae*	Gen. Rotavirus	gastroenteritis, respiratory infections	Ingestion of polluted water
Fam. *Adenoviridae*	Adenovirus	eye infections, respiratory disease, gastroenteritis	Bathing water
Bacteria			
Enterobacteriaceae	*Salmonella, Shigella, Yersinia enterocolitica, Yersinia pestis, Escherichia coli* (certain strains)	Typhoid, gastroenteritis, plague	Ingestion of polluted water or food
Campylobacteraceae	*Campylobacter*	gastroenteritis	Food
Vibrionaceae	*Vibrio cholera*	cholera	Ingestion of polluted water (cholera, typhoid)
Protozoa			
	Entamoeba histolytica	Amoebic dysentery	Ingestion of polluted water or food
	Giardia lamblia, Cryptosporidium	diarrhea	Ingestion of polluted water or food
Helminths			
Nematodes	*Ascaris lumbricoides, Trichuristrichiura*	Parasites of human intestine, lungs, and blood	Drinking water
	Ancylostoma spp.	Parasites of human intestine	Bathing water (penetration through skin or mucosa)
Trematodes	*Schistosoma mansoni*	Parasite of blood	Bathing water (penetration through skin or mucosa)
Cestodes	*Taenia saginata*	Parasite of human intestine	Drinking water

gastrointestinal system and frequently associated to poor hygienic conditions. None of them forms spores and have short persistence in natural waters. In contrast, the protozoan parasites of the genus *Giardia* and *Cryptosporidium* form cysts infective after ingestion that can survive for months in water and are more resistant to chlorination than enteric bacteria. The oocysts of *Cryptosporidium* have proved to be the most resistant enteric pathogen to inactivation by common disinfectants of any known so far. Concentrations of 1−2 mg/L of chlorine are not sufficient to kill the organism.[3]

Eggs of parasite nematodes (*Ascaris*, *Ancylostoma*, *Trichuris*, etc.), trematodes (*Schistosoma*) and cestodes (*Taenia*), excreted with the feces of infected individuals, are also very resistant to disinfection and environmental stresses, and may persist in both crude and treated sewage. Depending on the helminth species the infection may be transmitted by drinking waters or direct contact with skin and mucosa in bathing freshwaters.

Since it would not be feasible to try to identify in the environment every kind of pathogenic microorganism transmitted from human wastes, the most abundant enteric bacteria, *E. coli*, is universally used as **indicator** of fecal contamination. Each person liberates 300,000 million *E. coli* daily. Since this nonpathogenic symbiotic species has, in addition, a similar or even higher resistance to environmental conditions than disease-causing bacteria, its absence from water is an indication that water is bacteriologically safe. Conversely, the presence of this species, nonexistent as free-living form, is indicative of fecal contamination and thus potential presence of any of the other enteric organisms listed in Table 4.1 capable of causing disease.

> The enteric bacteria *Escherichia coli* is universally used as indicator of fecal pollution...

However, the use of *E. coli* as universal indicator has limitations. Namely, they do not specifically detect fecal pollution of human origin, and their environmental resistance is much lower than that of other important pathogens such as protozoans and viruses. Therefore, alternative fecal indicators have been proposed, including *E. coli* bacteriophages; viruses that infect only this species of bacteria; the pathogen bacteria *Clostridium perfringens*, a spore-forming bacteria with longer survival times in the marine waters; and other fecal bacteria such as enterococci (see Section 4.2). Coliphages and *C. perfringens* have lower decay rates than fecal coliforms, and are thus better indicators of remote fecal sources.[4] **Enterococci** are a more accurate indicator than coliforms because they are more closely associated with human rather than with animal fecal matter and survive longer in aquatic environments.[5] In addition, enterococci depuration rates from polluted mussels are more similar to those of pathogenic bacteria such as *Vibrio cholera* than *E. coli* depuration rates, which are more rapid.[6]

> ... but spore-forming bacteria, protozoans, and viruses are more persistent than *E. coli* in natural waters

Current water quality standards and mollusk depuration controls are only based on bacterial indicators. However, virus removal is known to be less effective than bacterial removal and the compliance with standards cannot guarantee the viral absence (see Section 4.3), a fact evidenced by the periodic outbreaks of hepatitis A and gastroenteritis following the consumption of depurated shellfish. Conventional water treatment methods reduce but do not eliminate human infectious viruses from wastewaters. Also, not all bivalve species are equally prone to depuration, and bivalves living in the sediments where viruses are known to accumulate, such as clams, showed slower depuration rates and higher contamination levels for viruses than mussels.[7]

4.2   MICROBIOLOGICAL ANALYSIS OF NATURAL WATERS; BATHING WATER REGULATIONS

For health issues, recreational waters are monitored for microbiological pollution every year

Waterborne (transmitted by ingestion through the fecal—oral route) or water-based (transmitted from pathogens that live in water and thus not necessarily of fecal origin) diseases have been traditionally abated by filtration and disinfection of drinking water with chemicals such as chlorine, a procedure introduced in large cities in the 1910s, with a resulting remarkable decline in the typhoid dead rates. However, transmission of infectious diseases can also be a problem for other uses typical of coastal waters such as recreational bathing (Section 4.2) and harvesting of shellfish (Section 4.3). Many microorganisms of intestinal origin can cause in bathers from mild infections of ear and upper respiratory system to serious illness. For that reason, bathing waters are monitored every year, especially during the tourist season.

Bacterial loads in water may be assessed using serial dilutions or membrane filter incubations

Quantification of enteric bacteria for monitoring natural waters may be conducted by two methods: the most probable number (MPN) test, normally applied to samples with high loads of coliforms, and the membrane filter test, with highest precision at low levels of microbial pollution. MPN begins with a presumptive test where serial dilutions of the problem sample are incubated in a prescribed liquid culture medium with lactose (e.g., lauryl sulfate tryptose broth) in test tubes (five per dilution) incubated at 35°C. After 24 and 48 h positive tubes are identified by production of gas trapped in a small inverted inner tube and acid, and a confirmed test is conducted using specific media. Official standard methods take advantage for the confirmative step of the β-glucuronidase activity typical of *E. coli*.[8] The number of positive tubes at each dilution is then translated into a statistical estimation of the MPN of bacteria per 100 mL using the appropriate tables.

Fecal enterococci are differentiated from other streptococci by their ability to grow in 6.5 percent NaCl and at high pH (9.6) and temperature (45°C). *Entero-coccus faecalis* and *Enterococcus faecium* are the species most frequently found in humans.[9] *Enterococci* are especially reliable as indicators of health risk caused by fecal pollution in marine environments and recreational waters.

Due to its higher precision detecting individual colony-forming units (CFU) the membrane filter test (Fig. 4.1) is more suitable for waters with low bacterial numbers. In this method a measured amount of water—100 mL for presumably clean natural waters or lower volumes otherwise—is passed through a membrane filter (0.45 microns pore size) later placed on an absorbent pad saturated with the appropriate selective and differential culture media and incubated at 35°C for 24 h. Chromogenic media incorporate substrates specific for enzymes typical of each bacterial group and thus total coliform and *E. coli* colonies can be distinguished by its color and independently counted with the aid of the filter grid.

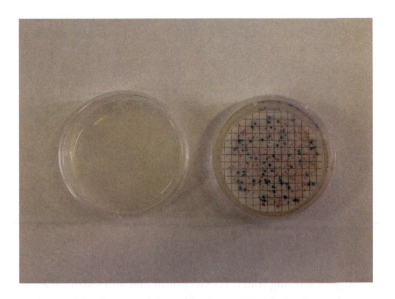

FIGURE 4.1

Sample of water for microbiological testing using the membrane filter method and a selective medium with bile salts and two chromogenic substances, one that develops salmon to red color in presence of β-galactosidase activity, present in all coliforms (total coliforms: reddish dots), and another that develops blue color in the presence of β-glucuronidase activity, specific for *Escherichia coli* (blue dots). *Source: Beiras R, Pérez S. Manual de métodos básicos en contaminación mariña costeira. Servizo de Publicacións, Universidade de Vigo; 2011.*

Traditional microbiological tests used the MPN method and targeted the presence of so-called total coliform bacteria that includes genus *Escherichia, Citrobacter, Enterobacter,* and *Klebsiella,* all easy to detect because they are Gram-negative, nonspore-forming, rod-shaped bacteria that produce gas upon lactose fermentation at 35°C within 48 h. Since total coliform counts include naturally occurring bacteria of no intestinal origin, the so-called fecal coliform counts were additionally conducted on a defined selective medium at 44°C. However, we know that certain nonintestinal thermotolerant bacteria, such as some *Klebsiella* biotypes, are incorrectly counted as fecal by this method. A systematic study conducted by US-EPA in weekend swimmers at recreational beaches found no correlation between both total or fecal coliforms and the incidence of gastrointestinal illnesses in the swimmers over 7–10 days after their visit to the beach. *E. coli* counts were positively correlated with illness rates (r = 0.54) but not significantly, and only Enterococcus showed a statistically significant correlation coefficient (r = 0.75; $P < .001$).[11] In addition, intestinal enterococci, which are Gram positive human enteric bacteria, were reported to be more persistent than *E. coli* in natural waters, especially in marine environments (see Section 4.1). As a consequence, the use of total coliforms and fecal coliforms MPN values was phased out, and current water quality criteria for recreational waters are based on *E. coli* and enterococci counts in selective chromogenic media.

Current tests of microbiological water quality are based on *E. coli* or enterococci counts

Table 4.2 Microbiological Quality Criteria for Coastal and Transitional Bathing Waters in the EU (Directive 2006/7/EC) and For Marine Recreational Waters in the United States (US-EPA, 2012)[12]

Directive 2006/7/EC	"Excellent"	"Good"	"Sufficient"
Intestinal enterococci (cfu/100 mL)	95th percentile < 100	95th percentile < 200	90th percentile<185
Escherichia coli (cfu/100 mL)	95th percentile < 250	95th percentile < 500	90th percentile<500
US-EPA, 2012	**"estimated illness rate": 32‰**	**"estimated illness rate": 36‰**	
enterococci (cfu/100 mL)	STV ≤ 110; 30-d geometric mean ≤ 30	STV ≤ 130; 30-d geometric mean ≤ 35	

STV, *statistical threshold value; it approximates the 90th percentile.*

According to Directive 2006/7/EC, European member states must yearly identify all bathing waters (both continental and marine), and classify them according to their microbial quality following the limits shown in Table 4.2 for transitional and coastal waters, slightly more stringent than for inland waters. Classification is conducted comparing percentiles of data accumulated from at least eight previous sampling dates to the criteria approved in the Directive. It should be noticed that waters may be classified as "Poor" (requirements for "Sufficient" not met) and bathing remains allowed if the competent authority so decides, as long as a warning signal alerting the public was shown. Apart from the microbial quality, when cyanobacterial proliferation or presence of solid wastes occur and a health risk had been identified or presumed, adequate management measures shall be taken, including information to the public. In the US, marine recreational waters are assessed on the basis of enterococci, while either enterococci or *E. coli* may be used for freshwaters. Geometric means of 30-d intervals are used and each state may choose between two levels of protection, associated to theoretical 32‰ and 36‰ prevalence of waterborne illness, to set their own standards (Table 4.2).

4.3   MICROBIAL SAFETY OF SHELLFISH AND OTHER FOOD OF MARINE ORIGIN

Molluskan shellfish safety is assured through microbiological tests compulsory prior to placing the shellfish in the market

Molluskan shellfish are one of the few animal food products frequently eaten raw or lightly cooked, and they are consumed whole, including gills and digestive tissues. In addition, they are active filter feeders, exposed to hundreds of liters of water per day, and able to concentrate from water all kinds of contaminants within their edible tissues: chemicals and biotoxins, but also parasites and infectious bacteria and viruses.[13] Outbreaks of infectious diseases associated with the consumption of oysters and clams in the US include typhoid fever, hepatitis A, cholera and other *Vibrio*-caused enteric pathologies, and Norwalk virus poisoning.[14] On the other hand, shellfish have remained a valuable source of protein and a highly appreciated and commercially important food in many coastal cultures worldwide. As a result, a wealth of regulations and controls involving the harvesting and commercialization of mollusks and other seafood were approved to avoid hazards to human health caused by these products. Seafood was the first food commodity in the United States to utilize a science-based system of preventive food safety controls. In Europe, current regulations include controls on the levels of marine biotoxins, chemical contaminants (Regulation EC 1881/2006, for Hg, Cd, Pb, benzo-*a*-pyrene, and other chemicals), and microbial contamination. According to Regulation EC 853/2004, live shellfish (bivalves, gastropods, echinoderms, and tunicates) may not be placed on the market for retail sale otherwise than via a dispatch center, where an identification mark must be applied to ensure these controls. The shellfish batch dispatched for commercialization or relying must be accompanied by a registration document indicating among other data: date of harvesting, location, and health status of the production area.

In the United States, shellfish growing areas are classified according to the microbiological quality of water into "approved" (geometric mean ≤ 70 total coliform or ≤ 14 fecal coliform MPN/100 mL), "restricted" (70—700 TC or 14 to 88 FC MPN/100 mL), and "prohibited" (>700 TC or >88 FC MPN/100 mL).[15] Shellfish from "restricted" areas cannot be harvested for direct human consumption but they may be placed on the market following depuration in land-based facilities, relaying, or heat processing. Depuration must achieve fecal coliform counts in the shellfish below 20 FC/100 g (geometric mean) for mussels, oysters, Manila clams, and Hard clams (*Mercenaria mercenaria*), and below 50 FC/100 g for Soft clams (*Mya arenaria*). If shellfish-growing areas meet the microbiological water quality standards only for certain periods because of predictable pollution events such as increased population during touristic season or increased runoff after heavy rainfall, authorities may classify them as "conditionally approved" or "conditionally restricted," and they may be harvested during periods when they meet the standards subject to a management plan.

In Europe, regulations focus on routine analysis of shellfish itself, including levels of *E. coli* and identification of *Salmonella*. Bivalve production areas are classified following Regulation EC 854/2004 in one of three categories (Table 4.3). Class A are areas from which live bivalve mollusks may be collected for direct human consumption because they meet the microbiological criteria established in Regulation EC 2073/2005, namely <230 *Escherichia coli* MPN in 100 g of flesh and intravalve liquid and absence of *Salmonella* in 25 g, provided a series of additional organoleptic requirements and biotoxin limits are also met. Class B are areas from which bivalves can be collected but need purification prior to be placed on the market to meet the above referred microbiological criteria. In these areas, mollusks must not exceed in a 5-tube, three

Bivalve producing areas are classified according to their microbiological quality

Table 4.3 Classification of Shellfish Growing Areas According to Their Microbiological Quality in Europe (Regulations EC 854/2004 and EC2073/2005) and the United States (U.S. FDA, 2015)

EU Regulations	"Class A"		"Class B"	"Class C"
Controls in shellfish meat	$\leq$230 *E. coli* MPN in 100 g Absence of Salmonella in 25 g		$\leq$4,600 *E. coli* per 100 g	$\leq$46,000 *E. coli* per 100 g
US Regulations	**"approved"**		**"restricted"**	**"prohibited"**
Controls in seawater	geometric mean $\leq$ 70 total coliform or $\leq$14 fecal coliform MPN/100 mL		70 to 700 TC or 14 to 88 FC MPN/100 mL	>700 TC or >88 FC MPN/100 mL

dilution MPN test 4600 *E. coli* per 100 g. Class C are areas from which bivalves can be collected but placed on the market only after relying over a long period so as to meet Class A requirements. In these areas mollusks must not exceed 46,000 *E. coli* per 100 g, otherwise harvesting is prohibited. Mollusks from Class B or C production areas can be sent to a processing establishment were they are boiled or sterilized by some other procedure.

The sampling program established to classify the production areas must be designed with a number of samples, a geographical distribution of the sampling points and a sampling frequency which must ensure that the results are as representative as possible for the area considered. Other legal imperatives for the classification of the production areas, sometimes overlooked by the competent authorities, include: making inventories of the sources of pollution, quantifying the organic chemical pollutants in the mollusks, and determining the physical patterns of circulation and distribution of pollutants in the production area.

Marine biotoxins are also monitored

The presence in the bivalves of marine biotoxins produced by planktonic species is also routinely monitored. In Europe Regulation EC 853/2004 imposes limits to the bivalve contents of paralytic shellfish poison (PSP), amnesic shellfish poison (ASP), okadaic acid, dinophysistoxins, pectenotoxins, yessotoxins, and azaspiracids. The US FDA imposes similar limits to PSP, ASP, okadaic acid, and neurotoxic shellfish poison (NSP). Biotoxins are synthetized by naturally occurring plankton species and are thus not considered as pollutants according to the internationally accepted definition, which states that pollutants must have anthropogenic origin (see Section 1.1).

Detection of viral contamination and complete disinfection of wastewaters remain unsolved issues

The applicability of the methods currently available for monitoring microbial food safety to the control of viral contamination has been challenged. No consistent correlation has been found between the presence of indicator microorganisms (i.e., bacteriophages, *E. coli*) and intestinal viruses.[16] Enterovirus and hepatitis A virus were detected in samples of bivalves with <10 *E. coli* UFC/100 g meat and, in contrast, samples that were heavily contaminated by *E. coli* were free of viruses.[17] A number of bacteriophages (viruses that infect bacteria and frequently a single bacterium species) have shown promise as indicators of microbiological quality of water. They are enumerated as plaque-forming units (PFU) like the membrane filtration method described earlier. Serial dilutions of the water sample are filtered through 0.45 µm pore size membranes later placed on an overlay plate culture of the host bacteria and incubated at 37°C overnight, and the number of PFU counted as "holes" (plaques) in the stained overlay.[18] More advanced techniques involve the identification of viral RNA from a specific species or group by using reverse transcriptase-polymerase chain reaction (RT-PCR).[19]

Microbial pollution originated from urban wastewater is currently one of the main environmental threats in the shellfish-producing areas.[20] Incomplete disinfection of fecal waters, even in effluents from wastewater treatment plant (WWTP) with tertiary treatment, is commonplace. Disinfection requires previous elimination of organic matter and suspended solids, which is not always achieved, for example, after strong rainfall events. Also, ultraviolet or ozone-based disinfection methods (see Section 5.4) are costly due to electricity consumption. However, these costs should be balanced against those generated by the intensive surveillance programs implemented to guarantee consumer health and by the depuration procedures needed for microbiologically polluted mollusks.

KEY IDEAS

- Advances in sanitation strongly decreased the incidence of waterborne infectious diseases such as typhus or cholera. In Western countries norovirus and hepatitis A virus are currently recognized as the most important human pathogens transmitted by the fecal—oral route in terms of number of people affected.

- Microbiological quality of water was traditionally assessed in terms of total and fecal coliform MPN values. Specific identification of *E. coli* is preferred because this species is of exclusively fecal origin. Use of enterococci shows the additional advantages of being a more selective indicator of human origin and showing longer environmental persistence than *E. coli* in seawater.

- Since some bivalves are consumed raw or lightly cooked they are a vector of infectious diseases to humans. To prevent this problem, shellfish growing areas are classified according to their microbiological quality, and shellfish from nonoptimal areas cannot be directly placed in the market but must be naturally or artificially depurated. The classification is based on the analysis of the shellfish meat in the EU but on water quality in the United States.

- Spore-forming bacteria, protozoans, and viruses are more resistant to disinfection methods than *E. coli*, and thus may pass unnoticed in standard microbiological quality tests based on the latter.

Endnotes

1. Koopmans M, Duizer E. Foodborne viruses: an emerging problem. Int J Food Microbiol 2004; 90:23–41.

2. Rao VC, Seidel KM, Goyal SM, et al. Isolation of enteroviruses from water, suspended solids, and sediments from Galveston Bay: Survival of poliovirus and rotavirus adsorbed to sediments. Appl Environ Microbiol 1984;48(2):404–409.

3. Gerba CP, Pepper IL. Chapter 11. Microbial contaminants. In: IL Pepper, C.L. Gerba, M.L. Brusseau (eds.) Environmental and Pollution Science. 2nd ed. London, Academic Press; 2006. pp. 144–169.

4. Lucena F, Araujo R, Jofre J. Usefulness of bacteriophages infecting *Bacterioides fragilis* as index microorganisms of remote faecal pollution. Water Res 1996;30(11):2812–2816.

5. Griffin DW, Lipp EK, McLaughlin MR, et al. Marine recreation and public health microbiology: quest for the ideal indicator. BioScience 2001;51(10):817-825.

6. Marino A, Lombardo L, Fiorentino C, et al. Uptake of *Escherichia coli, Vibrio cholera* non-O1 and *Enterococcus durans* by, and depuration of mussels (*Mytilus galloprovincialis*). Int J Food Microbiol 2005;99:281–286.

7. Romalde JL, Area E, Sánchez G, et al. Prevalence of enterovirus and hepatitis A virus in bivalve molluscs from Galicia (NW Spain): inadequacy of the EU standards of microbiological quality. Int J Food Microbiol 2002;74:119–130.

8. For waters see: ISO 9308-1:2014 Water quality - Enumeration of *Escherichia coli* and coliform bacteria - Part 1: Membrane filtration method for waters with low bacterial background flora. For food see: ISO TS 16649-3. Microbiology of food and animal feeding stuffs - Horizontal method for the enumeration of beta-glucuronidase-positive *Escherichia coli* - Part 3: Most probable number technique using 5-bromo-4-chloro-3-indolyl-beta-D-glucuronide.

9. Scott TM, Rose JB, Jenkins TM, et al. Microbial Source Tracking: Current Methodology and Future Directions. Appl Environ Microbiol 2002;68(12):5796–5803.

10. Beiras R, Pérez S. Manual de métodos básicos en contaminación mariña costeira. Servizo de Publicacións, Universidade de Vigo; 2011.

11. Cabelli VJ. Health effects criteria for marine recreational waters. EPA-600/1-80-031. Research Triangle Park, NC; 1983. 98 pp.

12. U.S. EPA. Recreational water quality criteria; 2012. Office of Water 820-F-12–058.

13. Richards GP. The evolution of molluscan shellfish safety. In: Villalba A, Reguera B, Romalde JL, Beiras R, editors. Molluscan shellfish safety. Xunta de Galicia-intergovernmental oceanographic comission of UNESCO; 2003. pp. 221–245.

14. US FDA. National shellfish sanitation program. Guide for the control of molluscan shellfish. 2015 Revision; 2015.

15. US FDA. (2015). op. cit.

16. Koopmans & Duizer (2004) op. cit.

17. Romalde et al. (2002). op. cit.

18. Sobsey MD, Schwab KJ, Handzel TR. A simple membrane filter method to concentrate and enumerate male-specific RNA coliphages. J Am Water Works Assoc. 1990;82(9):52–59.

19. Griffin et al. (2001) op. cit.

20. Fernández E, Álvarez-Salgado XA, Beiras R, et al. Coexistence of urban uses and shellfish production in an upwelling-driven, highly productive marine environment: The case of the Ría de Vigo (Galicia, Spain). Regional Studies in Marine Science 2016;8:362–370.

Suggested Further Reading

- Gerba CP, Pepper IL. Chapter 11. Microbial contaminants. In: Pepper IL, Gerba CL, Brusseau ML, editors. Environmental and pollution science. 2nd ed. Amsterdam: Elsevier; 2006. p. 144—69.
- Griffin DW, Lipp EK, McLaughlin MR, Rose JB. Marine recreation and public health microbiology: quest for the ideal indicator. Bioscience 2001;51(10):817—25.
- Romalde JL, Area E, Sánchez G, et al. Prevalence of enterovirus and hepatitis A virus in bivalve molluscs from Galicia (NW Spain): inadequacy of the EU standards of microbiological quality. Int J Food Microbiol 2002;74:119—30.

CHAPTER 5

Liquid Wastes: From Self-purification to Waste Water Treatment

5.1 URBAN SEWAGE AND SELF-PURIFICATION OF NATURAL WATERS

As discussed in Chapter 1, the impact humans have had on the environment is primarily a function of population density. The agglomeration of humans in large nuclei, the cities, associated to all civilizations, generated a change in the scale of human pressure per surface unit, and brought about the need to artificially provide basic services formerly supplied by nature, such as drinking water and sanitation. Ancient Eastern civilizations in Egypt, Mesopotamia, and the Indus Valley were all located near large rivers and exploited water courses to build up drinking water supplies and sanitation to houses, including covered sewer networks. Hygienic means of human waste disposal were also known in the Minoic culture (Crete) 3000 years ago, and the *cloaca maxima* (the "biggest sewer") in Imperial Rome had enough capacity to serve a city of 1 million people. The aquatic ecosystems carried away and diluted liquid wastes, and eliminated the deleterious components of wastes by a set of natural processes known as self-purification capacity, discussed later in the chapter.

> Sewerage substituted natural water courses as the means to get rid of liquid wastes

Ulterior Western cultures trailed way behind. By mid-19th century the Thames River was a pestilent open sewage, leading to frequent outbreaks of waterborne diseases such as cholera and typhoid. These outbreaks were thought to be transmitted by airborne "miasmas" until Dr. Snow associated one of them to a particular drinking water pump in Broad Street. These public health problems led to a build-up in London of some 21,000 km of pipes, one of the first modern **sewerage** systems, along with cities such as Hamburg, Brooklyn, and Boston.

Sewerage with direct discharge to surface waters without any treatment ameliorated public health in the large cities but translated the issue downstream. Modern population densities are too high to make the transference and natural dilution strategy a viable option to manage liquid wastes. The discovery of **activated sludge depuration** in 1912 and the generalization of wastewater treatment before discharge in **wastewater treatment plants (WWTP)** eventually succeeded in drastically reducing the incidence of waterborne diseases, in particular those of fecal−oral transmission (see Section 4.1).

> Sewerage led to the need of treatment before discharge

53

Marine Pollution. https://doi.org/10.1016/B978-0-12-813736-9.00005-2

Generalization of complete wastewater depuration in WWTPs allows recovery of receiving waters

Besides health issues, urban liquid wastes now concentrated in the sewerage systems of large cities caused on receiving waters poor environmental quality in terms of organic load and the resulting deficit in **dissolved oxygen (DO)** (see also Chapter 2). In the 1930s the DO values in New York City's Lower East River bottom water were as low as 20 percent saturation as a consequence of discharging c. 500 tons of Biological Oxygen Demand (BOD) daily from untreated sewage. By the end of the 1980s bottom water DO values had recovered to 40%−50% saturation as a result of construction of new WWTPs and generalization of secondary treatment, mandated by the Clean Water Act from 1972.[1] DO values continued to improve in the following decades (Fig. 5.1) to nearly reach preindustrial levels in association to complete wastewater treatment, including tertiary treatment for nutrient removal. As shown in Fig. 5.1, other water quality indicators, such as fecal microorganisms, also improved as a result of the elimination of illegal discharges, and a reduction of combined sewer overflows through increased capture and treatment.[2] Present-day generalization of wastewater treatment is responsible for the improvement of the surface water quality parameters in most Western countries. In European rivers, average BOD_5 decreased from 3.4 to 2.1 mg/L ($n = 588$ sampling stations) and NH_4^+ from 0.22 to below 0.1 mg/L (n = 902) over the period 1992 to 2004, reflecting the general improvement in sewage treatment as a result of the EU's Urban Waste Water Directive,[3] but also due to a decline in polluting manufacturers after the economic recession of the 1990s.[4]

Urban waste water treatment still faces many challenges. In most cities old facilities combine the fecal waters from the sewer pipes with the run-off water from the streets. In the event of storms rainfall increases the sewage water flow above the capacity of the WWTP, imposed by the volume of the biological reactors and the retention time needed for correct depuration, and outlets allow excess of untreated water to be directly released into receiving natural waters. The so-called **combined sewer overflows** have become a major source of microbial pollution. Holding tanks have been proved inefficient in avoiding this problem, and separation of wastewater from storm water in an entire city is frequently considered as too costly.

Natural waters are capable of self-purification

The treatment of urban sewage in a WWTP mimics the physical and biological processes that take place in natural waters. When liquid wastes are spilled into surface waters the functioning of the aquatic ecosystem changes, and if the amount of wastes is moderate compared to the water volume it recovers the original environmental characteristics downstream. **Self-purification** of a natural water body may be defined as the partial or complete restoration, by natural processes, of the original conditions following the introduction of anthropogenic matter that causes a change in the physical, chemical, and/or biological characteristics of the water body.[5] **Assimilatory capacity** is the extent to which a water body can receive wastes without permanent deterioration of the water quality, anthropocentrically defined as a function of the water uses (drinking, recreational, aquaculture, etc.), or just in terms of ecological criteria

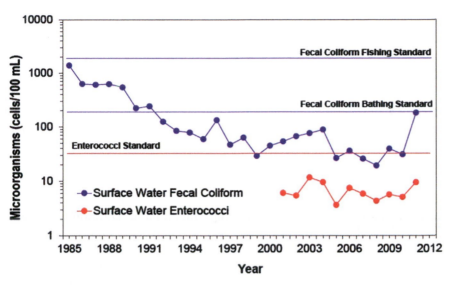

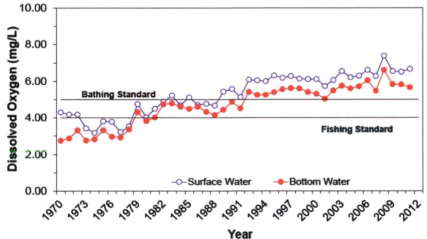

FIGURE 5.1

Improvement of surface water quality, in terms of reduction of fecal microorganisms and increase in dissolved oxygen, over the last decades in the New York City Inner Harbor. This improvement has allowed the City to open the Inner Harbor waters to most recreational activities. The progress is attributed to the cessation of raw sewage dumping through the full build-out of New York City's Wastewater Treatment Plants. *Reproduced from NYC Environmental Protection (2011).*

(diversity, ecosystem functioning). Therefore, self-purification of an aquatic ecosystem can be redefined as the set of physical, chemical, and biological natural processes that allow a water body to recover its original structure and functioning after the discharge of liquid wastes. This recovery can be recorded as a function of the distance from the point of discharge or, provided a discontinuous input of waste, time after the discharge takes place.

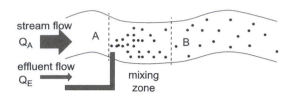

FIGURE 5.2

Dilution of an effluent discharge on a water stream. The dilution factor is the ratio between the effluent and the stream flow rates. The treatment is analogous in a coastal water mass provided prevailing currents replacing the stream flow were known.

Physical components of self-purification: dilution, sedimentation, and evaporation

Natural water bodies provide an appropriate means of dilution of liquid wastes, the most primary way of removing contaminants. Dilution is the result of two physical processes, diffusion and transport, the latter provided by hydrodynamics (currents, tides, waves). In the simplest possible scenario of a continuous and homogeneous point discharge to a constant flow water course, depicted in Fig. 5.2, the extent of dilution is determined by the **dilution factor**, or ratio between the effluent and the stream flow rates.

For the purpose of monitoring compliance with regulations, a mixing zone around the mouth of the effluent determined according to prevailing hydrodynamic conditions must be established and excluded from sampling. Original conditions and effluent effects can be recorded upstream (A) and downstream (B) of the **mixing zone**, respectively. The mixing zone can be defined as a limited area or volume of water where initial dilution of a discharge takes place and where numeric water quality criteria can be exceeded.[6] The size and shape of the mixing zone depend on local conditions, and different regulations may use different criteria to establish the mixing zone, but the rationale underneath is to exclude from sampling the limited area around the point of discharge where pollutant concentrations and environmental parameters may change unpredictably because of incomplete physical mixing and heterogeneous composition of the discharge, and thus be not representative of downstream conditions of the water body as a whole. Notice that in tidal estuaries, unlike the case of the stream depicted in Fig. 5.2, the direction of the mixing zone changes during ebb and rising tides.

Coming back to the stream scenario (Fig. 5.2), and considering a mass balance where contaminant materials in B come either from A or E, then,

$$C_B \times Q_B = (C_A \cdot Q_A) + (C_E \cdot Q_E) \qquad \text{(Eq. 5.1)}$$

where C are concentrations (g/L) and Q flows (L/h). If A corresponds to pristine conditions then $C_A = 0$ and

$$C_B = C_E(Q_E/Q_B) = C_E \cdot DF \qquad \text{(Eq. 5.2)}$$

where DF is the dilution factor, or ratio between the effluent and stream flows;

$$DF = Q_E/Q_B. \qquad \text{(Eq. 5.3)}$$

For nonpristine conditions, the concentration of the contaminant upstream must be considered. Then,

$$C_B = (C_A \times Q_A/Q_B) + (C_E \times Q_E/Q_B) \qquad \text{(Eq. 5.4)}$$

when the stream flow is not significantly affected by the effluent then $Q_A \approx Q_B$, and

$$C_B = C_A + C_E \cdot DF. \qquad \text{(Eq. 5.5)}$$

Downstream concentration can thus be predicted adding to Eq. (5.2) the upstream concentration of the pollutant of interest.

Additional physical processes that may be involved in self-purification of natural waters are **sedimentation** of solid waste particles or dissolved molecules that adsorb to natural particles, and **evaporation** of volatile compounds.

Chemical mechanisms of self-purification are heterogeneous. Dissipative pollutants such as **acids and bases** disappear from the water just by acid–base reactions that produce the corresponding salts and water. Mineral components of the water such as feldspar or limestone act as bases that neutralize acid discharges, while silica and bicarbonate can neutralize basic discharges. In the case of seawater the buffering capacity of the carbonate system is well known. The aqueous CO_2, a weak acid, can be transformed to bicarbonate and carbonate ions through the following equilibria:

Neutralization and other chemical components of self-purification

$$CO_2(aq) + H_2O \leftrightarrow H^+ + HCO_3^- \qquad \text{(Eq. 5.6)}$$

$$HCO_3^- \leftrightarrow H^+ + CO_3^{2-} \qquad \text{(Eq. 5.7)}$$

This carbonate system can buffer seawater pH at a value of about 8. CO_2 is about 200 times more soluble than O_2 in water, but since air has about 700 times more oxygen than CO_2, natural waters contain more oxygen (5–10 mg O_2/L compared to 0.5–1 mg CO_2/L).

Dissolved organic matter may bind metals and other chemical pollutants and render them unavailable to organisms. This issue is discussed in Section 10.4.

Biological mechanisms of self-purification are involved in the natural mineralization of **organic matter**, the most universal pollutant in human wastes. An input of biodegradable organic matter in natural waters stimulates rapid heterotrophic microbial growth, resulting in the consumption of the available dissolved oxygen according to a DO sag curve depicted in Fig. 5.3. Notice that, due to the current flow and microbial growth dynamics, maximum oxygen deficit, corresponding to the situation where deoxygenation rate (heterotrophic consumption) equals reaeration (oxygen diffusion and photosynthesis) does not takes place at the point of discharge but somewhere downstream.[7]

Biological degradation of anthropogenic organic matter is the basis of self-purification

The bacteria and fungi responsible for the mineralization of the organic matter are the base of a community of decomposers that also include protozoans

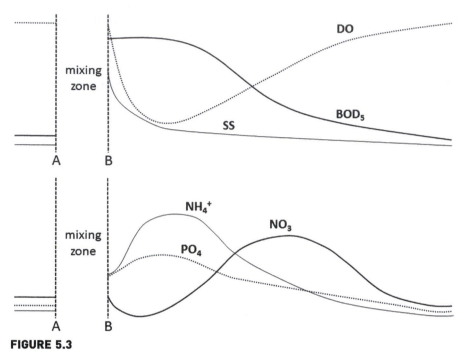

FIGURE 5.3

Effects of an effluent discharge on the suspended solids (SS), dissolved oxygen (DO), biological oxygen demand (BOD_5) and nutrients concentration in a water stream. The discharge takes place at A, and from A to B lies the mixing zone (see also Fig. 5.2). Notice that minimum DO levels and maximum NH_4^+ occur downstream the point of discharge.

(e.g., *Tetrahymena*, *Colpidium*, *Paramecium*) and saprobic invertebrates. The composition of the biological community thus is affected by the discharge, and this can be used as biological indicator of pollution (see Chapter 15).

Anthropogenic **nutrients** are considered as dissipating pollutants because they are readily taken up by aquatic microorganisms. They do not pose a risk provided the input does not exceed the assimilatory capacity of the receiving waters. Otherwise hypereutrophication takes place. At the point of minimum DO values nitrogen is chiefly present as ammonium, while subsequent reaeration restores the dominance of nitrates (Fig. 5.3).

Self-purification of **fecal microorganisms** involves physical, chemical, and biological mechanisms. Bacterial die-off is primarily controlled by sedimentation, solar radiation (particularly the ultraviolet spectrum) and nutrient deficiencies, with suboptimal thermic and salinity conditions, naturally occurring antibiotics, predators (protozoans, *Bdellovibrio*) and bacteriophages acting as additional factors. When the self-purification capacity of a natural water body is quantified sedimentation effects are discounted by using tracing particles of similar physical properties than the fecal bacteria.

It has been experimentally demonstrated that a dense microbial population of fecal origin has a higher die-off rate than a scarce population. Over this basis, wastewater effluents can be characterized according to their T_{90}, a parameter corresponding to the time required for a 90% reduction in number of the effluent's fecal microbiota, frequently *Escherichia coli* or enterococci. The T_{90} value is inversely related to the fecal microbial pollution of the effluent, so some environmental regulations set minimum T_{90} values to allow the discharge. The T_{90} is calculated adjusting the bacterial counts (C, MPN/100 mL) to an exponential decay with time (t, hours) according to the expression:

$$C_t = C_0 \cdot e^{-kt} \tag{Eq. 5.8}$$

that can be linearized using log scale,

$$\ln C_t = \ln C_0 - kt \tag{Eq. 5.9}$$

where k is the extinction rate of the microorganism. This parameter will be affected by the self-purification ability of the receiving waters. Taking $C_0 = 100$ and $C_t = 10$, and solving for t in Eq. (5.9),

$$T_{90} = (\ln 100 - \ln 10)/k = \ln(100/10)/k = 2.3/k. \tag{Eq. 5.10}$$

Therefore, upon standardization in laboratory of the physic-chemical factors that affect self-depuration, the T_{90} of an effluent can be measured according to Eq. (5.10).

> T_{90} quantifies the microbial quality of an effluent

5.2 CONVENTIONAL WASTEWATER TREATMENT: REMOVAL OF PARTICLES AND ORGANIC MATTER

Modern wastewater treatment is a modular process consisting of the steps depicted in Fig. 5.4. **Pretreatment or preliminary treatment** aims at the elimination of large objects by means of **grids** or **screens**, high-density particles (sand and grit removal) by sedimentation, and surface foams, oils, and floating objects by means of **skimmers**.

> Wastewater treatment mimics natural self-purification

Primary treatment aims at the removal of particulate matter, by the physic-chemical processes of sedimentation, coagulation, or flocculation. In all cases a settling tank is used (see Fig. 5.4), producing a primary sludge output. Particulate matter is gravimetrically quantified after filtration as **suspended solids** (**SS**), and this variable is used to characterize effluents and check compliance with regulations.

> Primary treatment aims at removal of particulate matter

Secondary treatment aims at the elimination of organic matter by biological oxidation, either aerobic or anaerobic, provided by heterotrophic microorganisms, mainly bacteria. Its efficiency and compliance with regulations is checked by measuring BOD. Low cost technologies include outdoors stabilization ponds that recreate a full ecosystem of heterotrophic microorganisms that liberate the nutrients for the photosynthetic microorganisms that provide the oxygen. The depth of the pond will determine prevailing aerobic (1–2 m depth) or

> Secondary treatment aims at mineralization of organic matter

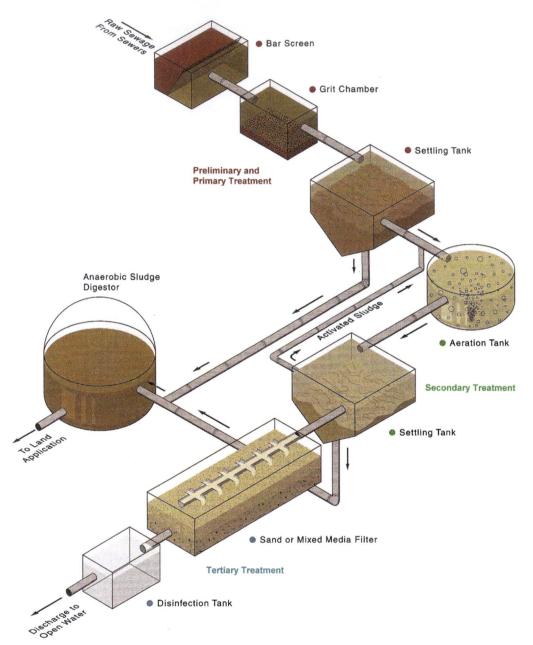

FIGURE 5.4

Depuration steps in a wastewater treatment plant based on the activated sludge process: primary, secondary, and tertiary treatment. Notice that the influent is divided into separated lines of water and sludge. The latter ends up in an anaerobic digestor whose contents are incinerated or used for land-filling or compost. *Modified from Gerba CP, Pepper IL. Chapter 26. Municipal wastewater treatment. In: Pepper IL, Gerba CL, Brusseau ML, editors. Environmental and Pollution Science. 2nd ed. Amsterdam, Elsevier; 2003. pp. 429–450.*

anaerobic (up to 10 m depth) conditions, the later need longer detention times (3—5 days for aerobic vs. 20—50 days for anaerobic).[8]

High capacity plants must use open-flow systems, mainly activated sludge reactors, trickling filters, or rotating biological contactors.[9] In activated sludge systems the decomposing microorganisms, technically referred to as mixed liquor volatile suspended solids (MLVSS), are maintained in liquid suspension by means of strong aeration. **Activated sludge** process is based on two serial steps: forced aeration (aerobic reactor) followed by decantation (clarifier tank) to return part of the settled biomass back to the reactor (see Fig. 5.4). Other portions of the decanted solids are removed as secondary sludge through the sludge line. The decomposition process in the reactor is enhanced because the inflow contains sludge that has already passed through the tank, and is "activated" or enriched with decomposers. This inoculum helps to maintain an active microbial biomass in the reactor, and provides stability to the community over a range of nutrient concentrations.[10] Among the heterotrophic microorganisms responsible for the mineralization of the organic matter the bacteria *Zooglea* spp. are especially abundant. They secrete extracellular polymers that form a gel responsible for the formation of clumps of bacteria and abiotic particulate matter that aid the clarification of the sewage. Because the MLVSS are returned to the reactor, their actual residence time is longer than that of the nonsettling fluids, increasing the efficiency of the BOD reduction. Controlling the sewage inflow rate allows controlling the residence time of the MLVSS and selecting the optimum age of the microbial community. Rapidly growing populations will quickly lower BOD but will also produce larger amounts of secondary sludge to be processed. Ciliates and other free-living (naked and testate amoebae, *Paramecium*) and attached (*Vorticella*) protozoans are also abundant in the reactors. They feed on free-swimming bacteria and thus play a role reducing turbidity in the secondary effluent. The dominant protozoan species in the activated sludge can be used as indicators of the correct functioning of the reactor.[11]

In **trickling filters** and **rotating biological contactors** (RBC) the heterotrophic microorganisms responsible for the mineralization of the organic matter grow as biofilms attached to an inert packing material (gravel, sand, plastics, etc.) of appropriate size to allow air circulation in the void spaces. In the trickling filters, wastewater is homogeneously distributed by dripping or prying over the packing materials in round tanks a few meters depth, frequently using rotating arms. An RBC consists of a series of closely spaced plastic disks of up to 3—4 m diameter that are partially submerged in wastewater and rotated through it.[12] As compared to activated sludge, the diversity of heterotrophic microorganisms in trickling filter systems is much higher, including not only bacteria but also algae, fungi, protozoans, and invertebrate grazers such as *Collembola*.

Anaerobic reactors are an interesting alternative with the advantages of lower operational costs, lower volumes of sludge generated, higher BOD loads admissible, and production of combustible biogas. Industrial effluents with chemical oxygen demand (COD) values above 1500 mg/L are especially suitable for anaerobic digestion, though mineralization is only partial and further treatment is needed to achieve admissible organic loads in the discharged effluent. The main limitations are the slower metabolic rate of anaerobic microorganisms, incomplete mineralization of the anaerobic routes, higher sensitivity of anaerobic microorganisms to toxic chemicals, and the higher initial investments derived from the control of oxygen absence and appropriate temperature in the reactor.[13] Methanogen archaea and other key microorganisms involved in the anaerobic digestion are mesophile, and their metabolic activity is negligible below 12–15°C, limiting the whole process. That is the reason why conventional anaerobic reactors are more promising in tropical countries.

Unlike aerobic mineralization, anaerobic digestion of organic matter is a complex sequential process involving different steps mediated by different strict and facultative anaerobic microorganisms: hydrolysis of biopolymers (carbohydrates, proteins, lipids), fermentation of sugars and aminoacids, anaerobic oxidation of long chain fatty acids and alcohols, oxidation of short-chain fatty acids to acetate and H_2, conversion of acetate to CH_4 (acetoclastic methanogenesis), and reduction of CO_2 to CH_4 (hydrogenotrophic methanogenesis). Different bacteria are associated to each step, the products of some steps limit the progress of the next, and only the simultaneous activity of all of them in the biofilms make the sequence proceed. For example the oxidation of ethanol to acetate,

$$CH_3-CH_2ON + H_2O^- \rightarrow CH_3-COOH + H_2; \Delta G_o = +6.3KJ \qquad (Eq.\ 5.11)$$

is thermodynamically unfavorable at atmospheric hydrogen partial pressures, and only takes place exothermically in the presence of hydrogenotrophic microorganisms such as methanogen archaea that act as a sink of H_2.[14]

The rate BOD/ microbial biomass is the key for a correct wastewater depuration

From a biotechnological standpoint, secondary treatment is the key step in a WWTP. Assuming complete mixing, the rate of mineralization of the organic matter is determined by the rate food to microorganisms (F/M). Food is estimated from the BOD (g/m^3) and microorganisms from the total microbial biomass in the reactor (MLVSS, g/m^3).

$$F/M = Q \cdot BOD/V \cdot MLVSS = BOD/\tau \cdot MLVSS \qquad (Eq.\ 5.12)$$

where Q is the influent flow rate (m^3/d), V is the aeration tank volume (m^3) and τ is the hydraulic retention time (d).[15]

Optimal F/M ratios are achieved by designing appropriate reactor volumes and manipulating flow rates and microbial loads. Other variables that affect the process are temperature, pH, and the presence in the raw effluent of any chemical (trace metals, chlorine products, phenol, etc.) toxic to the heterotrophic microorganisms. For this reason industrial effluents cannot be treated by municipal WWTPs without previous elimination of these substances.

The decantation phases in the primary and secondary treatments generate a **line of sludge** separated from the **line of water** (see Fig. 5.4). This sludge is further mineralized by anaerobic processes yielding biogas, mostly composed by methane (60%–80%), which can be used to produce energy (6000 kcal/m^3). The resulting sludge is later dried and either incinerated, or used for land-filling or compost.

Tertiary treatment, not included in conventional WWTPs, can consist of any process aimed at further improvement of the quality of the effluent, normally including chemical disinfection and biological elimination of inorganic nutrients (N, P). Industrial wastewaters may undergo special processes aimed at the removal of metals or other ions by specific methods such as reverse osmosis, electrodialysis or ultrafiltration, or elimination of organics by adsorption to activated carbon.

> **Tertiary treatment commonly targets nutrients and microorganisms**

5.3   REMOVAL OF INORGANIC NUTRIENTS

Secondary effluents of urban wastewaters show high levels of NH_3, originated from decomposition of organic nitrogen, mainly urea. Nitrifiers are slow growing bacteria that need long residence times. Primary and secondary treatments are frequently not sufficient for the effluent to meet environmental requirements intended to prevent hyper-eutrophication, and may need additional tertiary treatment with this aim. Some industrial wastewaters, such as those coming from fertilizer industry, explosive industry, or some pharmaceutical processes require nitrogen removal as well.

> **N removal can be achieved by bacterial nitrification (aerobic) plus denitrification (anaerobic)**

Biological elimination of nitrogen in wastewaters is conventionally achieved by the sequential combination of nitrification, an aerobic process conducted by chemolithotroph bacteria that oxidize NH_3 to NO_2 (*Nitrosomonas*) and NO_3 (*Nitrospira, Nitrobacter*) in the presence of O_2, and denitrification, an anaerobic process conducted by heterotrophic bacteria that use NO_3 and NO_2 as electron acceptors in the absence of O_2 to oxidize some organic source of carbon to CO_2 and produce N_2 gas (Fig. 5.5).[16]

An alternative treatment is the combination of a partial nitrification, where 50% of ammonia is oxidized into nitrite in an aerobic reactor, and a subsequent anaerobic ammonium oxidation (Anammox), where ammonia is oxidized by nitrite to N_2 in a second tank. This second step takes place at echimolar NH_3 and NO_2 concentrations by the bacterial phylum Planctomycetes. The alternative process avoids the requirement of an organic carbon source to denitrify, allows saving over 65% of the oxygen supply, and produces a lower amount of sludge. The limitation for this alternative is the optimum temperature range, from 30 to 40°C. [17]

> **Anammox allows efficient N removal at low C/N ratios**

Although removal of P from wastewater may be achieved by biological processes, most WWTP with tertiary treatment aimed at P removal resort to chemical methods, mainly precipitation of phosphates by means of chemical

> **P removal is achieved by chemical coagulation**

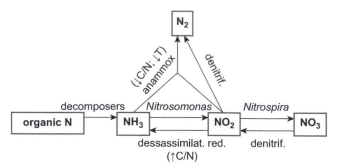

FIGURE 5.5

Nitrogen species and microorganisms relevant for the nitrogen cycle, including nitrifiers (aerobic chemo-litotrophic bacteria) and denitrifiers (anaerobic heterotrophic bacteria). The latter are useful for the elimination of nitrates from sewage. An alternative is the anammox (anaerobic ammonium oxidation) that oxidizes ammonium with nitrite to produce elemental nitrogen at low C/N ratios.

coagulants. The most commonly used coagulants are lime, $Ca(OH)_2$, alum, $Al^+(SO_4)_2$, or other Al and Fe salts. Insoluble metal phosphates precipitate and are eliminated as sludge in sedimentation tanks.

5.4 DISINFECTION OF WASTEWATERS

Disinfection refers to the partial (unlike sterilization that refers to total) destruction of pathogenic microorganisms. Physical and chemical oxidants damage cell walls, denaturalize proteins, and inactivate enzymes. UV light, in addition, cause DNA and RNA alteration, thus disrupting replication and protein synthesis.

Urban wastewaters show values of coliforms in the order of 100×10^6 MPN per 100 mL and a plethora of potentially infectious microorganisms (see Chapter 4). Conventional treatment (primary plus secondary) reduces the microbial load by about 95%, resulting still environmentally unacceptable levels $>1 \times 10^6$ MPN per 100 mL, which pose a risk to aquaculture and recreational uses of the receiving waters. The pathogenic bacteria *Salmonella*, for example, have been reported to resist conventional depuration.[18] Additional disinfection procedures are thus vital for safe discharge of wastewater.

Disinfection may be achieved by using chemical agents … Table 5.1 summarizes the advantages and limitations of the different methods available for the disinfection of waste waters. Because of its low cost, chlorination is the most common method of disinfection for both wastewater effluents and drinking water supplies. In addition, residual chlorine prevents the growth of pathogens after depuration along the distribution system, but this is not needed in the case of effluents. Chlorine gas (Cl_2) dissolves in water to form hypochlorous acid (HOCl), which is the actual oxidative agent responsible for killing microorganisms by inactivation of enzymes, denaturalization of proteins, or, at high concentrations, the destruction of cells. Other oxidizing

Table 5.1 Summary of the Main Advantages and Limitations of the Disinfection Methods Suitable for Application to Wastewater

Disinfectant	Advantages	Limitations
Cl_2	Cost-effective	Production of halomethanes and other by-side toxic products
ClO_2	Produces less halomethanes	More costly than Cl_2
O_3	Very effective against viruses and protozoans cysts, does not produce halomethanes	Costly; demands in situ production
UV	Does not produce halomethanes	Costly because of electricity consumption; suitable for low fluxes only due to limited penetration in water

chemicals suitable for disinfection through the same mechanisms include chlorine dioxide, bromine, iodine, and ozone.

However, Cl (and its alternative Br) chemicals react with organic matter and cause the formation of toxic substances such as trihalomethanes (chloroform, bromodichloromethane, dibromochloromethane, bromoform), haloacetic acids, and other undesirable disinfection byproducts.

Ozone and UV light lack this problem, and according to some sources may be cheaper than Cl for small plants (up to 4 million L per day). The efficiency of the UV radiation depends on the penetration of the UV rays into the water. Suspended matter, dissolved organics, and water itself adsorb the radiation and limits the efficiency of the process.

... or physical agents: UV light

The germicidal efficiency of chlorine compounds is reported in terms of $C_R \cdot t$ value, residual chlorine concentration present (titratable chlorine, mg/L) times contact time (min). For a 99.9% inactivation, Table 5.2 compares the germicide efficiency of chlorine, ClO_2, O_3, and UV. As already discussed in Section 4.1, viruses and protozoan cysts are particularly resistant to conventional disinfection methods.

Table 5.2 Germicidal Efficiency of Different Disinfectants Applied to Wastewater

	$C_R \cdot t$ (mg min/L) for 99.9 percent Inactivation (pH = 7; T = 20°C)		
Disinfectant	Bacteria	Virus	Protozoan Cysts
Cl_2	1.5–3	4–5	70–80
ClO_2	20–30	6–12	20–25
O_3	–	0.5–0.9	0.7–1.4
UV	60–80	50–60	15–25

Own elaboration. Data from Metcalf & Eddy, Inc. Wastewater engineering. Treatment and reuse. 4th ed. McGrawhill; 2004.

KEY IDEAS

- In large cities, sewerage substituted natural water courses as the means to get rid of liquid wastes. This practice deteriorated water quality in the receiving waters and forced the construction of wastewater treatment plants to depurate the effluents.

- Wastewater treatment mimics the self-purification capacity of natural water bodies. It commonly consists of preliminary elimination of large solids and floating oils and foams, removal of suspended particles (primary treatment), mineralization of the organic matter (secondary treatment), removal of nutrients, and disinfection (tertiary treatment).

- Despite generalized depuration of liquid wastes in WWTPs, occasional spillage of fecal microorganisms due to raw effluent overflow after rainy conditions remains an unresolved environmental issue.

- The most traditional form of secondary treatment is the activated sludge method that consists of an open-flow aerobic digestor in line with a decantation tank that recovers the microorganisms back to the digestor. The digestor must be designed in order to keep an optimum ratio BOD/microorganisms.

- Primary and secondary decantation originates a line of sludge that is commonly treated anaerobically with production of biogas. The resulting dried solids are incinerated or used for compost or land-filling.

Endnotes

1. Brosnan TM, O'Shea ML. Long-term improvements in water quality due to sewage abatment in the Lower Hudson River. Estuaries 1996;19(4):890−900.

2. NYC Environmental Protection. New York Harbor Water Quality Report; 2011. 20 pp.

3. Directive 91/271/EEC. Council Directive of 21 May 1991 concerning urban waste water treatment.

4. See Chapter 2, p. 111 in: EEA. Europe's environment. The fourth assessment. European Environment Agency, Copenhagen; 2007.

5. Benoit RJ. Self-purification in natural waters. In: Ciaccio LL, editor. Water and water pollution handbook. N.Y: Marcel Dekker Inc; 1971

6. US-EPA. Water quality standards handbook: Second edition. EPA 823-B-94-005a. United States Environmental Protection Agency, Washington; 1994.

7. Benoit (1971) op. cit.

8. Gerba CP, Pepper IL. Chapter 26. Municipal wastewater treatment. In: Pepper IL, Gerba CL, Brusseau ML, editors. Environmental and Pollution Science. 2nd ed. Amsterdam, Elsevier; 2003. pp. 429−450.

9. See p. 551 in: Metcalf & Eddy Inc. Wastewater engineering. Treatment and reuse. 4th ed. Boston, McGrawhill; 2004.

10. See p. 341 in: Beeby A. Applying ecology. London, Chapman & Hall; 1993.

11. See p. 343 in Beeby (1993) op. cit.

12. See p. 930 in Metcalf & Eddy (2004) op. cit.

13. Orozco Barrenetxea C, Pérez Serrano A, González Delgado MN, et al. Contaminación ambiental. Madrid, Thomson Editores; 2003.

14. Sanz JL. 10. Tratamiento de aguas residuales. In: Marín I, Sanz JL, Amils R (eds). Biotecnología y medioambiente, 2nd ed. Madrid, Ephemera; 2014. pp. 179-195.

15. See p. 599 in: Metcalf & Eddy (2004) op. cit.

16. Mosquera A, Campos, JL. 11. Eliminación biológica de nitrógeno de aguas residuales. pp. 197−213. In: Marín I, Sanz JL, Amils R, editors. Biotecnología y medioambiente. 2nd ed. Madrid: Ephemera; 2014.

17. Dosta J, Fernández I, Vázquez-Padín JR, et al. Short- and long-term effects of temperature on the Anammox process. J Hazard. Mater. 2008;154:688-693.

18. Leclerc H, Mossel DAA. Microbiologie: le tube digestif, l'eau et les aliments; 1989. In: Doin, editor. Paris. 468 pp.

Suggested Further Reading

- Benoit RJ. Self-purification in natural waters. In: Ciaccio LL, editor. Water and water pollution handbook. New York: Marcel Dekker Inc.; 1971. p. 223−61.

- Gerba CP, Pepper IL. Chapter 26. Municipal wastewater treatment. In: Pepper IL, Gerba CL, Brusseau ML, editors. Environmental and pollution science. 2nd ed. Amsterdam: Elsevier; 2006. p. 429−50.

CHAPTER 6

Plastics and Other Solid Wastes

6.1 SOLID WASTE MANAGEMENT

As a result of the concentration of human populations in large cities and the adoption of a hyper industrialized and technological lifestyle characterized by enhanced rates of consumption, the amount of urban or **municipal solid waste (MSW)** generated per capita has reached such a dimension that MSW management ranks among the main economic problems of local administrations. This is not exclusive of the most developed countries. According to the World Bank, it is common for municipalities of developing countries to spend 20%–50% of their budget on solid waste management, despite 30%–60% of all the urban solid waste in those countries remaining uncollected.[1] In fact, reduction of consumption rates and reutilization of packaging and other solid wastes ranks first in the hierarchy of waste management advocated by the United States Environmental Protection Agency (US-EPA)[2] (Fig. 6.1).

Managing municipal solid waste (MSW) became a challenge for local communities

Source reduction ranks first in the sustainable management of solid waste. In Europe, the same philosophy reflected in Fig. 6.1 inspires Directive 94/62/EC on packaging that aims at "limiting the production of packaging waste and promoting recycling, re-use and other forms of waste recovery," leaving disposal as a last resort solution.[3] According to this regulation, packaging volume and weight shall be limited to the minimum, and it must permit its reuse or recovery, including recycling. The Directive, modulated by subsequent amending acts, set a calendar to achieve quantitative targets for percentages of solid waste that must be recycled, recovered, or incinerated at waste incineration plants with energy recovery. For materials contained in packaging waste the percentages of recycling should be 60% for glass, paper, and board; 50% for metals; 22.5% for plastics; and 15% for wood.

Source reduction in the production of waste is the first priority

Reusing waste objects is preferable to recycling as it uses less energy and fewer resources. **Primary recycling** (conducted during the production process) is especially efficient in reducing the production of plastic waste compared to recycling at later stages of the plastic object life cycle. In the UK, 95% of the plastic solid waste arising from process scrap has been recycled in 2007 by reextrusion, i.e., reintroduction of scrap to the extrusion cycle to produce products of a similar

Reusing and primary recycling especially contribute to reduce waste

69

Marine Pollution. https://doi.org/10.1016/B978-0-12-813736-9.00006-4

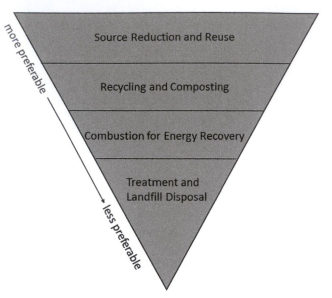

more preferable
less preferable

Source Reduction and Reuse

Recycling and Composting

Combustion for Energy Recovery

Treatment and
Landfill Disposal

FIGURE 6.1

Hierarchy, from most preferred to least preferred methods, in the management of municipal solid waste (MSW). *Modified from US Environmental Protection Agency (US-EPA).*

material. **Secondary recycling** (conducted after use and disposal of the plastics by consumers) demands selective waste collection and separation of the different recyclable components of waste from organic debris. In fact, postconsumer plastic recycling demands separation of the different thermolabile polymers; polyethylene (PE), polypropylene (PP), polystyrene (PS), etc. Excessively weathered plastic objects are not suitable for recycling and must be discarded. Common additional steps are washing, size reduction by milling, grinding, or shredding, density separation, extrusion, pelletization, and water cooling to produce the final product suitable for fabrication of new plastic objects.[4]

Composting of organic waste is preferable to incineration

For organic waste, composting is preferable to combustion, since it avoids atmospheric pollution and reduces the production of CO_2 and other greenhouse gases. Predisposal treatment is intended to reduce the volume and potential toxicity of the waste, and it may include physical (e.g., shredding), chemical (e.g., incineration), and biological (e.g., anaerobic digestor) processes. Landfill disposal is the least preferable option, and yet because of its short-term cost-effectiveness it is by far the most commonly chosen one. For example, according to a study dated from 2011, >69% of the total municipal waste in the United States, and >85% of the postconsumer plastic waste, ends up in landfills (Fig. 6.2A).[5] In Europe the figures are very variable, but in countries that did not ban landfilling between 40% and 85% of plastics are disposed by this method (Fig. 6.2B).

Environmental awareness of the society pushes for waste management policies inspired in Fig. 6.1. This is not limited to Western countries. In a study covering

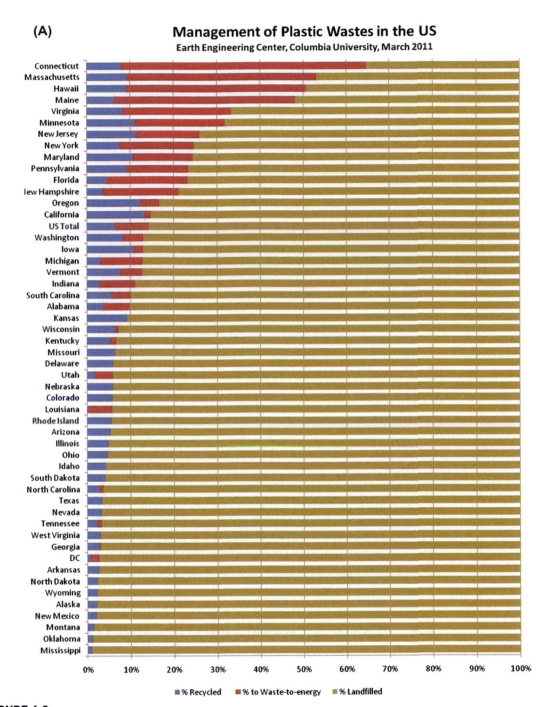

(A)

Management of Plastic Wastes in the US

Earth Engineering Center, Columbia University, March 2011

■ % Recycled   ■ % to Waste-to-energy   ■ % Landfilled

FIGURE 6.2

Percentage of plastic waste recycled, used for energy recovery, or landfilled in each US State in 2008 (A) and in each European country in 2012 (B). *Sources: Earth Engineering Center, Columbia University (A); Consultic/Plastics Europe (B).*

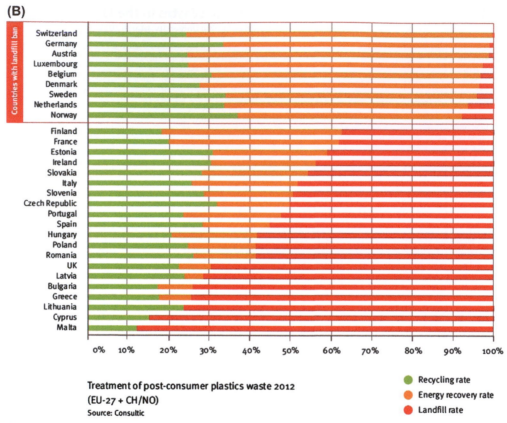

Treatment of post-consumer plastics waste 2012
(EU-27 + CH/NO)
Source: Consultic

- ● Recycling rate
- ● Energy recovery rate
- ● Landfill rate

FIGURE 6.2 cont'd.

cities from Asia, Africa, and Latin America, awareness of citizens and municipal leaders on the impacts of waste management systems ranked first and far above logistic factors such as availability of equipment and recycling machinery when explaining the variance in waste separation at household level.[6] It seems again that societal perception of the environmental problems contributes to a lower per capita impact of human activities, in line with the conceptual model enunciated in Section 1.1.

6.2   PLASTICS: CHARACTERISTICS AND TYPES

Plastic production increased during the last 40 years and does not level off

Plastics are **synthetic polymers** unknown to human kind prior to 1910, when the Belgian chemist Leo Baekeland in New York synthesized a thermosetting resin that could be used as insulator by the emerging electric industry. The more usual plastics such as PE were discovered 20 years later and not massively produced until the late 1950s. Since then its world production has been rising until the current 322 million tons per year (Fig. 6.3), nearly one-half now produced in Asia. Their low cost, excellent oxygen/moisture barrier properties,

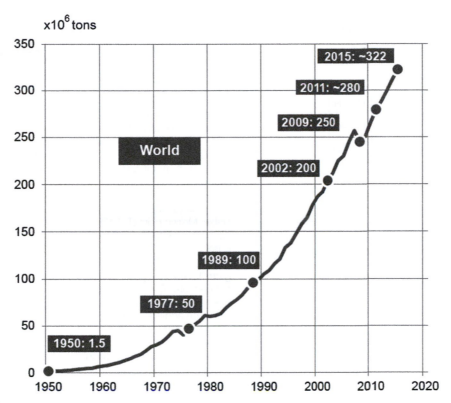

FIGURE 6.3
Increase of global plastic production from the 1950s to date. *Source: Plastics Europe.*

bio-inertness, and light weight make them excellent packaging materials. Packaging, with c. 40% of the total plastic market (mainly PE, PP, and polyethylene terephthalate (PET)) followed by building and construction (polyvinyl chloride (PVC), polyurethane (PUR)) automotive (PP), electrical-electronic, and agriculture are the main sectors demanding plastics.[7] This means that c. 40% of the over 300 million tons of plastic produced annually are intended for immediate disposal.

Plastic objects are composed by a matrix of polymer plus chemical additives. According to their thermal resistance, polymers can be classified as **thermosetting** and **thermolabile**. The first are also termed resins, and include PUR, polyester, and epoxy resins. Thermolabile polymers, the most common ones, can be repeatedly reshaped after heating and cooling, a characteristic that lend the plastic its name. According to the fabrication process they can be additional polymers such as PE, PP, PS, PVC, or condensation polymers such as PET, polycarbonates (PC), or polyamides (Table 6.1). Thermolabile polymers can be recycled, but only about 8%–26% of plastic waste is actually recycled.[8] It is interesting to note that recycling of plastic, in contrast to that for other litter components such as paper, did not succeed so far to level off global production,

Plastic objects are composed by inert polymers plus chemical additives

Table 6.1 Most Common Types of Polymers, Properties, and Examples of Uses in Plastic Products.

Polymer	Monomer	Typed and Properties	Examples of Uses
Addition Polymers			
Polyethylene (PE)	[structure: $-[CH_2-CH_2]_n-$ drawn as H–C–C–H repeating unit]	Types: LDPE (low density): Branched chains, low crystallinity. LLDPE (linear low density): linear chains. HDPE (high density): linear chains. Properties: Soft, flexible, and chemically stable. Melting point (T_m) 105–125°C. Odorless and no skin irritation. Floats (density 0.94–0.965 g/cm^3). Very low glass transition temperature ($T_g = -120$°C), i.e., does not need plasticizers. Common additives are colorants, flame retardants (cable insulation) and UV stabilizers (outdoor application).	LDPE and LLDPE: Packaging, wrapping foils, bags, cable sheaths. HDPE: Bottles, bottle crates.
Polypropylene (PP)	[structure: $-[CH-CH_2]_n-$ with CH_3 side group]	$T_m = 165$°C. Semicrystaline. $T_g = -10$°C. Floats (density 0.90–0.91 g/cm^3). All commercial PP is stabilized with antioxidants. Other common additives are colorants and (in cable applications and electronics) flame retardants.	Food packaging, household appliances, outdoor furniture.
Polystyrene (PS)	[structure: styrene repeating unit with benzene ring]	Types: PS, HIPS (high impact), EPS (expanded). PS, HIPS: Transparent, hard, and brittle. $T_g = 100$°C. Sinks (density: 1.04–1.07 g/cm^3). EPS: fabricated as foam. Expansion gases are pentane, CO_2 and (in the past) CFC. Flame retardants are used in building applications.	PS: CD and DVD cases, yogurt containers, disposable tableware. EPS: meat trays, thermal, and electrical insulation.
Polyvinyl chloride (PVC)	[structure: $-[CH_2-CHCl]_n-$ repeating unit]	Robust, insulating (very good air and water barrier), and fire-resistant, but rigid ($T_g = 82$°C) and needs phthalates (c. 30% in weight) as plasticizers. Other common additives are colorants and stabilizers. Sinks (density 1.38–1.53 g/cm^3)	Pipes, windows, seals, hoses, toys.

Table 6.1 Most Common Types of Polymers, Properties, and Examples of Uses in Plastic Products. *continued*

Polymer	Monomer	Typed and Properties	Examples of Uses
Condensation polymers			
Polyethylene terephthalate (PET) or polyester		Types: Amorphous (PET A) and semicrystaline (PET C). PET A Is transparent. PET C is opaque and white. Sinks (density 1.33–1.4 g/cm^3). Light, impact-resistant. Very good chemical resistance, thermal resistance ($T_m = 265°C$), and gar barrier. May leach traces of acetaldehydes and antimony, used in the fabrication.	PET A: Bottles PET C: Food and cosmetics packaging, household appliances, reinforcer in textiles, tires, belts.
Polycarbonate (PC)	Bisphenol A	High physical resistance. Sinks (density 1.2 g/cm^3).	Automotive and electronic industries, helmets.
Polyamides (PA)		High durability, strength and heat resistance ($T_m = 233–272°C$). Sinks (density 1.12–1.16 g/cm^3). May leach aromatic amines.	Textiles, automotive applications, carpets, fishing nets, and lines.
Reaction polymers			
Polyurethane (PUR)	isocyanates	High resilience, suitable for foams. It is a thermoset not mechanically recyclable. Common additives: flame retardants and biocides. Hg may be used as catalyst.	Insulation panels, tires, sponges.

which continues to show an upward trend[9]; reduced consumption and reutilization seem more efficient options to limit their increasing environmental ubiquity.

Raw polymers lack most of the properties demanded in the final plastic products. Additional components are used to satisfy the requirements of the end products or simply reduce cost. With this aim manufacturing plastic objects include a step of compounding in which different **additives** intended to save polymer (fillers), improve toughness (impact modifiers), enhance flexibility (plasticizers), reduce risk of ignition (flame retardants), give color (colorants), inhibit degradation (antioxidants and preservatives), stabilize against weathering (UV filters), give conductivity (antistatics) or improve the bond between the polymer matrix and the fillers (coupling agents) are added to the virgin polymer. In some cases chemical additives are used already as polymerization and processing aids, and may be present in the polymer before the compounding step. These include initiators, cross-linking agents, curing agents, blowing agents, or heat stabilizers. The most common plastic additives according to their volume of market in Europe and North America are reflected in Fig. 6.4.[10]

Conventional plastics are not biodegradable

The first environmental concern caused by plastics is their ubiquity in all environmental compartments, which derive from their nondegradability. Accumulation of plastic litter causes nuisance and lose of touristic value in many coastal environments, and problems of entanglement or clogging of the digestive system to marine fauna (see next Section). Some special plastics made from starch, cellulose, or aliphatic polyesters, such as the polylactic acid (PLA), are **biodegradable**; they can be microbially degraded and show environmental half-lives of months.[11] However, conventional plastics are nonbiodegradable and have persistence of hundreds of years. Plastics termed oxo-degradables are actually nonbiodegradable conventional polymers (PE, PP, PS) whose chains are treated to break up in shorter pieces by oxidation in the environment, thus accelerating their fragmentation into microplastics.

Plastic objects include many chemical additives in high quantities

The second problem of plastics is the suspected or demonstrated toxicity of some of the chemical additives present in their composition (See Section 8.3). These additives are not traces, but are normally added in large amounts to the polymers. Colorants, for example, account for up to 3% in weight of PE objects, and flame retardants up to 20%.[12] Flexible PVC requires up to 50% in weight of plasticizers, which represents approximately 85% of the total demand of plasticizers. Moreover, most additives are not bound to the polymer chain and can migrate to any medium in contact with the plastic material.

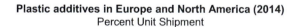

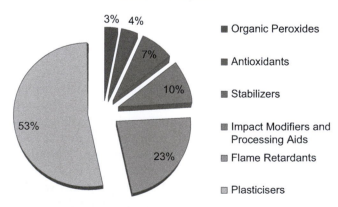

Plastic additives in Europe and North America (2014)
Percent Unit Shipment

- Organic Peroxides
- Antioxidants
- Stabilizers
- Impact Modifiers and Processing Aids
- Flame Retardants
- Plasticisers

3%   4%   7%   10%   53%   23%

FIGURE 6.4
Main plastic additives used in Europe and North America (United States and Canada) markets in 2014. Organic peroxides are used for polymerization. *Source: Frost & Sullivan (2015).*

While virgin polymers are biologically inert and thus nontoxic, many chemicals used as plasticizers, stabilizers, colorants, or flame retardants may be toxic, and preservatives (arsenic compounds, triclosan, etc.) are intended to be toxic since their role is as biocides. Cadmium is used as a pigment in many plastic products. Heat stabilizers, mainly used in PVC, may be lead or organo-tin based. Highly persistent halogenated organics are used as flame retardants. Polybrominated diphenyl ethers are teratogen at environmentally relevant concentrations for fish.[13] The ban of those compounds stimulated the use of phosphorus flame retardants, some of which already raise concern about their environmental persistence and risks to human health.[14]

About 70% of plasticizers are phthalates. Esters of the orthophthalic acid such as the di-2-ethylhexyl phthalate used as plasticizers in PVC are endocrine disrupters, and were also associated to selected allergies and asthma in children.[15] Plasticizers, like most plastic additives, are not bound to the plastic matrix and can dissolve through contact with liquids and fats. Therefore, the use of certain phthalates as plasticizers was banned for some applications such as toys and children-care products (for the United States, see Consumer Product Safety Improvement Act, 2008; for the European Union, see Directive 2005/84/EC). Plastic materials intended to be in contact with food are submitted to special regulations (e.g., Directive 2002/72/EC and its subsequent amendments summarized in EU Regulation No 10/2011).

Ortho-phthalates used as plasticizers cause effects similar to hormones

In addition, some polymers are synthetized from toxic chemicals, and unreacted residual monomers or polymerization aids may contaminate the end-product. Polyurethanes are synthetized from isocyanates, and the presence of free isocyanates in the PUR foam varies according to the curing time. Acrylonitrile, an acutely toxic and possibly carcinogenic chemical, is used in several polymers, including acrylonitrile-butadiene-styrene (ABS). Bisphenol A (BPA), the structural base of polycarbonates, is a well-known estrogenic chemical. BPA is also the main component of epoxy resins used as varnishes to coat the insides of drink and food cans, drink canisters, and drainage pipes. Leaching of BPA from certain food and beverage containers has been reported.

6.3 MARINE LITTER: DISTRIBUTION AND EFFECTS

According to United Nations, 6.4 million tons of anthropogenic litter end up in the oceans every year. Most of this (up to 83% according to some sources) is from plastics. This is mainly due to the disposable nature of some plastic objects, and mismanagement of solid wastes in many coastal countries, which caused 1.7%–4.6% of the total plastic waste generated in 2010 in those countries to end up in the sea.[16] In line with their main use in packaging, the most common plastic floating objects found in the sea are bottles (11%) and bags (10%).[17] Therefore, plastic pollution originates mainly from domestic consumers, rather than industrial activities. The first reports on plastic marine pollution identified plastic pellets, the bulk material used by the industry, as

Discarded plastic objects found their final destination in the sea

major components of the marine litter,[18] but currently fragments of consumer objects are the main component of the plastic litter (88% according to Cózar et al.[19]). Continental plastic litter enters the ocean largely through rivers, storm-water runoff, discharged into the shoreline during recreational activities or directly into the sea from ships.

Plastic debris is commonly identified by visual sorting, and classified according to basic properties such as size, shape (fibers, sheets, fragments, etc.), and color.[20] Separation may take advantage of its low density compared to mineral particles. Density separation may even contribute to identify plastic class, since polymers differ in density in the order PP (c. 0.9 g/cm^3), PE (0.92–0.97), PS (1.04–1.1 g/cm^3), polyester, or PVC (1.16–2.3 g/cm^3). A more rigorous classification is possible when using infrared spectrometry. Identification of the polymer and additives may be attempted through Fourier transform infrared spectroscopy (FT-IR), by comparison of the spectra of a given particle with libraries of spectra characteristic for each kind of material. However, in environmental samples weathering interferes with these spectra.

Despite increasing plastic production, plastics in the sea do not show increasing trends

Despite the mounting world plastic production and the last decades trend to increasing plastic discard in municipal wastes, plastic debris in the ocean's surface was not observed to increase.[21] The deep-sea floor is supposed to be the final sink of plastics. On the seabed plastic aggregates locally in response to local sources and bottom topography. The amount of plastic litter is so great in some areas with large amounts of shipping traffic that initiatives have been started to clean the seabed with trawls. Reported densities on the ocean's bottom range between 30 and 500 items per square km, but the few studies on temporal trends do not show increasing levels either.[22]

Global estimates of floating plastic debris in open ocean ranged from 10,000 to 269,000 ton, with a tendency toward fragmentation and loss from the sea surface by stranding on sea shore, sinking, and ingestion by biota.[23] The horizontal patterns of distribution agree with those expected from ocean surface circulation models,[24] with similar maximum values, up to one plastic fragment per 2 m^2, recorded in the large-scale convergence zone of the five main subtropical gyres: North and South Pacific, North and South Atlantic, and Indian Oceans (Fig. 6.5).

Effects of marine litter include entanglement of young seals and ingestion of nondigestible plastics by seabirds

Marine fauna get entangled in drifting plastics and ingest nondigestible plastic objects that may block their digestive system. The increasing use of plastic replacing natural materials in fishing gears that may be eventually lost or discarded in the sea contributes to the increasing abundance of these objects that affect medium and large size marine animals. The entangled animal may show a reduced ability to move and feed or even get trapped and die out. Pups and young seals are more likely to become entangled than adults.[25] Estimated entanglement rates (percentage of individuals of the population observed to get entangled) for different species of seals ranged from 0.1% to 1.9%,[26] with one of the higher values (0.7%) for the endangered Hawaiian

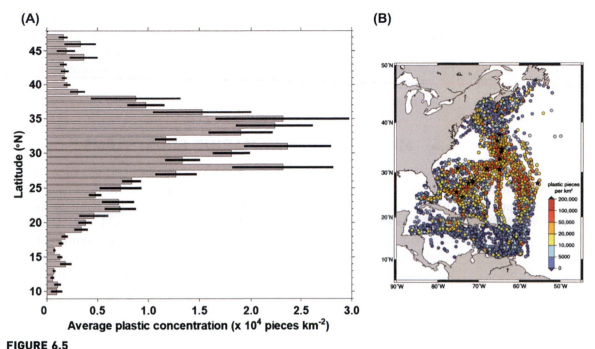

FIGURE 6.5

Distribution of plastic debris in the Northeast Atlantic Ocean collected in annual cruises from 1986 to 2008. Notice the accumulation in the subtropical gyre (latitudes 22—38°N) (A), at plastic densities above 200,000 pieces per km^2 (black stars) (B). *Source: Law et al. (2010).*

monk seal.[27] Despite the impact of those findings, information is frequently anecdotic and scientific studies on the effects on population dynamics are extremely scarce. One of these rare studies was reported by D.P. French and M. Reed in 1990. Although harvesting was the main driving factor explaining population dynamics of the Northern fur seal, the ban of harvesting in 1968 did not lead to the recovery of the population size, and the persisting decline was largely attributable to an observed entanglement rate of the young seals of 50,000 per year.[28] In this species, entanglement rates increased as a result of increased commercial fishing effort, and this explains the observed reduced survival at sea compared to the survival predicted from on-land records.[29] However, other factors likely contributed to the observed decline in the studied populations, namely the impact of climatic events and overfishing on depletion of fish species that make up food resources for the seals.[30]

Concerning ingestion of plastic items, many studies have focused on seabirds. Large differences among taxa, resulting from differences in size and feeding habits, have been described, with Procellariidae especially prone to ingest plastics.[31] The *Fulmarus* species have been proposed as biomonitors of oceanic plastic pollution (See Section 18.2) because they feed exclusively at sea, have vast habitats, and show a nonselective surface foraging.[32] In this species prevalence of plastics in stomachs reach for certain populations 97%, although total amounts per stomach in current studies range between 0.04 and 1.09 g,[33] well below the amounts and particle sizes reported to have killed individuals of other seabird species.[34] The albatross chicks, which feed at land, may show much higher plastic contents in their stomachs (9–24 g) and yet observed mortalities were not directly attributable to plastic ingestion.[35] However, evidence of sublethal effects on the affected chicks, including lighter body masses and lower fat indices, were reported.[36]

6.4   MICROPLASTICS IN MARINE ECOSYSTEMS

Small plastic fragments are termed microplastics

According to its size, plastic debris is commonly classified as **macroplastic** (>20 cm), **mesoplastic** (0.5–20 cm), and **microplastics** (<5 mm). The threshold between micro and mesoplastic derives from the habit of macroscopic examination of plastic debris in the field and the fact that fragments <5 mm are difficult to identify at naked eye, but a limit of 1 mm would be more consistent with the international system of units and with other fields of knowledge. Thus a separation between large (1–5 mm) and **small** (<1 mm) **microplastic** has been proposed.[37] Plastic particles below the 1 μm threshold are considered **nanoplastics**.

Weathering of plastic objects produces secondary microplastics while primary microplastics come from textiles, cosmetics, and other sources

Secondary microplastics originate in the environment from physical degradation of larger plastic objects. As mentioned, conventional plastics are nonbiodegradable, and they may persist in the environment for hundreds of years. However, throughout environmental exposures disposed plastics undergo weathering processes (leaching of plasticizers, thermo-oxidative breakage of the polymeric chains, photodegradation, physical abrasion) that make them more brittle and susceptible to fragmentation into small particles. In highly energetic environments such as exposed coasts, waves accelerate this fragmentation. The high temperature and light exposure in beaches also contribute to this process.[38] Plastic fragments became ubiquitous in the marine environment, and they can be found in increasing amounts in the water column or floating at the sea surface,[39] in plankton samples,[40] stranded in the upper intertidal zone of beaches,[41] or even in subtidal sediments. This is consistent with the trends observed over the last few decades, mentioned in the previous Section, which do not evidence an increase in macroplastic accumulation in marine environment while average size of plastic particles seems to be decreasing, and the abundance of microscopic plastic fragments have increased.[42] In addition, we can find in the environment **primary microplastics**, produced or released already as small particles such as micronized pellets used by the plastic industry, micro-beads in personal care products, or microfibers originated from washing of textiles.

Microplastics are ubiquitous in the marine compartments …

Floating microplastics are sampled using neuston nets with 200–333 μm mesh sizes, and may reach densities >100,000 per km², which are roughly equivalent to 1–10 μg/L (Table 6.2). Identification is conducted on the basis of the visual aspect under a binocular microscope, occasionally after density separation and organic matter digestion, but standard techniques for

Table 6.2 Mean (or Median) and Maximum Plastic Particle Abundance Reported (in Bold) and Estimated in Surface and Subsurface Marine Waters. Surface Particles are Collected Using Neuston Nets (Mesh Size Indicated), and Subsurface Particles by Filtering Pumped Water. Plastic is Identified Just by Visual Aspect, Except for Lusher et al. and Enders et al. Who Also Used Raman Spectra, and Frias et al. and Suaria et al. Who Used Fourier-Transformed Infrared Spectroscopy. Particle Number and Mass per Volume Were Estimated From the Reported Particle Number and Mass Per Area Data and Trawl Mouth Size, Assuming a Half of the Trawl Mouth Submerged in the Water. When Only Particle Number Is Reported, Mass of Particles Was Estimated From Mean Particle Size Assuming Sphericity and/or 1 Density.

Region	Mesh Size	Largest Particles	Part/km²		Part/m³		µg/L		Ref
			Mean	Maximum	Mean	Maximum	Mean	Maximum	
Surface									
Sargasso Sea (Northwest Atlantic)	330 µm	n.r.	**2579**	**12,080**	0.0066	0.0308	0.26	4.5	43
Northwest Atlantic Ocean	947 µm	1 cm	**855**	**166,991**	0.0017	0.3340	0.02	2.8	44
North Pacific Ocean	333 µm	>3 mm	**3370**	**96,100**	0.0080	0.2269	0.11	2.9	45
Coast of South Africa	900 µm	>60 mg	**3640**	**445,860**	0.0173	2.12	0.20	52.0	46
North Pacific Central gyre	333 µm	>4.76 mm	**332,545**	**969,777**	4.43	12.9	**34**	201	47
South California Coast	333 µm	>4.75 mm	n.r.	n.r.	**7.25**	**20.0**	**2.0**	**9.0**	48
Japanese Coast (Northwest Pacific)	330 µm	>11 mm	**174,000**	**3,520,000**	0.696	14.1	14.4	612	49
Northeast Pacific Ocean	505 µm	>10 mm	n.r.	n.r.	**0.0505**	**0.19**	**0.06**	**0.21**	50
Northwest Mediterranean Sea	333 µm	5 mm	**116,000**	**892,000**	1.16	8.92	2.02	22.8	51
Portuguese Coast	180, 280 µm	n.r.	n.r.	n.r.	**0.0235**	**0.036**	65	320	52
Central-West Mediterranean Sea	500 µm	5 mm	n.r.	n.r.	**0.18**	**0.35**	0.75	1.47	53
North Pacific Ocean	335 µm	n.r.	**1485**	**12,320,000**	0.0059	49.3	0.08	670	54
China Coast	500 µm[a]	>5 mm	n.r.	n.r.	**0.1168**	**0.455**	0.5	1.93	55
North Iberian Coast	333 µm	5 mm	**60,500**	**285,000**	0.605	2.85	0.21	0.72	56
Mediterranean Sea	200 µm	10 cm	**244,000**	**1,115,000**	**0.423**	**1.93**	1.69	7.74	57
West Mediterranean Sea	330 µm	5 mm	**130,000**	**420,000**	1.3	4.2	0.57	2.16	58
Northwest Sardinia	200 µm	5 mm	n.r.	n.r.	**0.17**	**1.69**	1.39	13.8	59
West Mediterranean Sea	200 µm	20 mm	**400,000**	**4,520,000**	**1**	**11.3**	**1.68**	**26.1**	60
Subsurface									
Northeast Atlantic Ocean	250 µm	1 mm	—	—	**2**	**20**	8.4	83.8	61
Northeast Pacific Ocean	62.5 µm	5 mm	—	—	**2080**	**9200**	231	1019	62
North Atlantic Ocean	10, 50 µm	10 mm	—	—	**c. 100**	**501**	0.05	0.25	63

n.r. = not reported.
[a]Mesh size 333 µm, but only particles >500 µm were numbered.

automatic quantification of microplastics are not developed yet. Maximum values are found in subtropical gyres (See Fig. 6.5). The distribution of microplastics in subsurface waters differs from that of floating plastic debris, and shows higher values in coastal compared to oceanic sites.[64] In beach sediments at the strand zone microplastic densities are much higher, with maximum values in the order of thousands of items per square meter,[65] although amounts of stranded debris is submitted to the interference of both natural (beach dynamics, storm events) and anthropogenic (touristic use, cleanup activities) factors.

... at levels that currently do not cause ecotoxicological concern

Microplastics are readily taken up by planktonic and benthic particle feeders and incorporated into trophic webs,[66] being present in both pelagic and demersal fish at concentrations that rarely exceed 1−2 particles per fish,[67] and in bivalves at levels below 1 particle per gram.[68] None of these figures support any potential ecotoxicological effect on marine organisms. Human health is not endangered either by the microplastic contents of marine food. FAO has recently reported that, even under the worst case scenario that assumes the highest concentrations of additives and sorbed contaminants, and complete transfer to the consumer, exposure to these substances derived from consumption of fisheries and aquaculture products is negligible compared to other dietary sources.[69] Despite lack of toxic effects at environmentally relevant concentrations and no risk for marine food safety, the fact that plastic microparticles are becoming ubiquitous should rise concern and prompt us to limit release of plastic into the environment.

The role of microplastics as vectors of hydrophobic pollutants to marine organisms is currently challenged by experimental evidence and modeling

Polychlorinated biphenyls (PCBs), polybrominateddiphenyl ethers (PBDEs), perfluorinated compounds, and many other organic pollutants have a very large polymer-water distribution coefficient, $K_{P/W}$ [L/kg], in favor of the plastic.[70] It has been hypothesized that ingestion of microplastics by aquatic species leads to increased exposure to both plastic additives and lipophilic compounds adsorbed to their surface. Concerning additives, biodynamic models for both invertebrates and fish do not give relevance to such exposure pathway.[71]

In addition to the potential toxicity caused by their additives, plastics offer a hydrophobic surface where many hydrophobic organic pollutants adsorb. Laboratory experiments at very high levels of exposure to microplastics have demonstrated for marine organisms with different feeding habits that the organics concentrated on the plastic surface can be assimilated and bioaccumulated, and may cause a number of biochemical and physiological changes in the consumers, from acetylcholinesterase inhibition to weight loss.[72] These results must be taken with caution. Thermodynamic considerations do not always support desorption of hydrophobic organics from ingested microplastics, but may even predict reduction in the organics body burden as a consequence of their high affinity for the polymers.[73] Also, adsorbed pollutants do not stay at the surface of the particles but diffuse among the polymer chains, reducing their availability. In addition, in environmentally realistic conditions, abundance of plastic items is low compared to that of other carrier media, such as suspended organic particulates or natural prey items, and their ingestion is not likely to increase the exposure to hydrophobic organics.[74] In summary, the most recent studies do not support a relevant role of microplastics as vectors of pollutants to the organisms, advocate a change of paradigm, or recommend taking into account the many other living and inert organic phases present in the seawater under environmentally realistic conditions.[75]

KEY IDEAS

- The reduction of consumption rates and reutilization of packaging is the most sustainable solution for the problem of solid waste management in the cities.

- Plastic objects have two components: a polymer matrix and a series of chemicals called additives used to give the object its physical properties.

- Hundreds of chemicals are used as plastic additives to provide flexibility (plasticizers), resistance to light (UV filters), low flammability (flame retardants), prevent oxidation or microbial attack, etc. A few of these chemicals (BPA, orthophthalates) are restricted because of their suspected or demonstrated endocrine disruption potential.

- Conventional polymers, such as PE, PP, or PVC are not biodegradable and once disposed persist in the environment for many years.

- More than 6 million tons of anthropogenic litter end up in the oceans every year. Most of this (up to 83% according to some sources) is plastics.

- In the marine environment plastics tend to leach their additives, become more brittle, and fragment into smaller particles called microplastics that become ubiquitous in all marine compartments.

- Floating marine litter accumulates in the central gyres of the global ocean, while subsurface microplastics are more abundant in coastal areas.

- Drifting plastic objects and particularly discarded fishing gear entangle marine mammals and other large marine organisms.

- Microplastics accumulate hydrophobic chemicals found at very low concentrations in the seawater, but they do not easily transfer them to organisms upon ingestion.

Endnotes

1. http://web.worldbank.org/WBSITE/EXTERNAL/TOPICS/EXTURBANDEVELOPMENT/ EXTUSWM/0,menuPK:463847~pagePK:149018~piPK:149093~theSitePK:463841,00. html.

2. https://www.epa.gov/smm/sustainable-materials-management-non-hazardous-materials-and-waste-management-hierarchy.

3. Directive 94/62/EC of the European Parliament and Council Directive of 20 December 1994.

4. Al-Salem SM, Lettieri P, Baeyenset J. Recycling and recovery routes of plastic solid waste (PSW): A review. Waste Management 2009;29:2625–2643.

5. https://journalistsresource.org/wp-content/uploads/2011/11/Report-from-Columbia-Universitys-Earth-Engineering-Center.pdf.

6. Abarca Guerrero L, Maas G, Hogland W. 2013. Solid waste management challenges for cities in developing countries. Waste Management 2013; 33:220–232.

7. Plastics Europe. Plastics - the Facts 2016. An analysis of European plastics production, demand and waste data; 2016.

8. OECD. OECD Series on emission scenario documents Number 3. Emission scenario document on plastic additives. ENV/JM/MONO(2004)8/REV1. Plastics Europe 2015 Plastics – the Facts 2014/2015 An analysis of European plastics production, demand and waste data; 2009.

9. See Endnote 7.

10. Frost and Sullivan. Strategic analysis of the European and North American plastics additives market; 2015.

11. http://www.plasticgarbageproject.org/en/plastic-garbage/solutions/bioplastic/.

12. See Endnote 8.

13. Mhadhbi L, Fumega J, Boumaiza M, Beiras R. Acute toxicity of polybrominated diphenyl ethers (PBDEs) for turbot (*Psetta maxima*) early life stages (ELS). Environ Sci Pollut Res 2012;19:708–717.

14. Van der Veen I, de Boer J. Phosphorus flame retardants: Properties, production, environmental occurrence, toxicity and analysis. Chemosphere 2012;88:1119–1153.

15. Bornehag C-G, Sundell J, Weschler CJ, et al. The Association between Asthma and Allergic Symptoms in Children and Phthalates in House Dust: A Nested Case–Control Study. Environ Health Perspect 2004;112(14):1393–1397.

16. Jambeck JR, Geyer R, Wilcox C, et al. 2015. Plastic waste inputs from land into the ocean. Science 2015;347(6223):768–771.

17. www.plasticgarbageproject.org.

18. Colton JB, Knapp FD, Burns BR. Plastic Particles in Surface Waters of the Northwestern Atlantic. Science 1974;185:491–497.

19. Cózar A, Sanz-Martín M, Martí E, et al. Plastic Accumulation in the Mediterranean Sea. PLoS One 2015;10:e0121762.

20. Hidalgo-Ruz V, Gutow L, Thompson RC, et al. Microplastics in the Marine Environment: A Review of the Methods Used for Identification and Quantification. Environ Sci Technol 2012;46:3060–3075.

21. Law KL, Morét-Ferguson SE, Maximenko NA, et al. Plastic Accumulation in the North Atlantic Subtropical Gyre. Science 2010; 329:1185–1188; Law KL, Moret-Ferguson SE, Goodwin DS, et al. Distribution of Surface Plastic Debris in the Eastern Pacific Ocean from an 11-Year Data Set. Environ Sci Technol 2014;48(9):4732–4738.

22. Galgani F. Marine litter, future prospects for research. Front Marine Sci 2(87):1–5.

23. Eriksen M, Lebreton LCM, Carson HS, et al. Plastic Pollution in the World's Oceans: More than 5 Trillion Plastic Pieces Weighing over 250,000 Tons Afloat at Sea. PLoS ONE 2014;9(12): e111913. Cózar A, Echevarría F, González-Gordillo JI, et al. Plastic debris in the open ocean. PNAS 2014;111(28):10239–10244.

24. Law et al. (2010) op. cit.

25. Hanni KD, Pyle D. 2000 Entanglement of Pinnipeds in Synthetic Materials at South-east Farallon Island, California, 1976–1998. Mar Pollut Bull 2000;40(12):1076–1081.

26. Page B, McKenzie J, McIntosh R, et al. Entanglement of Australian sea lions and New Zealand fur seals in lost fishing gear and other marine debris before and after Government and industry attempts to reduce the problem. Mar Pollut Bull 2004;49:33–42.

27. Henderson JR. A Pre- and Post-MARPOL Annex V summary of Hawaiian Monk Seal entanglements and marine debris accumulation in the Northwestern Hawaiian islands, 1982–1998. Mar Pollut Bull 2001;42(7):584–589.

28. French DP, Reed M. Potential impact of entanglement in marine debris on the population dynamics of the northern fur seal, Callorhinus ursinus, p. 431–452. In: Shomura RS, Godfrey ML editors. Second International Conference on Marine Debris. United States Department of Commerce. 1990.

29. Fowler CW. Marine debris and fur seals: a case study. Mar Pollut Bull 1987;18(6B):326–335.

30. Laws EA. Aquatic pollution. An introductory text. 4th ed. Hoboken, Wiley; 2018.

31. Furness RW. Ingestion of Plastic Particles by Seabirds at Gough Island, South Atlantic Ocean. Environ Pollut A 1985;38:261–272.

32. Avery-Gomm S, O'Hara PD, Kleine L, et al. Northern fulmars as biological monitors of trends of plastic pollution in the eastern North Pacific. Mar Pollut Bull 2012;64:1776–1781.

33. Bond AL, Provencher JF, Daoust P-Y et al. Plastic ingestion by fulmars and shearwaters at Sable Island, Nova Scotia, Canada. Mar Pollut Bull 2014;87:68–75.

34. Dickerman RW, Goelet RG. Northern Gannet Starvation After Swallowing Styrofoam. Mar Pollut Bull 1987;18(No. 6):293; Pierce KE, Harris RJ, Larned LS, et al. Obstruction and starvation associated with plastic ingestion in a Northern Gannet Morus bassanus and a Greater Shearwater Puffinus gravis. Marine Ornithol 2004;32:187–189.

35. 46 cc according to Sileo L, Sievert PR, Samuel MD. Causes of mortality of albatross chicks at Midway Atoll. J. Wildl. Disease 1990;26:329–38; and 18-24 g according to Auman HJ, Ludwig JP, Giesy JP, et al. Plastic ingestion by Laysan Albatross chicks on Sand Island, Midway Atoll, in 1994 and 1995. In: Robinson G, Gales R, editors. Albatross biology and conservation. Chipping Norton: Surrey Beatty & Sons; 1997. pp. 239–244.

36. Auman et al. (1997) op. cit.

37. Hanvey JS, Lewis PJ, Lavers JL, et al. A review of analytical techniques for quantifying microplastics in sediments. Anal. Methods 2017;2017(9):1369–1383.

38. Andrady AL. Microplastics in the marine environment. Mar Pollut Bull. 2011;62:1596–1605.

39. See Endnote 20.

40. Thompson RC, Olsen Y, Mitchell RP, et al. Lost at sea: where is all the plastic? Science 2004; 304:838.

41. Martins J, Sobral P. Plastic marine debris on the Portuguese coastline: a matter of size? Mar Pollut Bull 2011;62:2649–2653.

42. Barnes DK, Galgani F, Thompson RC, et al. Accumulation and fragmentation of plastic debris in global environments. Phil Trans R Soc B 2009;364:1985–1998.

43. Carpenter EJ, Smith Jr. KL. Plastics on the Sargasso Sea surface. Science 1972;175:1240–1241.

44. See Endnote 18.

45. Day RH, Shaw DG. Patterns in the abundance of pelagic plastic and tar in the North Pacific Ocean, 1976–1985. Mar Pollut Bull 1987;18(6B):311–316 and cites therein.

46. Ryan PG. The characteristics and distribution of plastic particles at the sea surface off the southwestern Cape Province, South Africa. Mar Environ Res 1988;25:249–273.

47. Moore CJ, Moore SL, Leecaster MK, Weisberg SB. A comparison of plastic and plankton in the North Pacific central gyre. Mar Pollut Bull 2001;42:1297–1300.

48. Moore CJ, Moore SL, Weisberg SB, et al. A comparison of neustonic plastic and zooplankton abundance in Southern California's coastal waters. Mar Pollut Bull 2002;44:1035–1038.

49. Yamashita R, Tanimura A. Floating plastic in the Kuroshio Current area, western North Pacific Ocean. Mar Pollut Bull 2006;54:464–488.

50. Doyle MJ, Watson, W, Bowlin NM, et al. Plastic particles in coastal pelagic ecosystems of the Northeast Pacific Ocean Mar Environ Res 71, 41–52. Mar Environ Res 2011;71:41–52.

51. Collignon, A. et al. Neustonic microplastic and zooplankton in the North Western Mediterranean Sea. Mar. Pollut Bull 2012;64:861–864.

52. Frias PGL, Otero V, Sobral P. Evidence of microplastics in samples of zooplankton from Portuguese coastal waters. Mar Environ Res 2014;95(2014):89–95.

53. de Lucia GA, Matiddi M, Marra S, et al. Neustonic microplastic in the Sardinian coast (Central-Western Mediterranean Sea): amount, distribution and impact on zooplankton. Mar Environ Res 2014;100:10–16.

54. Law et al. (2014) op. cit.

55. Zhao S, Zhu L, Wang T, et al. Suspended microplastics in the surface water of the Yangtze Estuary System, China: First observations on occurrence, distribution. Mar Pollut Bull 2014;86:562–568.

56. Gago J, Henry M, Galgani F. First observation on neustonic plastics in waters off NW Spain (spring 2013 and 2014). Mar Environ Res 2015;111:27–33.

57. See Endnote 19.

58. Faure F, Saini G, Potter G, et al. An evaluation of surface micro-and mesoplastic pollution in pelagic ecosystems of the Western Mediterranean Sea. Environ Sci Pollut Res 2015:22:12190–12197.

59. Panti C, Gianneti M, Baini M, et al. Occurrence, relative abundance and spatial distribution of microplastics and zooplankton NW of Sardinia in the Pelagos Sanctuary Protected area, Mediterranean Sea. Environ Chem 2015;12:618–626.

60. Suaria G, Avio CG, Mineo A et al. The Mediterranean Plastic Soup: synthetic polymers in Mediterranean surface waters. Sci Rep 2016;6:37551. DOI: 10.1038/srep37551.

61. Lusher AL, Burke A, O'Connor I, et al. Microplastic pollution in the Northeast Atlantic Ocean: validated and opportunistic sampling. Mar Pollut Bull 2014;88:325–333.

62. Desforges JPW, Galbraith M, Dangerfield N, et al. Widespread distribution of microplastics in subsurface seawater in the NE Pacific Ocean. Mar Pollut Bull 2014;79:94–99.

63. Enders K, Lenz R, Stedmon CA, Nielsen TG. Abundance, size and polymer composition of marine microplastics $\geq$ 10 μm in the Atlantic Ocean and their modelled vertical distribution. Mar Pollut Bull 2015;100:70–81.

64. Desforges et al. (2014) op. cit.; Enders et al. (2015) op. cit.

65. Van Cauwenberghe, et al. Mar Environ Res 2015;111:5–17.

66. Setälä O, Fleming-Lehtinen V, Lehtiniemi M. Ingestion and transfer of microplastics in the planktonic food web. Environ Pollut 2014;185:77–83.

67. Lusher AL, McHugh M, Thompson RC. Occurrence of microplastics in the gastrointestinal tract of pelagic and demersal fish from the English Channel. Mar Pollut Bull 2013;67:94 Neves et al. 2015; Bellas et al. MPB 2016.

68. Van Cauwenberghe L, Jansen CR. Microplastics in bivalves cultured for human consumption. Environ Pollut 2014;193:65−70.

69. Lusher AL, Hollman PCH, Mendoza-Hill JJ, 2017. Microplastics in fisheries and aquaculture: status of knowledge on their occurrence and implications for aquatic organisms and food safety. FAO Fisheries and Aquaculture Technical Paper. No. 615. Rome, Italy.

70. Andrady. (2011) op. cit.

71. Koelmans AA, Besseling E, Foekema EM. Leaching of plastic additives to marine organisms. Environ Pollut 2014;187:49−54.

72. Besseling E, Wegner A, Foekema EM, et al. Effects of Microplastic on Fitness and PCB Bioaccumulation by the Lugworm *Arenicola marina* (L.). Environ Sci Technol 2013;47: 593−600.
Brown MA, Niven SJ, Galloway TS, et. al. Microplastic Moves Pollutants and Additives to Worms, Reducing Functions Linked to Health and Biodiversity. Current Biology 2013;23: 2388−2392.
Oliveira M, Ribeiro A, Hylland K, et al. Single and combined effects of microplastics and pyrene on juveniles (0+ group) of the common goby *Pomatoschistus microps* (Teleostei, Gobiidae). Ecol Indicators 2013;34:641−647.
Bakir A, Rowland SJ, Thompson RC. Enhanced desorption of persistent organic pollutants from microplastics under simulated physiological conditions. Environ Pollut 2014;185:16−23.
Avio CG, Gorbi S, Milan M, et al. Pollutants bioavailability and toxicological risk from microplastics to Pollutants bioavailability and toxicological risk from microplastics to marine mussels. Environ Pollut 2015;198:211−222.

73. Gouin T, Roche N, Lohmann R, et al. Thermodynamic Approach for Assessing the Environmental Exposure of Chemicals Absorbed to Microplastic. Environ Sci Technol 2011;45: 1466−1472.

74. Koelmans AA, Bakir A, Burton GA, et al. Microplastic as a Vector for Chemicals in the Aquatic Environment: Critical Review and Model-Supported Reinterpretation of Empirical Studies. Environ Sci Technol 2016;50:3315−3326.

75. Herzke D, Anker-Nilssen T, Nøst TH, et al. Negligible Impact of Ingested Microplastics on Tissue Concentrations of Persistent Organic Pollutants in Northern Fulmars off Coastal Norway. Environ Sci Technol 2016;50:1924−1933.
Ziccardi LM, Edgington A, Hentz K, Kulacki KJ, Driscoll SK. Microplastics as vectors for bioaccumulation of hydrophobic organic contaminants in the marine environment: A state-of-the-science review. Environ Toxicol Chem 2016;35:1667−1676.
Beckingham B, Ghosh U. Differential bioavailability of polychlorinated biphenyls associated with environmental particles: Microplastic in comparison to wood, coal and biochar. Environ Pollut 2017;220:150−158.
Lohmann R. Microplastics are not important for the cycling and bioaccumulation of organic pollutants in the oceans—but should microplastics be considered POPs themselves? Integrated Environmental Assessment and Management 2017;13(3):460−465.

Suggested Further Reading

- Galgani F. Marine litter, future prospects for research. Frontiers in Marine Science 2015;2(87): 1−5.

- Cózar A, Echevarría F, González-Gordillo JI, et al. Plastic debris in the open ocean. Proc Natl Acad Sci Unit States Am 2014;111(28):10239−44.

- Hansen E. Hazardous substances in plastic materials. COWI-Danish Technological Institute 2013. 149 pp. Available on line, http://www.miljodirektoratet.no/old/klif/publikasjoner/3017/ta3017.pdf.
- Koelmans AA, Bakir A, Allen Burton GA, Janssen CR. Microplastic as a vector for chemicals in the aquatic environment: critical review and model-supported reinterpretation of empirical studies. Environ Sci Technol 2016;50:3315−26.
- Page B, McKenzie J, McIntosh R, et al. Entanglement of Australian sea lions and New Zealand Fur seals in lost fishing gear and other marine debris before and after Government and industry attempts to reduce the problem. Mar Pollut Bull 2004;49:33−42.

CHAPTER 7

Hydrocarbons and Oil Spills

7.1 SOURCES OF OIL IN THE SEA

Gas (propane, butane) and liquid fuels (gasoline, diesel, kerosene), plastics, lubricants, asphalt, solvents, and a myriad of other chemical products are obtained from underground reserves of fossil fuels: crude oil or petroleum and natural gas. Fossil fuels are originated at high temperature and pressure conditions from dead organisms buried under sedimentary rocks. They are obtained from the underground by drilling, and processed in refineries by the petrochemical industry to yield the petroleum products that are used for heating, transport, plastic synthesis, and other industrial and domestic applications. Billions of barrels of oil a day are shipped in oil tankers to various destinations all over the world. This is why accidental oil spills (see Section 7.5) account for some of the most feared marine pollution events worldwide.

Crude oil is the source of fuels, plastics, and many chemicals

However, at a global scale, tanker accidents are not the main source of oil in the sea. Estimates of total amount of oil entering the world's oceans vary more than one order of magnitude among sources, but tend to fall between 1 and 3 million tons per year.[1] The amount accounted for by natural sources (submarine oil seeps) is roughly estimated in 0.6 million T/y.[2] Concerning anthropogenic inputs, more than a half comes from land-based sources (urban run-off and industrial discharges), one-fourth from maritime transport (either intentional or accidental), and the rest from atmospheric sources, mainly (Table 7.1). The number of accidental oil spills has decreased over the last 20 years, partly as a result of implementation of measures for ship safety and prevention of maritime accidents.

Globally land-based sources are the main input of oil in the sea, but locally accidental oil spills caused high ecological impact

7.2 CHEMICAL COMPOSITION OF OIL; HYDROCARBONS

Crude oil is mainly composed of hydrogen and carbon that normally account for c. 97 percent of the oil total mass,[4] in addition to minor components such as sulfur (up to 6 percent), oxygen (up to 3.5 percent), nitrogen (up to 0.5 percent), and trace metals such as vanadium (39 ppm) and nickel (11 ppm).[5] That is because the major components of oil are nonpolar organic

Oil is a very complex mixture of organic molecules mainly composed of H and C

89

Table 7.1 Relative Importance of Sources of Oil Entering the World's Oceans

	Oil Entering Global Oceans. Total: 1—3 million Tons Per Year
50%	Land-based sources (urban run-off, discharges from industry, etc.)
18%	Operational ship discharges
13%	Atmospheric sources (petrochemical industry, vehicle exhaust, etc.)
10%	Natural sources (underwater seeps)
6%	Accidental spills
3%	Offshore extraction

Source: EEA (2007).

molecules entirely constituted by H and C called **hydrocarbons**. This is the only feature hydrocarbons share. In that complex mixture of molecules with molecular weights ranging from 16 (methane) to >1500 (asphaltene) there are compounds that are gaseous, liquid, and solid at atmospheric pressure. Some are very stable and environmentally persistent and others readily undergo transformation through physical, chemical, or biologically mediated pathways. The most simple hydrocarbon molecules are **alkanes**: saturated (with all available carbon valences linked to hydrogen atoms) molecules that can be linear (paraffins), branched, or cyclic (cycloalkanes) (Fig. 7.1). Waxes are high molecular weight linear alkanes (C18 and up). Thus, for industrial use, this complex mixture must be resolved during the process of refining, which is essentially a distillation that separates oil fractions according to their boiling point. Liquefied petroleum gases such as propane and butane are obtained at 30°C, gasoline (5—9 C) between 30 and 140°C, naphtha (10—12 C) at 150—200°C, kerosene (13—17 C) between 170 and 250°C, gasoil (17—20 C) up to 320°C, and fuel oil (20—35 C) up to 500°C. The solid residual fraction, tar, can be used for waterproofing or road surface.

Unsaturated hydrocarbons include **alkenes** or olefins, with at least one carbon—carbon double bond, such as ethylene and propylene, which are the source of most plastics, and **alkynes**, with at least one triple bond, such as acetylene. The double and triple bonds give these molecules more reactivity than that of alkanes. Olefins are not obtained directly from crude oil but synthetized from alkanes or naphtha by the petrochemical industry. Significant amounts of olefins are found in refined products only.

Aromatic hydrocarbons account for 1—20 percent of total hydrocarbons in crude oil.[6] They are structurally characterized by benzene rings with delocalized electrons shared by the entire six-carbon ring. The more common aromatic compounds found in crude oil are often referred to as BTEX (benzene, toluene,

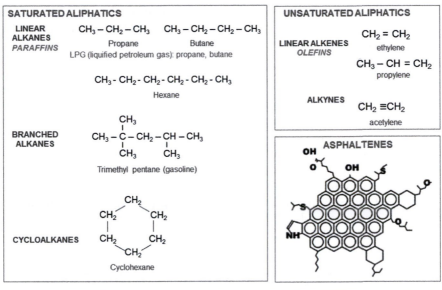

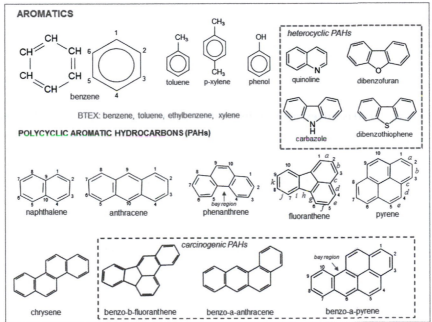

FIGURE 7.1

Chemical structures of the main types of hydrocarbons: saturated (linear, branched, and cyclic) and unsaturated aliphatics, aromatics, and asphalthenes. Alkanes are saturated molecules with all single C—C bonds, alkenes show some double bonds and alkynes some triple bond. Linear alkanes are commonly termed "paraffins," alkenes "olefins," and the most common monocyclics "BTEX."

ethylbenzene, and xylene). Aromatic hydrocarbons with two or more benzene rings are termed polycyclic aromatic hydrocarbons. They bear particular ecotoxicological interest and will be dealt with in Section 7.2.

Polar components of crude oil bear molecular charges in O, N, or S groups and are divided in **resins**, with the lowest molecular weights, and **asphaltenes**, very large molecules with many condensed aromatic rings. For analytical purposes, oil components were traditionally separated according to the SARA classification into "saturated," "aromatics," "resins," and "asphaltenes." The separation is based on the use of different solvents. Asphaltenes typically precipitate in *n*-pentane. The other components are obtained by loading the *n*-pentane dissolved fraction onto a silica gel column and eluting with solvents (and solvent mixtures) of increasing polarity such as hexane for saturates, benzene and dichloromethane for aromatics, and dichloromethane and methanol for resins.[7]

7.3   POLYCYCLIC AROMATIC HYDROCARBONS

Minoritary, highly stable components of oil provide "finger-prints" that help in investigating its origin

Polycyclic aromatic hydrocarbons (PAHs) are hydrocarbons containing two or more fused rings, commonly benzene rings (see Fig. 7.1), but occasionally 5C-rings or heterocycles containing N (indols, carbazoles, quinolones), S (mercaptanes, thiophenes), or O (fatty acids, phenols, furanes). "True" hydrocarbon and heterocyclic PAHs are detected with the same analytical techniques and share similar molecular weight-dependent properties. Most PAHs are exclusively formed by fused benzene rings, and they are responsible for most of the ecotoxicological issues caused by oil. However, some minority PAHs with more complex structures such as hopanes and steranes are important because they are highly stable and microbiologically resistant, and their structures have suffered minor changes compared to the original biological substrate. Thus, their ratios can be used as signatures of petroleum from a certain origin, a technique known as *fingerprinting*. This technique provides information on the biological precursors and geological conditions of the oil formation, and it has an important applied relevance in forensic investigations since it facilitates the identification of the origin of spilled oil.

Polyaromatic hydrocarbons are responsible for most ecotoxicological problems caused by oil

Polyaromatic hydrocarbons and their metabolites are responsible for most of the toxicological issues caused by oil spills. However, their environmental fate and biological interactions, including carcinogenesis, are greatly affected by structure and molecular weight. The latter strongly influences volatility, solubility in water, bioaccumulation potential, and environmental persistence.

Water solubility decreases as molecular weight increases, from $>1\,g/L$ for benzene until $<1\,\mu g/L$ for the largest PAHs, and volatility decreases from c. $12\,Pa$ for benzene to $10^{-8}\,Pa$ for large PAHs. K_{ow} in turn increases more than five orders of magnitude, and bioaccumulation has been shown to be a direct function of this parameter at least within certain ranges (see Section 11.3).

The effects of temperature and salinity on water solubility are in comparison less remarkable. A 10°C increase in temperature increases roughly twofold the solubility of PAHs.[8] Ionic strength (and thus salinity) reduces water solubility of organic chemicals according to the empirical model:

$$\log S/S_o = -K_{salt} C_{salt} \qquad \text{(Eq. 7.1)}$$

where S and S_o are the solubility in salt and fresh water respectively, C_{salt} is the molar salt concentration and K_{salt} is a constant (called Setschenow constant) typical for each chemical. K_{salt} values for PAHs typically range from 0.22 to 0.34 L/mol. Ni and Yalkowski[9] found that for a wide range of organic chemicals K_{salt} was a direct linear function of K_{ow} according to the empirical equation:

$$K_{salt} = 0.040 \log K_{ow} + 0.114 \qquad \text{(Eq. 7.2)}$$

i.e., the higher the K_{ow} of a chemical the more remarkable is the effect of salinity on its water solubility.

Dissolved organic matter can sorb PAHs and increase their solubility. For higher molecular weight PAHs environmental concentrations are significantly correlated with suspended solids, and uptake via particles or food is more relevant than the waterborne fraction (Table 7.2).

Table 7.2 Concentrations of Polycyclic Aromatic Hydrocarbon (PAH) (μg/Kg Dry Weight) in Sediments (Fraction <2 mm) From Clean (CA) and Polluted (VN) sites At Ría De Vigo (NW Iberian Peninsula), Sediment Quality Criteria (TEL and PEL), and Enrichment Factor at VN, Calculated as Concentration at VN Divided by Concentration at CA. Notice the Regularity in EF_{VN} Values Despite Large Variability in Individual PAH Concentrations

	Ref. (CA)	TEL	PEL	Polluted Site (VN)	EF_{VN} (C_{VN}/C_{CA})
Phe	40.6	86.7	544	1038	26
Ant	8.8	46.9	245	290	33
Flu	84.6	113	1494	2175	26
Pyr	69.6	153	1398	1957	28
BaA	46.5	74.8	693	1258	27
Chry	48.8	108	846	1289	26
BaP	50.7	88.8	763	1749	34
DBA	12.0	6.2	135	190	16
IPy	59.8	—	—	1334	22

Data Beiras et al. Ecotoxicology 2012;21:9–17.

Petrogenic PAHs, originated from oil, are more bioavailable than pyrogenic ones

Although PAHs can be synthesized by plants, bacteria, and fungi (biogenic), they are normally originated from partial combustion of organic matter (**pyrogenic**) or as components of fossil fuels (**petrogenic**). Natural sources in aquatic environments include underwater cold seeps or surface run-off after forest fires. However, the majority of PAH released into the sea are derived from anthropogenic activities such as tanker operations, deliberate and accidentals spills, municipal and industrial effluents, and atmospheric deposition into surface waters. Pyrogenic PAHs may be tightly bound to particles and thus be much less bioavailable than petrogenic PAHs.

Considering their hydrophobic nature and their tendency to adsorb to organic particles, sediments are a sink for PAHs. Moreover, since the environmental persistence of hydrocarbons within sediments is longer than in the water column, sediments are the ideal matrix to monitor marine pollution by these compounds.

PAHs accumulate in sediments and invertebrates, while fish can metabolize them.

Concerning biota, bivalve mollusks and, to a lesser extent, other invertebrates, tend to accumulate PAHs with BCF in the order of 10,000 or even higher.[11] The kinetics of uptake is dependent on the exposure route. Uptake of PAHs from the dissolved phase is much quicker than uptake from sediments,[12] and PAH accumulation in suspension feeders reached equilibrium more rapidly than in deposit feeders. However, since PAH tend to sorb to particles and accumulate in sediments, final bioaccumulation is higher in the latter, particularly for the larger PAH, since BCF increases as the number of aromatic rings increases.[13]

Fish have a more efficient biotransformation capacity that leads to their excretion (see later in the chapter), and thus show BCF much lower than those from invertebrates (Table 7.3).

A wide variety of bacteria, fungi, and algae have the ability to metabolize aromatic hydrocarbons and are responsible for their biodegradation in the aquatic environments (see Section 10.3). Biodegradability decreases with number of benzene rings, and no microorganisms have been isolated capable to completely degrade to carbon dioxide PAHs with more than three rings.[16]

Biotransformation of PAHs produce carcinogenic intermediate metabolites

In vertebrates (Fig. 7.2), biotransformation of PAHs is accomplished by cytochrome P-450 dependent monooxygenase enzymes called **mixed function oxidases (MFO)**, such as benzopyrene hydroxylase and other aryl hydrocarbon hydroxylases, capable of oxidation of parent PAHs and production of polar metabolites that accumulate in the fish bile (See Section 12.3). Both MFO activity and bile metabolites can be used to trace exposure of fish to these compounds (See Section 15.3).

The mechanism of carcinogenicity in fish by high molecular weight PAH such as BaP has been reviewed by Baumann.[17] It involves activation of the parental

Table 7.3 Concentrations of PAH (µg/Kg Dry Weight) in Wild Mussels From Clean (BA) and Polluted (Vigo Harbor) Sites at Ría De Vigo (NW Iberian Peninsula). OSPAR BAC Values are Also Shown

	Mussel BAC[15]	Clean Site (BA)	Vigo Harbor
Phe	11	11.1	15.9–295
Ant	–	2.4	2.0–91.0
Flu	12.2	15.3	34.9–127
Pyr	9.0	11.1	34.8–564
BaA	2.5	3.4	7.6–148
Chry	8.1	6.5	15.3–409
BbF	-	5.0	21.5–99.1
BaP	1.4	1.0	5.7–75.5
BghiP	2.5	2.8	15.1–24.2
IPy	2.4	2.3	10.7–23.9
SUM13[a]	–	67.7	220–2066

[a]Phe, Ant, Flu, Pyr, BaA, Chry, BeP, BbF, BkF, BaP, BghiP, DBA, IPy.
Data Beiras et al. Ecotoxicology 2012;21:9–17

compound through MFO enzymes. MFO catalyze the oxidation of BaP to form **epoxides**, toxic metabolites that undergo nonenzymatic rearrangement to form phenols or are the substrate of epoxide hydrolases (EH) which rend dihydrodiols (metabolites with two contiguous hydroxyl groups). However, dihydrodiol epoxydes subsequently formed by MFO are poor substrates for EH and can only be detoxified by conjugation with glutathione, a reaction catalyzed by the Phase II biotransformation enzyme GST. Certain isomeric forms of dihydrodiol epoxides such as the 7,8-dihydrodiol-9,10-epoxides are the ultimate carcinogenic agents responsible for the mutagenic and carcinogenic effects of BaP, because they covalently bind to DNA (mainly to guanine bases) to form DNA adducts and impair correct gene expression, triggering neoplasias. Fish liver is a target tissue of PAH-induced carcinogenesis, in contrast to mammal liver, which is relatively resistant to neoplasia. Elevated levels of high

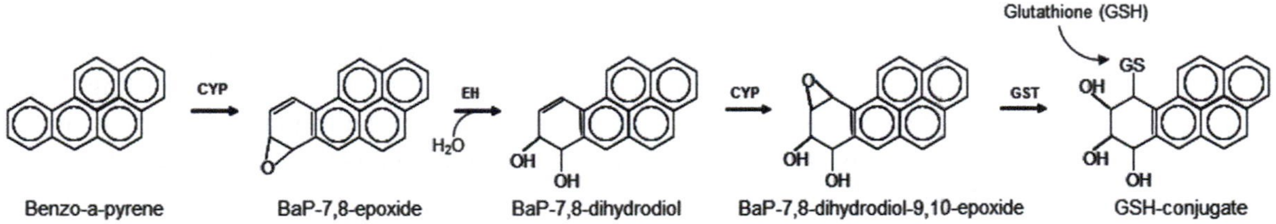

Benzo-a-pyrene   BaP-7,8-epoxide   BaP-7,8-dihydrodiol   BaP-7,8-dihydrodiol-9,10-epoxide   GSH-conjugate

FIGURE 7.2

Metabolism of BaP in vertebrates. *EH*, epoxide hydrolase; *GST*, glutathion transferase; *MFO*, mixed function oxydase. MFO and EH are Phase I biotransformation enzymes, while GST is a Phase II biotransformation enzyme (see Chapter 12). Dihydrodiol epoxides are responsible for the carcinogenic properties of BaP.

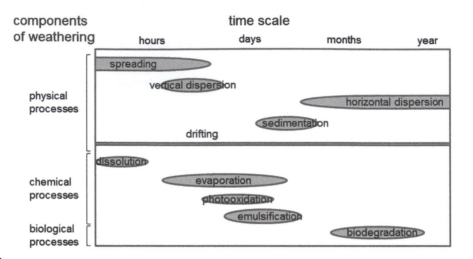

FIGURE 7.3

Schematic representation of the time-course of events, globally termed "weathering," that modify the composition and properties of oil once spilled in the sea.

molecular weight PAHs in sediments have been repeatedly associated to enhanced occurrence of liver and skin neoplasia in bottom-dwelling fishes.

Because of their carcinogenic properties, in Europe maximum levels for the sum of BaA, Chry, BbF, and BaP in foodstuffs are regulated to a maximum of 30 μg/Kg WW in bivalves and a maximum ranging from 12 to 35 μg/Kg WW (depending on the species) in smoked fish and shellfish.[18]

7.4   WEATHERING OF OIL IN THE SEA

The "weathering" process changes the composition and fate of the spilled oil slick

When oil is spilled in the marine environment a set of physical, chemical, and biological processes termed "weathering" begin immediately to move it, distribute it among the environmental compartments, and transform its composition, properties, and environmental fate. Weathering can be mimicked in the laboratory by increasing temperature (up to 250°C) and/or irradiating with PAR or UV light. The time scale of the different processes that include weathering is summarized in Fig. 7.3. Oil is less dense than water and it spreads to form a thin film, the oil slick, that drifts on the sea surface according to the atmospheric and hydrodynamic conditions. **Spreading** takes place first, and it is primarily dependent on the viscosity of the oil, with lighter crudes spreading faster.

Dispersion is a mixing process with horizontal and vertical components, and it is primarily determined by the turbulence caused by waves, currents, and other hydrodynamic factors. **Drifting** of the oil slick is mainly determined by winds, with oil often assumed to travel at one-tenth of the wind speed. All these physical processes can be modeled in order to predict the distribution and displacement of the oil in the sea, which is important to plan efficient measures of response to an oil spill.

Within a few days following the spill light crude oils can lose up to 75 percent of their initial volume and medium crudes up to 40 percent as a consequence of **evaporation**.[19] Heavy or residual oils will not lose more than 10 percent. The rapid loss by evaporation of the lighter aromatics (BTEX) explains the initial decrease in toxicity of the water-accommodated fraction of weathered oil.[20] **Dissolution** accounts for only a small proportion of oil loss, but it is important because soluble components, particularly the smaller aromatic compounds, are the most toxic to aquatic species. Energy from sunlight radiation, particularly from the shorter wave-length spectrum, can cause the formation of highly reactive singlet oxygen that react with hydrocarbons causing **photooxidation**. This process is unimportant from a mass balance consideration but very relevant ecotoxicologically since the oxidized derivatives may be much more toxic than the parental compounds. Aliphatic and aromatic fractions are oxidized in sunlight to more soluble ketones, aldehydes, and other polar compounds.

Evaporation and dissolution of lighter components contribute to an increase in the density of the weathered oil slick that may eventually sink. **Sedimentation** is also promoted by the adsorption of oil to particles either in suspension or after stranding onshore, or even by the ingestion and egestion of oil droplets by the water column fauna.

Emulsification of spilled oil may form very stable water-in-oil emulsions, often called "chocolate mousse" with increased viscosity, decreased rate of evaporation, and longer environmental half-life. In contrast, oil in water emulsions increase the rate of biodegradation since oxygen diffusion, the process that often limits the degradation of oil by heterotrophic microorganisms, is enhanced. Aerobic heterotrophic bacteria and fungi are able to obtain energy from the oxidation of hydrocarbons. Bacteria of the genus *Pseudomonas* are capable of complete oxidation of small alkanes to CO_2 and water. As the molecular weight increases and the molecular structure includes unsaturation and branches degradation pathways get more complex and consortia of multiple microorganisms need to be present. In that case, and particularly in pristine environments, natural populations of the required microorganisms may not be present and longer time lags may occur between the oil spill and the onset of significant biodegradation. There is a hierarchy for rates of biodegradation of hydrocarbons from more rapid to slower: saturated alkanes, small aromatics, branched alkanes, multiring and substituted aromatics, and polycyclic compounds.[21]

PHOTOGRAPH 7.1

Persistent water-in-oil emulsion, technically called "'chocolate mousse,'", formed after the *Prestige* fuel -oil spill in As Furnas beach (Galicia, Northwest Iberian Peninsula.) *Photograph: R. Beiras.*

Among the environmental factors, oxygen is the most important in determining the **biodegradation** rate of oil. This can be illustrated with figures; to mineralize 1 Kg of oil, the decomposing biota need 2.6 Kg of oxygen.[22] This is the oxygen dissolved in 320,000 L of saturated seawater at 16°C. Although anaerobic degradation of hydrocarbons by sulfate and iron-reducing bacteria is also possible, oxygen availability greatly limits the velocity of recovery of oil polluted sediments also. A direct correlation has been described between grain size and recovery rate (in terms of reduction of hydrocarbon content in the sediments with time) in areas of the Brittany coast affected by the *Amoco Cadiz* oil spill, and while medium sands reduced their hydrocarbon content by a factor of 100 in 1 year, in fine sands the reduction was tenfold and in silt no marked reduction was observed.[23]

Since hydrocarbons are very deficient in N and P, biodegradation rates can be enhanced by "fertilizing" the oil with the appropriate amounts of N (nitrates, ammonium) and P (phosphates) sources, a technique termed bioremediation (see in the next section).

7.5   OIL SPILLS: LESSONS LEARNED

Table 7.4 lists the largest accidental oil spills in the sea. Intended war actions were not included, but the Gulf War in 1991 caused by far the largest oil spill, with estimations exceeding 1 million tons. The two largest accidental spills were caused by drilling platforms in the Gulf of Mexico, and these incidents caused prolonged spillage lasting several months. The other oil spills reported concern accidents of tankers, mostly by collisions or fires and explosions. Although the locations affected spread throughout the world, at least 14 from the 23 tanker accidents listed in the table occurred in the main global route of oil transportation that joins the Persian Gulf (main global producers) to Western Europe (main consumers) harbors. Bad weather was involved in many of the incidents (notice that most of them in both the Northern and Southern hemispheres took place in winter), and the places particularly dangerous for maritime traffic (English Channel, Brittany, Northwest Spain, South Africa) were particularly affected. The three traditional European's *Finisterrae* or "land's end" in Galicia, Brittany, and Cornwall were termed as the hottest oil spill hotspots worldwide (Fig. 7.4).

Major oil pills were caused by blowouts in drilling platforms and tanker accidents associated to bad weather conditions and dangerous navigation "hot spots"

The reader will miss in Table 7.4 some cases particularly mediatized, such as the *Exxon Valdez* in Mar 1989 in Alaska (38,000 T of crude oil spilled) and the *Erika* in Dec 1999 in Brittany (28,000 T of fuel oil spilled). Although minor events in terms of amount spilled, both cases gave origin to thorough scientific studies on the ecological impact, and the earlier raised unprecedented public awareness that prompted the most expensive spill response to date and legislative initiatives in the United States—followed by other countries—intended to enhance maritime safety, prevent tanker accidents (e.g., double hull), and improve liability and economic compensations to affected stake-holders (see also Section 18.5).

As a result of the measures taken worldwide the amount of oil spillages due to tanker accidents showed a significantly decreasing trend, particularly remarkable from mid 1990s on.[26] Earlier initiatives focused also on prevention of operational oil pollution. The MARPOL 73/78 convention adopted regulations to limit the amount of oil that may be discharged into the sea to 1/30,000 of the total cargo volume, require larger tankers to have segregated ballast tanks, completely separated from the cargo oil, and restrict routing of ballast piping through cargo tanks, and vice versa.[27] However, in 2003 the US NRC estimated that one-third of the crude oil tankers still carried ballast in cargo tanks.

Preventive regulations succeeded in decreasing the number of accidental oil spills

Despite prevention, accidental spills still happen. When oil is spilled offshore the main goal is to avoid the oil slick to strand on the coast, where economic and ecological consequences are far more costly than in open sea. To achieve

Table 7.4 World Largest Accidental Oil Spills in Marine Environment Ranked by Amount Spilled.[25] Note that the Two Most Important Correspond to Accidents in Drilling Platforms, and that Most Tanker Accidents Took Place in Winter Time of the Corresponding Hemisphere

Estimated Amount Spilled (T)	Name	Cause	Location	Date
400–700,000	BP Deepwater Horizon	Drilling blowout	Louisiana, US	Apr-2010
476,000	Ixtoc I well	Drilling blowout	E Mexico	Jun-1979
2,86,000	Atlantic Empress	Ship collision	Tobago	Jul-1979
260,000	ABT Summer	Explosion	Open sea 700 miles off Angola	May-1991
254,000	Castillo de Bellver	Fire	W South Africa	Aug-1983
230,000	Amoco Cadiz	Ran aground	France	Mar-1978
144,000	Haven	Fire	Genoa, Italy	Apr-1991
132,000	Odyssey	Broke in two in bad weather	Canada	Nov-1988
121,000	Sea Star	Ship collision	Gulf of Oman	Dec-1972
40–122,000	Torrey Canyon	Ran aground	W Cornwall	Mar-1967
103,000	Irenes Serenade	Explosion	SW Greece	Feb-1980
100,000	Urquiola	Ran aground	NW Spain	May-1976
98,000	Independentza	Ship collision	Bosphorus, Turkey	Nov-1979
88,000	Jakob Maersk	Ran aground	N Portugal	Jan-1975
85,000	Braer	Ran aground	NE Scotland	Jan-1993
77,000	Prestige	Hull damage	NW Spain	Nov-2002
74,000	Aegean Sea	Ran aground	NW Spain	Dec-1992
72,000	Sea Empress	Ran aground	SW Wales	Feb-1996
72,000	Katina P	Storm damage	Mozambique	Apr-1992
70,000	Khark 5	Fire	Open Sea, 400 miles N of Canary Islands	Dec-1989
70,000	Nova	Ship collision	S Iran	Dec-1985
68,000	Yuyo Maru	Ship collision	Tokyo Bay, Japan	Nov-1974
52,000	Assimi	Fire	Gulf of Oman	Jan-1983
50,000	Metula	Ran aground	S Chili	Aug-1974
50,000	Andros Patria	Fire	NW Spain	Dec-1978

that goal the means of combat against oil spills must be managed and deployed according to operational oceanography models that predict the trajectory of the oil slick as a function of the oil properties, and the prevailing winds and currents. Tourism, fisheries, and shelf harvesting activities are particularly threatened by oil spills. From an ecological standpoint not all types of coast are equally vulnerable to oil pollution. The degree of exposure and type of bottom are important characteristics. Exposed rocky headland will be less vulnerable, with wave reflection and hydrodynamics contributing to natural clean up. On the other end, fine-grained soft bottoms and sheltered locations

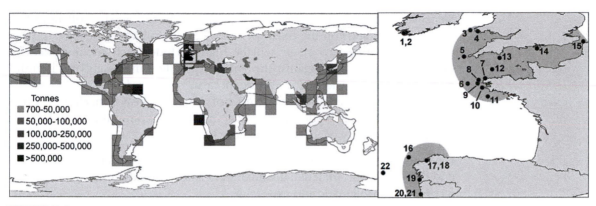

FIGURE 7.4

Left: Worldwide distribution of oil spilled in the seas by maritime transport from 1965 to 2002, with the hottest spot marked in red. Right: Distribution of principal oil spill accidents in the world's hottest spot, the Atlantic European waters. Numbers identify accidents. Those spilling over 10,000 tons are: 1 Betelgeuse; 4 Sea Empress; 5 Torrey Canyon; 6 Gino; 7 Amoco Cádiz; 11 Erika; 12 Tanio; 15 Olympic Alliance; 16 Andros Patria; 17 Urquiola; 18 Aegen sea; 19 Polycommander; 20 Jakob Maersk; 22 Prestige. *From Vieites et al. (2004).*

will delay natural recovery. Sheltered tidal flats and salt marshes are particularly vulnerable to oil spills.[28]

Three types of means are available to fight oil spills and try to minimize their ecological and economic impact.

Containment means: oil at sea can be contained by air or foam floating booms with a weighted skirt below the water line. These booms are particularly effective in sheltered areas such as ports or inland waters. Large oceanic booms of up to 1.5 m freeboard are available, but their efficiency under rough weather conditions is doubtful.

Recovery means: once contained and accumulated the oil can be taken out of the water by a variety of mechanical apparatus such as slick-lickers, based on absorbent materials or, if not too viscous, pumped from the surface by skimmers. In all cases these methods are more effective in sheltered waters and they can only recover limited amounts of oil.

Elimination methods: chemical dispersants can be sprayed over the oil slick to enhance the natural process of emulsification of oil in the water. Dispersants are not effective against heavy or long weathered oils. The first generation of oils dispersants was highly toxic, and there was motivated reluctance to use them, but low toxicity dispersants are now available. Still, a priori risk assessment must be conducted before large-scale use of dispersants in natural environments. Emulsification enhances the biodegradation rate but also makes oil more available to aquatic biota.

Methods to combat oil spills include containment plus recovery in the sea, and bioremediation in the shoreline

Shoreline cleaning: If these means designed to fight the oil slick offshore fail, then the oil strands on the coast, where the ecological and economic impact, and also the media coverage, will be higher. Under such circumstances of public pressure to react against the spill, frequently unnecessary and harmful clean-up activities take place. Although clean up intends to speed recovery, drastic methods using high-pressure hot water or heavy machinery often increase the damage made by the oil to the benthic communities and delays the natural recovery of the affected shore line.[29]

Cleaning oiled beaches generates high volumes of oil-contaminated debris, and disposal of this material, unsuitable for conventional dumping nor energy recovery, is problematic and expensive.

Bioremediation is the use of techniques intended to mitigate the consequences of an oil spill and speed up the elimination of the oil by resorting to biological processes. Hydrocarbons are organic molecules susceptible to microbial degradation to CO_2 and water. Following an oil spill, oil-degrading bacteria proliferate. However, only a fraction of the hydrocarbons present, the lowest molecular weight alkanes, are easily mineralized. **Biostimulation**, also termed passive bioremediation, intends to create the ideal conditions to speed up this process. This may be achieved in first instance by fertilizing the oil slick with sources of N and P in the C:N:P ratios more suitable for microbial proliferation (approximately 100: 10: 1 in weight). Since O_2 limits the process, aeration of the substrate (e.g., by raking) and or surfactant addition may also be necessary. **Bioaugmentation**, also termed active bioremediation, goes a step further and includes the addition of cultured bacteria capable of degrading the more refractory constituents of oil, such as asphalthenes and resins. However, some field trials showed that the addition of commercial mixtures or enriched cultures of indigenous oil-degrading bacteria do not significantly enhance the rates of oil biodegradation over that achieved by nutrient enrichment alone.[30]

Low energy shorelines are the most suitable for bioremediation, while in high energy types of coast natural cleaning may be fast enough to make bioremediation unnecessary.

7.5.1 Mortalities of Seabirds and Mammals

Air breeding and sessile animals are specially vulnerable to oil spills

The most noticeable effect of oil pollution is the mortality of seabirds and marine mammals, animals that require routine contact with the sea surface. The deaths are not caused by chemical intoxication but are instead due to the physical properties of oil that dissolve the fat layer which isolates the seabirds plumage or the otters fur. Without this isolation, the feathers get dumped in contact with the water, causing death by drowning or hypothermia. Indirect effects of oil spills on reproduction and other fitness-related traits are less well known. Other animals such as cetaceans and seals are killed by stranding on the coast due to disorientation.

7.5.2 Mortalities on Benthic Organisms

Since the oil slick prevents oxygen diffusion, sessile organisms in the most impacted sites may be killed by smothering. Acute toxicity of oil is moderate to low. The most toxic components, with LC_{50} values ranging between 1 and 100 mg/L are the lowest molecular weight aromatics (BTEX), and thus the first to disappear from the oil slick during weathering. Early life stages of marine organisms are far more sensitive than adults. Invertebrate larvae show LC_{50} values for PAHs that are a function of the K_{ow}, ranging from 1 to 100 µg/L. Toxicity increases with exposure to light, likely due to the formation of oxidized metabolites more reactive than the parental PAHs.[31]

7.5.3 Unbalances in Community Structure

Ecological effects of oil spills show some common traits. High mortalities on grazing macrofauna (gastropods, echinoderms) cause blooms of fast-growing macroalgae and some opportunistic invertebrate species. The ecological imbalance is always reversible, and recovery time depends on the intensity of impact, hydrodynamics and geomorphology of the area, and reproductive potential of the affected species, but typically full recovery takes several years.

Ecological effects of oil spills are reversible, but recovery takes several years

In the *Tampico Maru* fuel oil spill (Pacific coast of Mexico, 1957), the disappearance of the dominant herbivores *Haliotis* and *Strongylocentrotus* caused proliferation of the giant kelp *Macrocystis*, and recovery of this ecological imbalance lasted 5 years.

The *Torrey Canyon* spill in 1967 caused mortalities on *Patella* in the affected coasts of Cornwall and invasion of *Fucus* algae. The stages in the succession were: oiled rocks, green algae proliferation, brown algae dominance and an eventual recovery of the herbivorous population 4 years after. The recovery of limpet and barnacle populations was slowed down by the use of toxic emulsifiers.

The *Florida* fuel oil spill in 1969 in Massachusetts caused a huge proliferation of *Capitella* the first year and another capitellid (*Mediomastus*) the second year, whereas ampeliscid amphipods virtually disappeared. As a result of these and other faunistic changes, species richness and Shannon diversity values took more than 2 and a half years to recover in heavily affected sites and 1 year in lightly affected sites. Sublethal effects on the fiddler crab were reported to decrease their populations in Wild Harbor Marsh for at least 7 years after the spill.[32]

The *Bahia Las Minas* spill in Panama in 1986 was caused by the rupture of a storage tank that released 14,000 T of crude oil into a tropical ecosystem that included mangroves and coral reefs. Oil trapped in mangrove soils showed little degradation and chronic re-oiling occurred for at least 5 years. Subtidal reef flats and seagrass beds (*Thalassia*) showed moderate impact and recovered within 6 months, but where oil was in direct contact with the reef flat during low tides extensive mortality was recorded and the effects persisted over 5 years, and estimates of recovery for some coral species extended to 10—20 years.[33]

The *Exxon Valdez* spill in Alaska (1989) caused heavy mortalities of limpets, periwinkles, whelks, and barnacles, the colonization of the upper intertidal by an ephemeral green macroalga and an opportunistic *Chthamalus* species, and the substitution in the lower intertidal of red algae by *Fucus*. *Laminaria* populations in the rocky shores recovered 1 year later. In contrast in soft-bottom communities, mortalities of amphipods and (partially as a consequence of cleaning) clams, were recorded. Oligochaetes and deposit feeding polychaetes took over and full recovery took 6 - 8 years. Other species whose abundance decreased were seastars and *Zostera*.[34]

After the *Sea Empress* oil spill in 1996 in Milford Haven (Wales) limpet mortalities were very high and resulted in a dramatic green phase of *Enteromorpha* spp., followed by a bloom of *Porphyra* spp. and ultimately a brown phase of *Fucus vesiculosus*. The structure of the benthic community in the lightly oiled sites recovered after 1 year, but in the highly impacted sites it took up to 5 years.[35]

After the *Prestige* fuel oil spill, a decrease in the cover of *Chthamalus montagui* in the upper intertidal zone of affected rocky coasts and a decrease in the abundance of *Patella vulgata* was recorded, and the *Paracentrotus lividus* sea-urchins completely disappeared from the most affected areas.[36] Demersal fisheries were also affected; abundance of *Lepidorhombus boscii* and *Nephrops norvegicus* (Norwegian lobster) markedly decreased after the spill but recovered 1—2 years after.[37]

KEY IDEAS

- Crude oil has a paramount economic importance because it is the main source of transport and heating fuels, plastics, solvents, and many other chemicals.

- Maritime navigation transports crude oil from the producer to the consumer countries, and tanker accidents, associated to dangerous navigation areas and bad weather conditions, caused some of the most relevant marine pollution issues worldwide.

- PAHs are more persistent and toxic than alkanes. Those with lower molecular weight are more soluble and acutely chronic to marine fauna, while those with higher molecular weight may be carcinogens. The mechanism of carcinogenicity involves activation by the detoxification metabolism of vertebrates.

- After a spillage in the aquatic environment, oil forms a slick whose composition and properties change as a result of the weathering processes, which cause an increase in density and an enrichment in the most persistent components of oil: those with higher molecular weight and a branched, unsaturated, and cyclic structure.

- In the event of oil spilled in open sea it is a priority to contain and recover the oil to avoid it reaching the coastline, where more important ecological and economic impacts are expected.

- Oil spills typically cause large mortalities of seabirds, and changes in benthic community structure characterized by mortalities of herbivores, excessive proliferation of macroalgae, and loss of ecological diversity. This ecological unbalance usually takes several years to recover.

Endnotes

1. EEA. Europe's environment. The fourth assessment. Copenhagen: European Environment Agency; 2007. 452 pp.
2. See p. 9 in: Fingas M, editor. Oil spill science and technology. Amsterdam, Elsevier; 2011.
3. EEA (2007) op. cit.
4. NRC. Oil in the sea III. Inputs, fates, and effects. Washington, The National Academies Press; 2003.
5. Merian E, Anke M, Ihnat M, Stoeppler M. editors. Elements and Their Compounds in the Environment. 2nd ed. Weinheim, Wiley-VCH; 2004.
6. NRC (2003) op. cit. (p. 20).
7. Fingas M. Oil spill science and technology. Amsterdam, Elsevier; 2011. (pp. 74–77).
8. See pp. 2-39 in: Varanasi U, editor. Metabolism of polycyclic aromatic hydrocarbons in the environment. Boca Raton: CRC Press; 1989.
9. Ni N, Yalkowski SH. Prediction of Setschenow constants. Int J Pharm 2003; 254: 167–172.
10. Beiras R, Durán I, Parra S, et al. Linking chemical contamination to biological effects in coastal pollution monitoring. 2012;21:9–17.
11. Bleeker EAJ, Verbruggen EMJ. Acumulation of polycyclic aromatic hydrocarbons in aquatic organisms. RIVM Report 601779002/2009; 2009.
12. Obana H, Hori S, Nakamura A, et al. Uptake and release of polynuclear aromatic hydrocarbons by short-necked clams (Tapes japonica). Wat Res 1983;17(9):1183–1187.
13. Roesijadi G, Anderson JW, Blaylock JW. Uptake of hydrocarbons from marine sediments contaminated with Prudhoe Bay crude oil: Influence of feeding type of test species and availability of polycyclic aromatic hydrocarbons. J Fish Res Board Can 1978;35:608–614.
14. Beiras et al. (2012) op. cit.
15. OSPAR Commission. CEMP 2011 assessment report. Publication Number: 563/2012; 2012.
16. See p. 48 in: Varanasi U, editor. Metabolism of polycyclic aromatic hydrocarbons in the environment. Boca Raton: CRC Press; 1989.
17. See pp. 269-89 in: Varanasi U, editor. Metabolism of polycyclic aromatic hydrocarbons in the environment. Boca Raton: CRC Press; 1989.
18. EC Regulation 835/2011.
19. NRC (2003) op. cit. (p. 90).

20. Neff JM, Ostazeski S, Gardiner W, et al. Effects of weathering on the toxicity of three offshore Australian crude oils and a diesel fuel to marine animals. Environ Toxicol Chem 2000;19(7): 1809–1821.

21. NRC (2003) op. cit. (pp. 96–97).

22. CEDRE. Guidelines for decisión making and implementation of bioremediation in marine oil spills. Centre de Documentation de Recherche et d'Expérimentations sur les pollution accidentelles des eaux 2001.

23. Amoco Cadiz. Conséquences d'une pollution accidentelle par les hydrocarbures. Centre national pour l'exploitation des océans Paris 1981.

24. Vieites DR, Nieto-Román S, Palanca A, et al. European Atlantic: the hottest oil spill hotspot worldwide. Naturwissenschaften 2004;91:535–538.

25. http://earth.tryse.net/oilspill.html.

26. See p. 45 in Fingas (2011) op. cit.

27. See p. 204 in: NRC (2003) op. cit.

28. Irvine, GV. Persistence of spilled oil on shores and its effects on biota. Chapter 126. In: Sheppard C, editor. Seas at the millennium. 2000;3:267–281.

29. See p. 80 in: Clark RB. Marine Pollution 5th ed. Oxford University Press; 2001.

30. See p. 25 in: CEDRE (2001) op. cit.

31. Bellas J, Saco-Álvarez L, Nieto O, et al. Ecotoxicological evaluation of polycyclic aromatic hydrocarbons using marine invertebrate embryo–larval bioassays. Mar Pollut Bull 2008;57: 493–502.

32. Sanders et al. Anatomy of an oil spill: long-term effects from the grounding of the barge Florida off West Falmouth, Massachusetts. J Mar Res 1980;38:265–324.

33. See p. 142 in: NRC (2003) op. cit.

34. CH Peterson, "Exxon Valdez" Oil Spill in Alaska: Acute, Indirect and Chronic Effects on the Ecosystem: Adv Mar Biol 2000;39:3–84.

35. Crump RG, Morley HS, Williams AD. West Angle Bay, a case study. Littoral monitoring of permanent quadrants before and after the *Sea Empress* oil spill. Field Studies 1999;9:497–511.

36. Penela-Arenaz M, Bellas J., Vázquez E. Effects of the prestige oil spill on the biota of nw spain: 5 years of learning. Adv Mar Biol 2009;56:365–396.

37. Sánchez F, Velasco F, Cartes JE, et al. Monitoring the Prestige oil spill impacts on some key species of the Northern Iberian shelf. Mar Pollut Bull 2006;53:332–349.

Suggested Further Reading

- Varanasi U, editor. Metabolism of polycyclic aromatic hydrocarbons in the environment. Boca Raton: CRC Press; 1989.

- National Research Council. Oil in the sea III. Inputs, fates, and effects. Washington: The National Academic Press; 2003. Available on line, https://www.nap.edu/catalog/10388/oil-in-the-sea-iii-inputs-fates-and-effects.

- Peterson CH, Rice SD, Short JW, Esler D, Bodkin JL, Ballachey BE, Irons DB. Long-term ecosystem response to the Exxon Valdez oil spill. Science 2003;302:2082–6.

CHAPTER 8

Persistent Organic Xenobiotics

8.1 ORGANOCHLORINE PESTICIDES, POLYCHLORINATED BIPHENYLS AND DIOXINS

In 1874 a young Austrian chemist, Othmar Zeidler, synthetized dichlorodiphe-nyltrichloroethane (DDT) as part of his doctorate research at the University of Strasbourg. The new molecule passed largely unnoticed until 65 years later, when the Swiss Paul Mueller discovered its **insecticide** properties, a finding that granted him the Nobel prize in 1948. By then the DDT had been extremely successful in controlling typhus, spread by lice, and mosquito-transmitted diseases such as yellow fever and malaria, and had saved millions of lives in tropical countries were the latter malady caused high mortalities. Just in Sri Lanka the cases of malaria decreased from 2.8 million when the DDT spraying campaigns began, down to 17 in 1963.

However, the massive use of DDT resulted in unexpected effects on nontarget species. The best known case is the detrimental effect on reproduction in birds due to **reduction of egg-shell thickness**, which leads to hatching failure due to cracking of eggs during incubation.[1] Inverse correlations between egg-shell thickness and DDT and DDT-metabolite residues have been reported in pelicans, gulls, and other carnivore bird species. Several underlying mechanisms were proposed. DDT-induced detoxification enzymes decrease the levels of estrogens, which leads to **delayed breeding** and reduced storage of a supply of calcium in the bone-marrow needed for the formation of the egg-shell.[2] The DDT metabolite DDE was shown in vitro to inhibit the Ca-ATPase, an enzyme controlling the calcium metabolism essential to promote egg-shell growth.[3] As a result of this scientific evidence, and also after the impact of Rachel Carson's book on the environmental problems of synthetic pesticides "Silent Spring," published in 1962, the use of DDT began to be restricted worldwide. In Sri Lanka the antimalaria DDT spraying campaign was terminated in 1964. The result was a rapid increase in the incidence until c. 0.5 million cases recorded in 1969. The spraying campaigns were resumed but they never were as much effective, suggesting that mosquitos had developed resistance to the pesticide, and the cases of malaria in Sri Lanka stabilized to about 250,000 per year. Only

Organochlorine (OC) pesticides: the first synthetic persistent pollutants

107

the expansion of insecticide-treated bed nets ultimately succeeded to reduce the prevalence of malaria to below 100 cases in 2012.

The history of DDT illustrates many common traits in the management of the use of environmentally unfriendly synthetic chemicals. On the one hand, useful chemicals are massively used before environmental consequences were foreseen. On the other hand discontinued use with no effective alternative may be a clear mistake, and replacement products may be even worse than the original ones. DDT and other organochlorine (OC) insecticides were initially replaced by **organophosphate insecticides**, but the former never caused a human casualty whereas the latter, much more toxic to humans, caused accidental poisoning in exposed workers.

Twelve other OC pesticides were widely used from the 1940s until their ban or restrictions from the 1970s on, including the insecticides aldrin, dieldrin, endrin, heptachlor, chlordane, toxaphene, mirex, endosulfan, kepone and hexachlorocyclohexane, and the fungicides hexachlorobencene and pentachlorobence. All of them are considered by the Stockholm Convention (see Box 8.1) as **persistent organic pollutants (POPs)** whose production must be eliminated—or restricted to sanitary uses in the case of DDT—and all share in their structure the typical feature of chlorine atoms bond to cyclic hydrocarbons. They are also lipophilic compounds that accumulate in the fat tissues of organisms, and when comparing levels among individuals of different age or species concentrations should be expressed on a lipid weight basis.

BOX 8.1 STOCKHOLM CONVENTION

In 1995 the UN Council requested an international assessment of 12 synthetic chemicals: aldrin, chlordane, dieldrin, endrin, heptachlor, hexachlorobenzene (HCB), mirex, PCB, toxaphene, DDT, PCDD, and PCDF, preliminary identified as persistent in the environment, since their half-life is typically higher than 2 months in water and 6 months in sediments. As a result, the need was identified for an internationally legally binding instrument to minimize the risks posed by the 12 **POP** through measures to reduce and/or eliminate their emissions. After a long process of international negotiations, a Convention was adopted for signature by more than 100 countries at a conference held in May 2001 in Stockholm, Sweden. Parties must take measures to **eliminate** the production and, with some exceptions, use of aldrin, chlordane, dieldrin, endrin, heptachlor, HCB, mirex, PCB, and toxaphene (Annex A of the Convention), **restrict** to disease vector control the production and use of DDT (Annex B of the Convention), and **reduce** the unintentional releases of HCB, PCB, PCDD and PCDF (Annex C of the Convention). The use of preexisting electric equipment with PCB, and recycling items containing polybromodiphenylethers, is allowed. Some specific uses such as ectoparasiticide and termiticide are also allowed for aldrin, chlordane, dieldrin, heptachlor, HCB, lindane, and mirex.

In 2009 hexachlorocyclohexane isomers (including lindane), chlordecone, pentachlorobenzene, and polybrominated diphenylethers of 4–8 bromine atoms were added to Annex A (some of them with specific authorized exemptions), PFOS to Annex B (production restricted to a list of uses from metal plating to textile, rubber and plastic additive), and pentachlorobenzene to Annex C. In ulterior meetings endosulfan, hexabromocyclododecane, hexachlorobutadiene, pentachlorophenol, polychlorinated naphthalenes and decabrominated diphenylethers were added to Annex A (some of them with specific exemptions), and hexachlorobutadiene and polychlorinated naphthalenes to Annex C. To date 181 countries have adopted the Convention. The United States signed the original Convention but has not adopted it.

Under sunlight and in the presence of photosensitizers that provide reactive oxygen species, DDT readily photooxidizes and shows a short half-life of just hours.[4] However, it tends to bind to organic matter and sink to the sediments where it can persist for years. Disregarding photolysis, DDT does not degrade in seawater unless some microorganisms were present, and as a result of reductive dechlorination the main metabolite formed is dichlorodiphenyldichloroethane (DDD) (Fig. 8.1), where one of the three chlorine atoms bond to the ethyl group is replaced by a hydrogen atom. DDD is also the predominant metabolite found in marine sediments. The microbial biodegradation route further proceeds through oxidation of that group to form more hydrosoluble products, but the bond between Cl and C from the phenyl groups, which confers OCs their **persistence**, is not broken.[5] Chlorine is an electronegative substituent, and when present on an aromatic ring shifts electron density toward itself and hence alters resonance across the ring. As a result, chlorine makes the aromatic ring less susceptible to oxidative attack.[6]

In marine fauna the biotransformation of DDT frequently begins with a dehydrochlorination that yields dichlorodiphenyldichloroethylene (DDE) (Fig. 8.1), which accounts from 50% to 75% of the total DDT metabolites accumulated in seabirds. Marine fish can biotransform DDT to both DDE and DDD,[7] while mussel biotransformation metabolism apparently only produces DDD.[8] In both sediments and mussels, %DDD is directly related to total amount of DDTs (Fig. 8.2), which is consistent with enhanced biodegradation, or biotransformation in the mussel, as environmental or accumulated DDT levels increase.

In marine mammals, DDE may be further metabolized to methyl sulfones (MeSO$_2$-DDE), formed by oxidation of DDE followed by conjugation with glutathione.[9] The ratios of persistent metabolites such as DDE and MeSO$_2$-DDE to the parental compounds may inform on the environmental age of the pollutant.[10]

As a result of its persistence, but also occasionally due to unexpected sources, DDT and its metabolites are still universally found in coastal sediments and

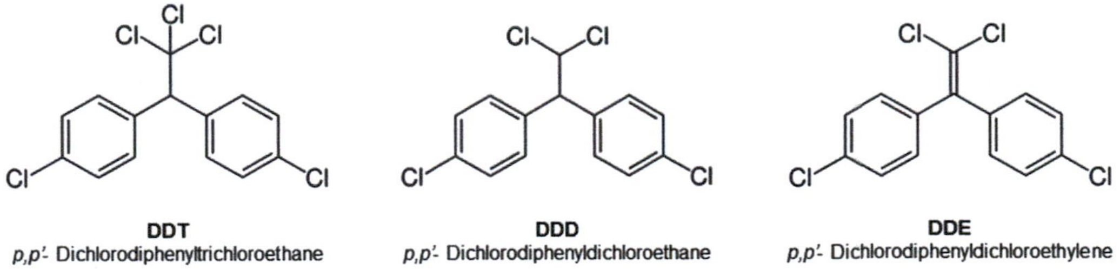

DDT
p,p'- Dichlorodiphenyltrichloroethane

DDD
p,p'- Dichlorodiphenyldichloroethane

DDE
p,p'- Dichlorodiphenyldichloroethylene

FIGURE 8.1

Molecular structures of DDTs. The commercial product already contained residues of both DDD and DDE, but in the environment they are mainly produced as metabolites of the DDT. Minoritary components are o,p′-isomers where Cl is located in ortho position in one of the rings. DDD is originated by reductive dechlorination, the first step in microbial degradation, while DDE is originated by dehydrochlorination, the major biotransformation route in birds and mammals.

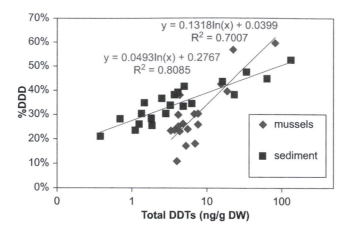

FIGURE 8.2

Levels of total DDTs (p,p'-DDT + p,p'-DDD + p,p'-DDE) and percentage of p,p'-DDD measured in marine sediments and wild mussels from clean and polluted sites at the Ría of Vigo (NW Iberian Peninsula). Notice the increase of %DDD in both matrices at elevated total DDT levels, consistent with a concentration-dependent reductive dechlorination of DDT in sediment microbiota and mussels *Data from: Beiras R, Durán I, Parra S, et al. Linking chemical contamination to biological effects in coastal pollution monitoring. Ecotoxicology 2012; 21:9–17.*

biota, despite having been discontinued many decades ago, and current levels of these and other OCs at industrial harbors may still exceed environmental quality criteria (Table 8.1). In a recent monitoring of the Spanish coast concentrations up to 45.1 ng/g WW were found in wild mussels,[11] and the Asian Mussel Watch reported even higher concentrations in some hot spots of the Chinese coasts (10–58 μg/g LW, roughly equivalent to 100–580 ng/g WW).[12] Even in remote

Table 8.1 Concentrations of Organochlorine Compounds in Marine Sediments and Wild Mussels From Clean (CA: Cangas, BA: Baiona) and Polluted Sites (Vigo Harbor) in Ría de Vigo (NW Iberian Peninsula). Environmental Assessment Criteria (OSPAR BAC/EAC, ERL/TEL, ERM/PEL, EU Regulations) are Also Shown. Concentrations are ng/g DW. Notice the High Concentration Still Found in Polluted Places Despite the Use of Most of These Chemicals was Phased out Decades Ago

	Sediments					Mussels		
	Clean Site (CA)[c]	ORPAR BAC	ERL	ERM	Vigo Harbor[c]	Clean Site (BA)[c]	Vigo Harbor[c]	OSPAR BAC
ΣDDT	0.7	n.r.	1.58	46.1	16–134	3.7	15–82	0.63[a]
γHCH	0.1	n.r.	0.32[(2)]	99 [(3)]	0.12–1.19	0.31	0.19–0.37	0.97
ΣPCB	10.6[b]	1.09[b]	22.7	180	58–251[b]	28[b]	61–217[b]	4.6[b]
Aldrin	0.01	n.r.	n.r.	n.r.	0.14–1.03	n.m.	n.m.	n.r.
HCB	0.03	16.9 [(1)]	n.r.	n.r.	0.13–0.32	n.m.	n.m.	0.63
chlordanes	0.01	n.r.	0.5	6	0.11–5.13	n.m.	n.m.	n.r.

[(1)] *OSPAR EAC;* [(2)] *TEL;* [(3)] *PEL; n.r.: not regulated; n.m.: not measured.*
[a]*DDE_{p,p'}.*
[b]*sume of 7 ICES congeners.*
[c]*Beiras et al. (2012) op. cit.*

areas, marine mammals continue to show high levels of DDT and other OCs in their tissues, especially in the blubber (see later in the chapter).

The case of DDT was also a paradigm of a novel feature in environmental science, the **biomagnification** of a chemical across a trophic web, a concept dealt with in Section 11.4. Biomagnification is due to enrichment in pollutant concentration in the consumer compared to its food at each step of the food chain. As a result, when we measure the concentration of the pollutant in the different organisms of the food web, we find a direct relationship between tissular concentration and the trophic level occupied by the organism. Thus, levels in herbivores are higher than in producers, and in carnivores higher than in herbivores, with top predators showing the highest concentrations. One of the first reports of this feature in a coastal ecosystem was provided by Woodwell et al.[13] who found DDT residues in algae and plankton ranging from 0.04 to 0.08 μg/g WW, from 0.16 to 0.26 in invertebrates, from 0.17 to 2.07 in fish, and from 1.48 to 75.5 in seabirds. Plants (*Spartina patens*) showed highest DDT contents than expected according to their trophic position (0.33−2.8 μg/g WW).

Levels of DDT and other OCs are elevated in top predators

The highest DDT concentrations reported in marine organisms correspond to the blubber (fat layer under the skin) of marine mammals, including endangered species such as beluga (up to 827 μg/g LW[14]) or monk seal (up to 950 μg/g WW[15]). The fact that Arctic species such as beluga were affected illustrates the potential **long-range transport** of OCs relative to local sources. On the other hand adult females show lower levels of DDT and other OCs than males and juveniles, which led to the hypothesis that fat-stored POPs are **mobilized during pregnancy** and transferred to the offspring. Frouin and co-workers[16] studied seal mother-pup pairs and reported a rise of POP concentrations in maternal blubber, blood serum, and milk as lactation progressed. High levels of OCs in marine mammals have been associated to immunosuppression and reproductive impairment, which may be triggered by the seasonal or physiological mobilization of the lipid tissues, but the toxicological implications of these processes are still today poorly understood.[17]

A second class of synthetic chemicals that share similar structural properties and environmental fate than OC pesticides are **PCB**. PCBs are not pesticides; they were synthetized for a myriad of industrial applications in electric apparatus, fluids, paints, inks, adhesives, and plastics due to their dielectric, coolant, and flame-retardant properties. That includes insulating material in transformers and capacitors, additive to motor oil to keep the viscosity at high temperatures, lubricant in hydraulic fluids and other high duty equipment, flame retardant in plastics, and plastifier in synthetic resins. The molecular structure of PCBs is illustrated in Fig. 8.3. Thus the general formula is: $C_{12}H_{10-x}Cl_x$, where x, the number of chlorine substitutions, varies from 1 to 10. Their hydrophobicity, half-life, and other properties are directly related to the number of chlorine atoms, which results in a total of 209 possible forms, although only some of them are represented in the commercial mixtures and thus in the environment. Fig. 8.3 also depicts an abundant hexachlorine congener, the

Polychlorinated biphenyls (PCBs) are OCs with many industrial applications

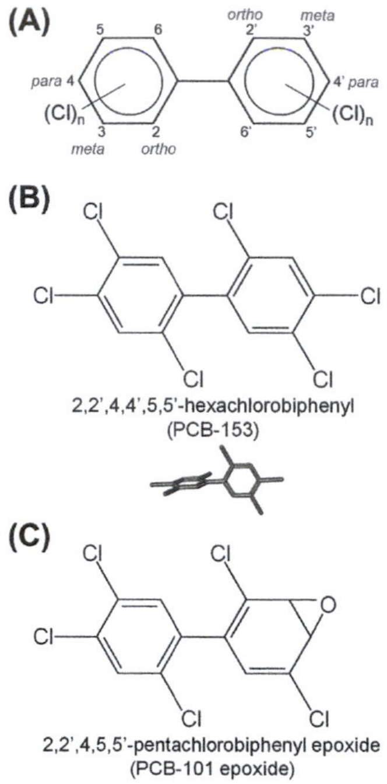

FIGURE 8.3
(A) Generic structure of polychlorinated biphenyls (PCB) indicating the carbon positions where hydrogen can be substituted by chlorine. Carbon 1 in each ring is not available for substitutions, thus n ranges from 1 to 5. Substitutions at carbon numbers 2, 3, and 4 are termed ortho, meta, and para, respectively. **(B)** an example of hexachlorinated congener, the PCB-153, common in marine samples, including its tridimensional configuration. Notice that the phenyl rings are twisted c. 90 degrees. **(C)** the first step in the route of degradation of PCBs requires two vicine nonsubstituted C atoms, in the case of PCB-101 the 3′ and 4′, to form an epoxide. *3D-image source: US National Library of Medicine.*

2,2′,4,4′,5,5′-hexachlorobiphenyl (PCB-153). PCB biodegradation pathways begin with an initial oxidation of two contiguous C atoms where H was not substituted by Cl to form an epoxide (Fig. 8.3). Thus, environmental persistence of PCB congeners increases as the degree of Cl substitutions increases. Biotransformation further proceeds through hydroxylation and conjugation with glutathione (see Chapter 12).

As a result of the increased levels of PCB found in human foods, the presence in breast milk, and the demonstrated toxic (including carcinogenic and teratogenic) effects, Monsanto, the main PCB producer in the United States, voluntarily terminated all manufacture of PCB in 1977, but they are still present in electric equipment now in use, an exemption of their ban specifically accepted in the Stockholm Convention. Unfortunately, there is no universal agreement on a definite set of PCBs to be routinely monitored, which hinders comparisons among data sets. In an effort to overcome this difficulty, ICES recommends the measurement in marine samples of 7 congeners: PCB-28, 52, 101, 118, 138,

153, and 180. In foodstuff, the European legislation places emphasis on the dioxin-like congeners. PCBs in fact should raise a higher environmental concern that OC pesticides. The toxicity of the dioxin-like PCB isomers is much higher than that of noncoplanar PCBs or OC pesticides. Besides, the environmental persistence of highly chlorinated isomers in water can be 300 times that of DDT.

Despite the efforts to eliminate PCB inputs to the environment and the decreasing trends generally found in coastal ecosystems, these OCs are still universally detected in marine sediments and biota. More than 30 years after the adoption of the first regulations for PCBs, these remain a focus of environmental concern. In a recent study in the Spanish Atlantic coastline by J. Bellas and coworkers,[18] the hexachlorinated congeners 153 and 138 dominated the profiles in all wild mussel populations, and sites located at industrial ports showed values above OSPAR EAC (sum of 7 ICES PCB congeners of 10 µg/kg WW). Similarly, sediments from industrial harbors show PCB concentrations above ERL and even ERM values, reflecting probable toxic effects on the local fauna. Frignani et al.[19] analyzed both ICES-recommended and coplanar congeners in mussels and sediments, and found that maximum concentrations of the most potent dioxin-like PCBs (congeners 126 and 169) occurred in the same sites showing maximum values for the ICES Σ^7PCB.

Finally, the third class of OC chemicals that rise environmental concern consists of **polychlorinated dibenzo-p-dioxins** (PCDD) and **polychlorinated dibenzofurans** (PCDF) (Fig. 8.4). Unlike pesticides and PCBs, those are by-products unintentionally produced during the synthesis of other chemicals or by incomplete combustion of chlorinated compounds. Dioxins such as tetrachlorodibenzodioxin (TCDD) are among the most toxic chemicals known. They are teratogenic and carcinogenic, and became sadly known because of the serious health issues suffered by millions of Vietnamese and thousands or US veterans exposed to the Agent Orange defoliant massively used by the US army during the Vietnam war. This warfare chemical consisted of two OC herbicides that contained PCDD originated as side products during their synthesis.

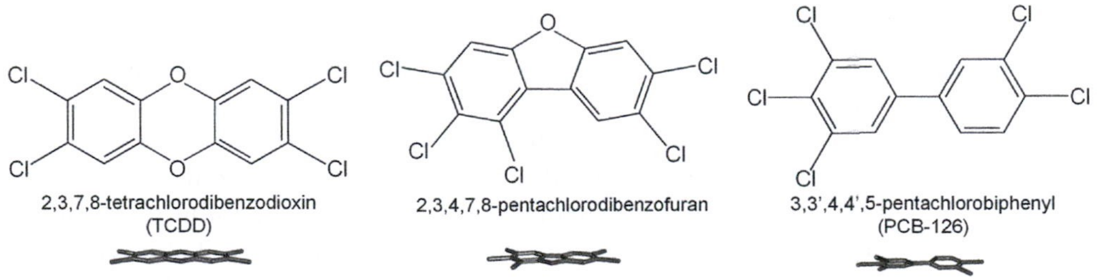

Polychlorinated dibenzo-dioxins and furans are among the most toxic chemicals

2,3,7,8-tetrachlorodibenzodioxin
(TCDD)

2,3,4,7,8-pentachlorodibenzofuran

3,3',4,4',5-pentachlorobiphenyl
(PCB-126)

FIGURE 8.4

Molecular structure of the tetrachlorodibenzodioxin (TCDD), one pentachlorodibenzophuran, and one nonortho dioxin-like PCB (PCB-126). Notice the coplanar structure of the three molecules and compare to the structure of the di-ortho PCB-153 in Fig. 8.3. *3D-image source: US National Library of Medicine.*

Nowadays the main source of PCDD and PCDF (up to 95% according to some reports) are solid waste incinerators, and minor sources include smelting, bleaching of paper pulp, and production of PVC and some pesticides. Curiously, nonanthropogenic sources of PCDDs have also been reported.[20]

PCDD and PCDF share most of the properties typical of OCs, such as environmental persistence ($T_{1/2}$ from 0.45 to 9 years in water and 0.63 to 148 years in sediments[21]), lipophilicity, and bioaccumulation. However, despite their chemical structure and high persistence they do not biomagnify up the food chain to an appreciable degree, and very moderate BMF values or even BMF < 1 have been reported in aquatic food webs.[22] Their environmental levels are extremely low, which challenges their analytical determination, and they are usually not targeted in conventional monitoring networks. (10^{-12} g = pg per gram). Values for total PCDD from 1 to 1000 pg/g DW in sediments and from 0.1 to 100 pg/g WW in marine biota have been reported.[23]

Following WHO criteria, the maximum level allowed in fish for the sum of PCDDs, PCDFs and dioxin-like PCBs is 8 pg/g WW, after applying to the actual concentration of each compound a **toxic equivalency factor** (**TEF**). The TEF takes a value of 1 for the two most potent dioxins (TCDD and 1,2,3,7,8-pentachlorodibenzo-p-dioxin) and values ranging from 0.1 to 10^{-5} for the remaining PCDD, the PCDF, and the dioxin-like PCB.[24] This approach assumes that the toxicity of these chemicals follow a common mechanism triggered by the binding of the chemical to the aryl-hydrocarbon receptor, and that the combined effects are additive. The **dioxin-like PCB** are molecules with none (PCB-77, 81, 126, 169) or a single ortho substitution (PCB-105, 114, 118, 123, 156, 157, 167, 189), which confers the double ring a coplanar tridimensional configuration similar to that of PCDD (see Fig. 8.4).

8.2   POLYBROMINATED COMPOUNDS

Polybrominated diphenyl ethers (PBDEs) and other polybrominated compounds have been massively used over the past decades as flame retardants in polyurethane and other plastics, resins, textiles, wood, and electronic equipment to meet the fire safety regulations imposed to the final commercial products.[67] Halogens, especially I and Br, released from organohalogen compounds at high temperatures are very effective in capturing free radicals, hence removing the capability of the flame to propagate. However, iodinated compounds are not stable, so the most effective flame retardants are organobromine substances. The amounts used vary from 0.8% to 28% in mass of the final product. Reactive flame retardants, such as tetrabromobisphenol A, are chemically bonded into the plastics, while additive flame retardants, which include PBDEs and hexabromocylododecane, are simply blended with the polymers, and are more likely to leach out of the products.

The basic structure of PBDEs consists of two phenyl rings linked by an ether bond. Thus, like PCBs, they can have from 1 to 10 halogen atoms substituting

H in the biphenyl skeleton, and the water solubility and vapor pressure decrease with increasing degree of bromination whereas hydrophobicity increases. However, unlike PCBs, the two phenyl rings can twist around the oxygen atom and do not form coplanar congeners.[68]

The actual composition of commercial PBDE mixtures is not disclosed, which hinders risk assessment and regulations. They are produced in formulations with varying degrees of bromination loosely termed penta-PBDE, octa-PBDE, or deca-PBDE. For thermodynamic reasons during their synthesis, these mixtures are thoroughly dominated by certain congeners. For example >70% in weight of penta-PBDE consists of congeners 2,2',4,4'-tetrabromodiphenyl ether (BDE-47) and 2,2',4,4',5-pentabromodiphenyl ether (BDE-99) (Fig. 8.5).

As a consequence of their massive use and environmental persistence, they have been detected at increasing levels in the environment (sediments,[69] sludge,[70] biota[71]), and in human breast milk.[72] In contrast with OCs, PBDEs showed increasing temporal trends during the last decades of the 20th century.[73] On the other hand, some in vitro and in vivo studies attributed to PBDEs antiandrogenic effects and interference with normal thyroid function. All this rose environmental and health concerns that led in Europe to a restriction in their use (max 0.1% in mass for the penta- and octa-BDE mixtures) (Directive 2003/11/EC). This was apparently effective in leveling the temporal trends of PBDEs in the European environment,[74] including foodstuff.[75] The European Court of Justice annulled the exemption of deca-BDE in electrical and electronic goods placed on the European Union market after June, 2008. In the United States, regulation occurred at the state level with some states banning the manufacture or distribution of the penta- and octa-BDE formulations. By the end of 2004 the Great Lakes Chemical Corporation, the only producer of the commercial penta- and octa-BDE formulations in the United States, voluntarily discontinued their production. PBDEs are currently listed as POPs by the Stockholm Convention (see Box 8.1).

Focusing on the marine environment, concentrations of BDE-47 reported in polluted marine waters range from 0.03 to 0.18 ng/L, and in sediments from <1 to 368 ng/g DW (Table 8.2). The most abundant congener in sediments, the decabromodiphenyl-ether (BDE-209),[76] does not bioaccumulate due to its extreme hydrophobicity (log K_{ow} > 10). The most abundant

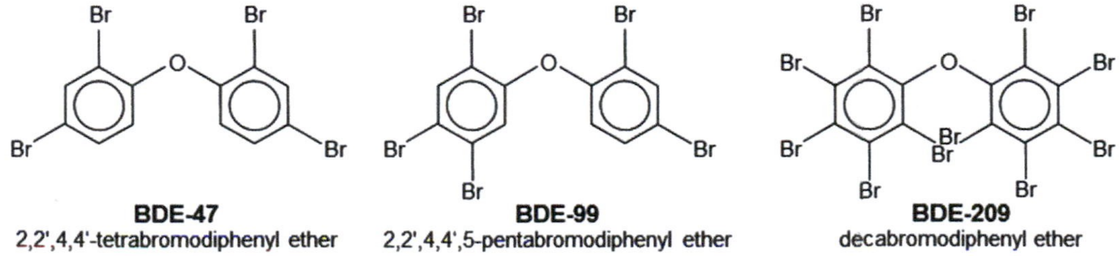

BDE-47	BDE-99	BDE-209
2,2',4,4'-tetrabromodiphenyl ether	2,2',4,4',5-pentabromodiphenyl ether	decabromodiphenyl ether

FIGURE 8.5

Molecular structures of common polybrominated diphenyl ethers. PBDEs do not form coplanar molecules.

Table 8.2 Maximum Concentrations (C_{max}) of Household Chemicals Reported in Marine Waters, Sediments, and Mussels, Mussel Bioconcentration Factor (BCF) Measured In Laboratory, Acute Toxicity to Marine Invertebrate Larvae, Reported Endocrine Disruption Effects on Marine Organisms, and EU Environmental Standards for Coastal Waters

Chemical	C_{max} Seawater (ng/L)	C_{max} Marine Sed (ng/g DW)	C_{max} Mussel (ng/g DW)	BCF Muss (L/Kg WW)	Acute Toxicity High/Low/None 48 h EC_{50} (µg/L)	EDC on Marine Organisms	EU Standard (Directive 2013/39/EU)
Bisphenol A	58-72-93-145-146-249-330-408-1700-2470 [1]	5.0–118 [2]	13.7–197 [3,4]	900 [5]	Low; 2085[b]	Estrogenic: increased VTG in fish, increased egg production in copepods	Not regulated
Nonylphenol	73-84-95-170-172-207-211-269-337-370-416-497-547-915-934-2760-4100-5200 [1]	13.6-86.3-192-640-1642 [7,8]	75.6-240-643-3000 [4,7]	1370 [13]	High; 140[b]	Estrogenic: increased VTG in fish	300 ng/L (AA); 2000 ng/L (MAC)
Octylphenol	18-72-1060 [16]	10 [8]	6.4–16 [3]	no data	High; 225[c]	Estrogenic: increased VTG in fish	10 ng/L (AA)
Triclosan	7-14-40-310 [1]	9.6-27-80 [9,10,14]	1.5–9.9 [6,15]	129 [15]	High; 173[b]	No effects reported	Not regulated
DEHP	400–2110 [17,49]	90-2860-3390 [18,19,49]	c. 170 [20]	2500 [21]	Above solubility limit	Estrogenic: increased VTG in fish	1300 ng/L (AA)
BDE-47	0.03-0.08- 0.18 [22,23,26]	0.13-0.6 -0.83-1.12-7.1-100-368 [19,24–28,39]	2.45 -c. 3.5-4.3-10-16.5-17-28-68 [12,25–30,38]	2180 [31]	Above solubility limit	No effects reported	14 ng/L (MAC)[d]; 0.0085 ng/g WW[d]
PFOS	25-58 [32,47]	0.11 [33]	<1.5 [33]	no data	None 20,000[c]	No effects reported	0.13 ng/L (AA); 7200 ng/L (MAC); 9.1 ng/g WW
Benzophenone	136 [11]	9.7[34]-110[11]	<3[11]	no data	no data	No effects reported	Not regulated
Benzophenone 3	69-216-2013-3300 [35,45]	<0.1[34]-3[39] - 39.8 [35]	<28[46] –63[36]	No accum. [48]	Low; 3473[b]	No effects reported	Not regulated
Benzophenone 4	<1-54-164 [35,45]		87 [36]	181 [48]	None; >10,000[b]	No effects reported	Not regulated

4-MBC	7-85-799 [35,45]	7.90 [35] – 17.2 [39]	49 [36]	160 [48]	High; 587[b]	No effects reported	Not regulated
EMHC	53-86-92-264 [43,45]	c. 10 [40] – 39 [39], 16.4 [42]	256[37], c. 1700[41]	no data	High; 284[c]	No effects reported	Not regulated
Octocrylene	1324–7301 [44]	15.6[35]-65 [40] - 82.1 [42]	20 [46] - 3500 [41] –7122 [37]	442 [48]	High; 737[c]	No effects reported	Not regulated
OD-PABA	111–182 [44,50]	4[40]–150 [35]	<2 [37] – 800 [41]	No accum.	High; 130[b]	No effects reported	Not regulated

[1] Reviewed by Tato et al.[25]. [2] reviewed by Careghini et al.[26]. [3] Isobe et al.[27]. [4] Staniszewska et al.[28]. [5] Gatidou et al. (2010); [6] Álvarez-Muñoz et al. (2015); [7] reviewed by Careghini et al. (2015); [8] Isobe et al.[29]. [9] Pintado-Herrera et al.[30]. [10] Fernandes et al.[31]. [11] Pojana et al.[32]. [12] Dodder et al.[33]. [13] Vidal-Liñán et al.[34]. [14] Miller et al.[35]. [15] Kookana et al.[36]. [16] reviewed by Salgueiro-González et al.[37]. [17] Jornet-Martínez et al.[38]. [18] Muñoz-Ortuño et al.[39]. [19] Klamer et al. (2005); [20] Muñoz-Ortuño et al.[40]. [21] Brown and Thompson[41]. [22] Kim et al.[42]. [23] Wurl et al.[43]. [24] Eljarrat et al.[44]. [25]a Christensen and Platz[45]. [26] Oros et al.[46]. [27]a de Boer et al.[47]. [28] Allchin et al.[47]. [29] Gama et al.[49]. [30]a Ramu et al. (2007); [31] Vidal-Liñán et al.[50]. [32] Yamashita et al.[51]. [33]a Nakata et al.[52]. [34] Jeon et al.[53]. [35] Sánchez-Quiles and Tovar-Sánchez[54]. [36] R. Rodil pers. comm.; [37] Bachelot et al.[55]. [38] Johansson et al.[56]. [39] Barón et al.[57]. [40] Amine et al.[58]. [41] Picot Groz et al.[59]. [42] Langford et al.[60]. [43] Goksøyr et al.[61]. [44] reviewed by Giraldo et al.[62]. [45] reviewed by Paredes et al.[63]. [46] Negreira et al.[64]. [47] EC (2011) PFOS EQS dossier; [48] L. Vidal-Liñán et al. (2018); [49] Naito et al.[65], 95% percentile estimated from the Japanese monitoring dataset; [50] Bratkovics and Sapozhnikova[66].

a Original data for mussels expressed on wet weight basis, conversion assuming 80% water.
b Bivalve larvae.
c Seaurchin larvae.
d Sum of the concentrations of PBDE congener numbers 28, 47, 99, 100, 153, and 154.

congeners in marine biota are BDE-47 and BDE-99.[77] Seafood is considered the major source of dietary intake of PBDEs in humans, and fish were reported to show higher PBDE concentrations than invertebrates.[78] In laboratory experiments, mussels accumulate BDE-47 with a BCF of 2180 L/Kg WW.[79] In wild mussels maximum BDE-47 concentrations found in different polluted areas ranged from 1.2 to 68 ng/g DW (Table 8.2). For the sum of PBDE congeners, the Asian Mussel Watch reported a range of values from <0.1 ng/g WW in clean sites to >1 ng/g WW in polluted sites, with maximum values of 10 ng/g WW approximately equivalent to 50 ng/g DW.[80]

In the EU, the EQS for marine waters is 0.014 μg/L, and for fish 0.0085 ng/g WW. These values refer to the sum of congeners 28, 47, 99, 100, 153, and 154.[81]

8.3   HOUSEHOLD CHEMICALS: PLASTIC ADDITIVES, CLEANERS, PHARMACEUTICALS AND PERSONAL CARE PRODUCTS

Contaminants of emerging concern include plastic additives, pharmaceuticals, and personal care products

We have dealt in Chapter 7 with naturally occurring hydrocarbons, including polyaromatic hydrocarbons, and in the previous sections of this chapter with synthetic chlorinated and brominated hydrocarbons, characterized by a bond between a chlorine (or bromine) atom and a C atom from a benzene ring, which give them their characteristic environmental persistence. PAHs, OCs, and to a lesser extent polybrominated chemicals are considered in most environmental legislations, and their levels are routinely assessed in marine pollution monitoring networks. We will briefly review now other synthetic organic chemicals generally not included in environmental regulations, and thus frequently termed **contaminants of emerging concern**. They are becoming ubiquitous in the environment because, although many of them are not persistent, they are increasingly used to manufacture everyday items, like plastics, cosmetics, electronics, textiles, cleaning, or pharmaceutical products. They are thus common **household chemicals** present in all urban wastewaters that reach the aquatic environments mainly through the urban wastewater effluents. For some of these household chemicals that frequently partition into the particulate phase eliminated by decantation, conventional sewage treatments efficiently remove ca. 90%.[82] Still the remaining 10% may constitute a continuous and ecologically relevant input to surface waters, where these chemicals reach maximum concentrations generally ranging from 100 to 1000 ng/L (Table 8.2). Therefore, the inputs to natural waterbodies of these household chemicals are universal and continuous, which in practice turns their presence and effects in the environment equivalent to that of nonbiodegradable persistent pollutants.

From the very beginnings of the **plastics** industry, it has been necessary to add materials to a basic polymeric matrix to improve its processing and end-use performance, and to achieve the desired mechanical, chemical, optical, and electrical properties in the final plastic product.[83] Ranked by total volume used, chemical **additives** may be classified as follows: reinforcing fibers, fillers, and coupling agents; plasticizers; colorants; stabilizers (antioxidants, ultraviolet absorbers, and biological preservatives); processing aids; flame retardants; peroxides; and antistats. In addition, albeit classical polymers are made from natural hydrocarbons, more recent polymers intended to manufacture high performance objects directly use as monomers synthetic chemicals.

Pharmaceuticals and personal care products used in everyday life in large amounts as medicines and cosmetics reach the natural waters through the wastewater effluents. Depending on their environmental persistence and the efficiency of removal in the urban wastewater treatment processes, they can be detected in surface waters at concentrations from 1 to 100 ng/L, although levels as high as 1 µg/L for certain common analgesics and antiinflammatories,[84] UV filters, and other cosmetics components[85] have been reported.

Perfluorooctane sulfonate (PFOS), and related compounds such as perfluorooctane sulfonamide (PFOSA) and perfluorooctanoic acid (PFOA), are synthetic chemicals with a hydrophobic chain where all the H atoms have been replaced by F, bond to a polar group (Fig. 8.6). This molecular structure gives them the useful property to repel both water and oils, and thus they are used as impregnation agents for protecting the surfaces of textiles, leather, paper, and metals. They are also used in wax, polishes, paints, varnishes, cleaning products, firefighting foams, and semiconductor industry. PFOA is also used as a processing aid in the synthesis of fluoropolymers, including Teflon.

Perfluoroalkyl compounds

Perfluorooctane compounds are highly persistent due to the extreme stability of the C—F bond. PFOS does not hydrolyze, photolyze, or biodegrade in any environmental condition tested. The half-life of PFOS was set to be greater than 41 years.[86] PFOS is a bioaccumulative compound with BCF from 1000 in invertebrates to 2800—3100 in fish. However, it does not follow the classical pattern of partitioning into fatty tissues, but does instead bind preferentially to proteins in the plasma and liver. In marine ecosystems PFOS and PFOSA showed remarkable biomagnification, particularly in air-breeding animals.[87] Maximum PFOS concentrations were measured in the liver of large fish (up to 50 ng/g WW), sea birds (up to 100 ng/g WW), and marine mammals (up to 1 µg/g WW).[88] In contrast PFOA does not biomagnify and it is more abundant than PFOS in organisms from low trophic levels.[89]

The toxicity of perfluorooctane compounds to aquatic organisms is moderate to low, with EC_{50} values > 1 mg/L. Available data suggest that PFOS has the potential to induce adverse effects on the endocrine system of animals,

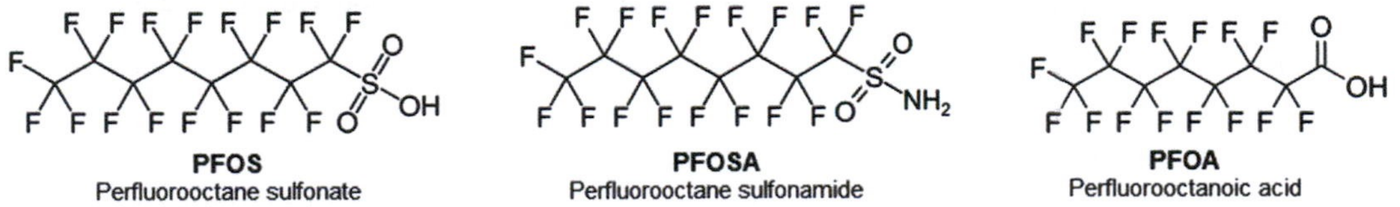

PFOS
Perfluorooctane sulfonate

PFOSA
Perfluorooctane sulfonamide

PFOA
Perfluorooctanoic acid

FIGURE 8.6
Molecular structures of common perfluorinated compounds.

including rats and fish. However, the data indicate that endocrine effects occur at concentrations higher than those causing effects on growth, reproduction, and mortality in standard toxicity tests.[90]

The major global producer of PFOS, 3M Company, phased out its production in 2000. Directive 2006/122/EC placed restrictions on the marketing and use of PFOS. Limits were set to the quantity of PFOS allowed in preparations and on finished products. The Stockholm Convention (see Box 8.1) added PFOS to the list of persistent organic pollutant in 2009.

Alkylphenols   Among the thousands of chemicals used as plastic component, phenolic compounds raise special environmental and human health concern. **Bisphenol A (BPA)** (Fig. 8.7) is mainly used as a monomer for the production of polycarbonate, epoxy resins, phenol resins, polyesters and lacquer coatings, and

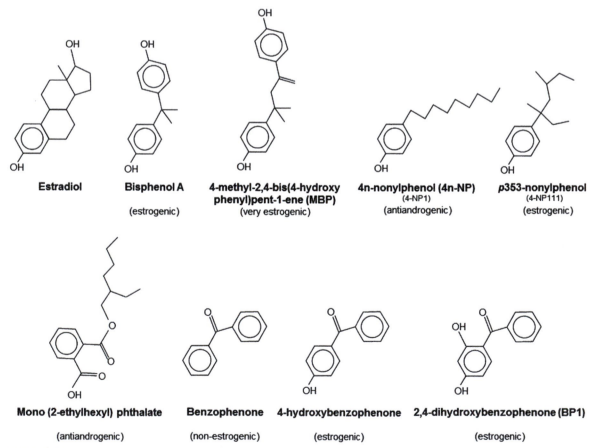

FIGURE 8.7

Molecular structures of the natural female hormone estradiol and structurally related synthetic endocrine disrupting compounds (EDCs): bisphenol A, a highly estrogenic bisphenol A metabolite, the linear isomer (4nNP) and a branched isomer of 4-NP (with two different nomenclature systems indicated), a monoester of the phthalic acid, benzophenone, a benzophenone metabolite (4-hidroxybenzophenone), and a synthetic derivative used as plastic additive (BP1). The EDC activity of each compound according to in vitro and in vivo tests is also indicated.

even food cans.[91] However, in those major applications BPA is immobilized in the polymeric chain and the main environmental source of BPA is its marginal use as plasticizer in PVC.[92] This chemical is classified as an endocrine disrupting compound (EDC) because of its detectable estrogenic and/or antiandrogenic potency in humans and wildlife, and its deleterious effects on mammary glands, brain, and behavioral development were demonstrated in different organisms.[93] In fact some BPA metabolites such as 4-methyl-2,4-bis(4-hydroxyphenyl)pent-1-ene (see Fig. 8.7) are several orders of magnitude more estrogenic than the parental compound,[94] apparently due to a higher affinity for the estrogen receptor (ER).

Despite the large global production of 4.5 million T/year, its environmental ubiquity and its toxicological effects, including concerns on human health effects,[95] BPA is not considered as a priority contaminant in the European Directive 2013/39/EU. Nevertheless, the use of BPA in infant plastic feeding bottles and toys manufacture was restricted by Directive 2011/8/EU and Directive 2014/81/EU, respectively. Furthermore, its use as additive in plastics in contact with food was also regulated in Europe with a specific migration limit of 0.6 mg/kg.[96] On the other hand, the European Food Safety Authority has recently concluded BPA poses no health risk to humans because current exposure to this compound via ingestion is too low to cause damage.[97] Therefore the ecotoxicological status of BPA is under strong debate and the environmental risk posed by this compound deserves further investigation.

Maximum BPA concentrations reported in marine waters range between 58 and 2470 ng/L,[98] with a BCF in mussels of 900 L/Kg WW.[99]

Nonylphenols (NP) are by far the most important alkylphenols, constituted by an alkyl chain with 9C located at either ortho- (2-NP), meta- (3-NP), or mainly para- (4-NP) position on the phenolic ring. These compounds are the degradation products of nonylphenol ethoxylates (NPEs), one of the most common nonionic surfactants used in detergents and cleaning products, with a total production of 500,000 T/year.[100] Moreover, NPs are used as pesticides, as a monomer in phenol/formaldehyde resins and mainly as plasticizers for high density polyethylene, polyethyleneterephthalate, and polyvinylchloride.[101] Other uses are intermediate in the production of tri-(4-nonylphenyl) phosphite, a plastic antioxidant, and as a catalyst in the curing of epoxy resins.[102]

Among NPs, only the 4-NP has been classified as an EDC because it mimics the female hormone 17-β-estradiol (estrogenic effects) and inhibits the aromatase enzyme essential for the synthesis of testosterone (antiandrogenic effects), according with a variety of both in vitro and in vivo assays.[103] In fact, the linear form 4n-NP (Fig. 8.7), frequently used as nonylphenol standard despite it is not present in technical mixtures, is not estrogenic, and from the >100 branched isomers identified only a few are considered as EDCs according to in vitro tests, although their potency is always several orders of magnitude lower than the natural hormone, the estradiol.[104]

The differential estrogenic potential of the 4-NP isomers is related to the tridimensional configuration, which determines their binding affinity to the ER protein. Fig. 8.8 depicts the conformation of a nonestrogenic isomer, the linear 4nNP, one weakly and one highly estrogenic isomers according to their ER affinity quantified by means of in vitro tests.[105]

Due to its persistence in the environment ($T_{1/2}$ 30–58 days in seawater), its moderate bioaccumulation in organisms, and its toxicity, 4-NP has been included in different EU and international regulations to preserve the environment and protect human health. Thus, product formulations marketed in the EU cannot contain more than 0.1% of NPE or NP (Directive 2003/53/EC). This applies to many industries, including the textile and leather industries, except in the case of closed application systems where no release into waste waters occurs. NPEs have been replaced by more expensive alcohol ethoxylates, which degrade more quickly in the environment. The EU has also included 4-NP on the list of priority hazardous substances for surface water and its chronic and acute environmental quality standards are 0.3 and 2 µg/L respectively (Directive 2013/39/EU). In North America, the United States Environmental Protection Agency (US-EPA) also set criteria which recommend that 4-NP concentration should not exceed 6.6 µg/L in fresh water and 1.7 µg/L in saltwater.[107] To meet these criteria, the US-EPA is encouraging a voluntary phase-out of 4-NP in industrial laundry detergents.[108]

NPs and NPEs do not biomagnify in marine food webs.[109] Maximum 4-NP concentrations reported in marine waters range between 16 and 5200 ng/L, with a BCF in mussels of 1370 L/Kg WW (see Tab. 8.2).

Octylphenol (OP) is mainly used to make phenolic resins (>90%), and in rubber processing to make tires. Minor uses include production of ethoxylates for surfactants, and a component in printing inks and electrical insulation varnishes. It can also be present as an impurity in NP, typically up to around 5%. Similarly to NP, it is used in the industry as a mixture of isomers mostly

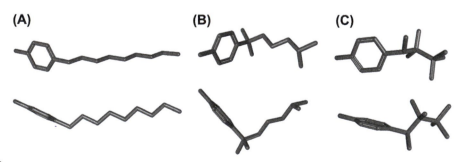

FIGURE 8.8

Tridimensional structure of three 4-NP isomers of very different estrogenic potency: (A) the linear isomer 4nNP, nonestrogenic; (B) 4-(2,6-dimethyl-2-heptyl)phenol, weakly estrogenic; (C) 4-(3,3,4,4-tetramethylpentan-2-yl)phenol, highly estrogenic. *3D image source: US National Library of Medicine.*[106]

in para position (tert-octylphenol), although the isolated linear isomer (4n-OP) is commercially available.

The environmental and toxicological properties of OP are analogous to those of NP. It is moderately persistent since half-life in seawater is 30 d, and no significant degradation takes place in anaerobic sediments. It is found in surface waters at concentrations <1 µg/L and shows a moderate BCF of a few hundreds in fish. It has been reported as an estrogenic and anti-androgenic EDC, inducing VTG synthesis in males and reducing reproductive success.[110]

Phthalates are the main plasticizers used by the plastic industry (see Section 6.2). Flexible PVC may contain more than 30% in weight of phthalates. They are diesters of the orthophthalic acid, where the carboxylic groups are esterified by linear or cyclic hydrocarbons between 2 and 10C atoms (Fig. 8.9). Orthophthalates, with the carboxylic groups in neighbor C atoms, (but not meta- or para-phthalates, the latter also known as terephthalates) are endocrine disruptors, particularly those with hydrocarbon chains between 4 and 8C. Benzyl-butyl-phthalate, Dibutyl-phthalate, and Diethylhexyl-phthalate demonstrated testicular toxicity and infertility in male mammals.[111] The endocrine mode of action likely involves inhibition of the synthesis of testosterone, and the monoester metabolites (such as the mono (2-ethylhexyl) phthalate illustrated in Fig. 8.7) were described as more active than the original diester.[112] Laboratory tests with rodents and epidemiological studies in humans raise concern on the potential effects of DEHP on immune system and its involvement in immunological disorders.[113]

Phthalates

After the ban on some polybrominated flame retardants (see Section 8.2) organo-phosphorus chemicals were proposed as an alternative, including the chlorinated chemicals tris(2-chloroethyl) phosphate (TCEP), tris(chloropropyl) phosphate (TCPP), and tris(1,3-dichloro-2-propyl)phosphate (TDCPP). They are used as flame retardants, plasticizers, and viscosity regulators in polyurethanes, polyester resins, polyacrylates, and other polymers.[114] For TDCPP, >85% is used in PU foam. TDCPP is more expensive than other P flame retardants such as the most common TCPP but shows a better performance. These

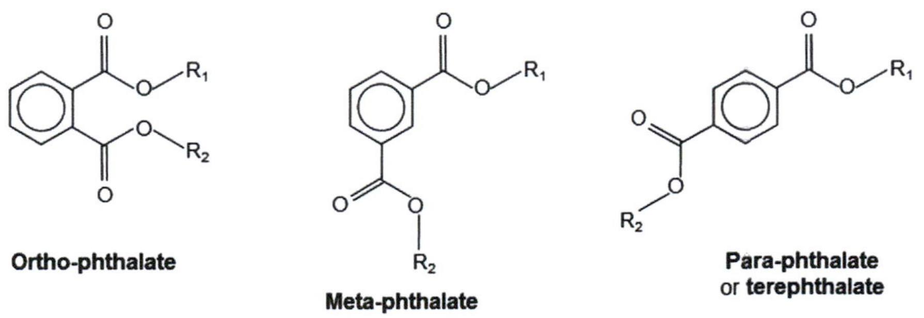

Chlorinated organo-phosphorus flame retardants

Ortho-phthalate   Meta-phthalate   Para-phthalate or terephthalate

FIGURE 8.9
Molecular estructure of phthalates; esters of the orthophthalic acid where R1 and R2 are linear or cyclic hydrocarbons between 2 and 10C atoms. Meta and Para-phthalates are also shown.

chemicals have low biodegradability and remarkable environmental persistence, but they show a limited bioaccumulation in marine organisms and do not biomagnify across the trophic webs.[115] However, due to their persistence they have been detected in fish even from remote areas such as the Arctic at concentrations reaching 10 ng/g WW.[116] The most abundant in seawater and marine biota from European coasts is TCPP, with up to 28 ng/L[117] in water and 1300 ng/g LW in mussels.[118] However, much higher concentrations were reported in Chinese coasts, with maximum values of 618, 170 and 378 ng/L for TCEP, TCCP, and TDCPP respectively.[119] In contrast in laboratory experiments the BCF was, as expected, higher for the most lipophilic of the three, the TDCP, with BCF values c. 30 L/Kg WW in invertebrates and fish. Toxicological information on these chemicals is still scarce, but TCEP is considered as carcinogenic for animals and toxic for aquatic organisms.[120]

Triclosan   **Triclosan** (TCS) is a halogenated biphenyl ether with the chemical name 2,4,4'-trichloro-2'-hydroxy-diphenyl ether (Fig. 8.10). It has been mainly used as broad spectrum antimicrobial in pharmaceutical and personal care products.[121] Because of its thermal stability, TCS is increasingly used also as an

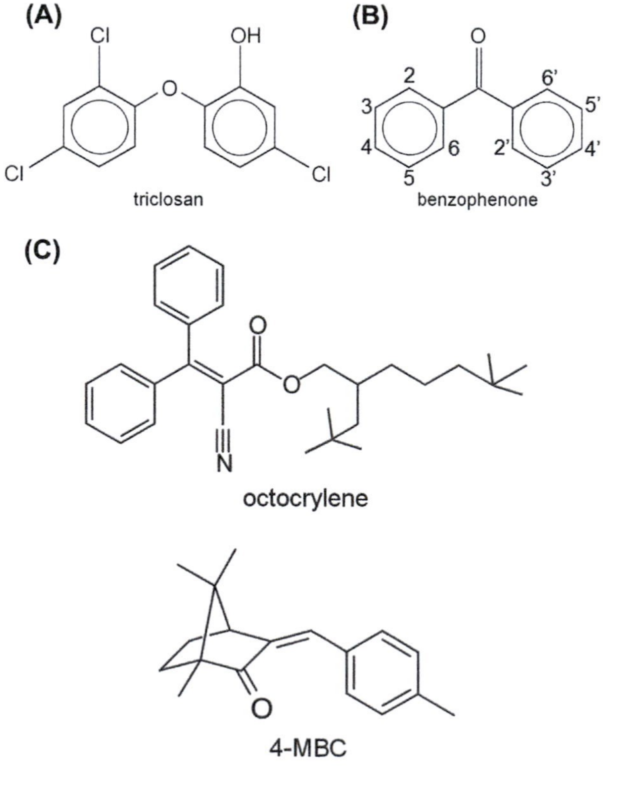

FIGURE 8.10
Molecular structures of aromatic compounds used in cosmetics and plastics: triclosan (biocide), benzophenone, octocrylene and 4-MBC (UV filters). For benzophenone, hydroxy and methoxy radicals in carbons 2, 4, 2' and 4' generate derivatives of industrial use.

antimicrobial material preservative in plastics and textile fibers.[122] Despite its moderate production (1500 T/y), and ready biodegradation, TCS has been widely detected in natural waters and sediments[123] (see Table 8.2). Maximum TCS concentrations reported in marine waters range between 7 and 310 ng/L,[124] and in sediments from 10 to 80 ng/g DW, with a BCF in mussels of 129[125] L/Kg WW. As the occurrence of TCS in the environment and human body (including breast milk) increased, emerging health concerns related to its use have raised. Trace levels of TCS have been suggested to promote the development of cross-resistance to antibiotics among bacteria.[126] Furthermore, TCS can be potentially transformed into more toxic chlorinated compounds such as dioxins in the environment.[127]

In Europe, TCS is regulated in Annex V of the Regulation (EC) No 1223/2009, to a maximum concentration of 0.3%, and it is currently under evaluation in the context of both REACH and the Biocides Directive. Currently Canada and Japan restrict the use of TCS in cosmetics,[128] and in 2010 TCS was removed from the EU list of provisional additives for use in plastic food-contact material.[129] Several European countries, including Denmark, Sweden, Norway, and Finland, have issued national consumer advisories for the use of TCS.[130] TCS is not regulated in the United States, however, EPA is reevaluating the risks and the agency may consider new regulatory action if warranted.[131]

Benzophenone (BP) is a diphenyl ketone (Fig. 8.10) used in plastics and cosmetics as UV filter, in inks and varnishes as curing photoinitiator, and as food additive in the United States. The parental molecule is not an EDC[132] but hydroxylized metabolites such as 4-hydroxy-BP and synthetic derivatives of industrial use in sun screens such as **benzophenone-1 (BP-1)** (2,4-dihydroxybenzophenone), **benzophenone-2 (BP-2)** (2,2′,4,4′-tetrahydroxybenzophenone), **benzophenone-3 (BP-3)** (2-hydroxy-4-methoxybenzo phenone, or oxybenzone), and **benzophenone-8 (BP-8)** (2-hydroxy-4-methoxy-2′hydroxybenzophenone, or dioxybenzone) are considered as estrogenic and antiandrogenic compounds.[133] 4-hydroxy-BP is an agonist of the human ER. BP-1 inhibits in vitro an enzyme involved in the synthesis of testosterone and it is a potent strogen agonist in vitro.[134] BP-2 was the most estrogenic benzophenone according to a cell proliferation in vitro test based on the activation of the human ER.[135] Both BP-1 and BP-2 at concentrations far above those found in the environment induced the synthesis of vitellogenin in juvenile fathead minnows, a freshwater fish. BP-3, widely used in cosmetics and as plastic additive, was found to be estrogenic in vitro and in rats. It is readily biodegradable and does not accumulate in mussels, but showed remarkable acute toxicity to phytoplankton and mysids.[136] Finally, BP-8 showed estrogenicity in a yeast test.[137]

Other organic UV filters found at concentrations reaching 1 μg/L in seawater are **4-Methylbenzylidene camphor (4-MBC)**, **Ethylhexyl methoxycinnamate (EHMC)**, the **octocrylene**, and **octyl dimethyl-paraaminobenzoic acid (OD-PABA)** (see Table 8.2). All of them include in their molecular structure aromatic

Benzophenones and other UV filters

rings that adsorb UV radiation. Hence, they are used as light stabilizers in a wide range of products including plastics, adhesives, rubber, and cosmetics (such as creams, shampoos, sunscreen lotions), and are intended to protect those products (or skin, in the case of sunscreens) from damage caused by the UV component of sunlight. Concentrations of these chemicals in cosmetics can reach 10% weight in the commercial product. The use of sunscreen lotions by swimmers represents a direct input into the seawater. Not surprisingly, octocrylene, EHMC, and OD-PABA have been found at concentrations up to 7000, 1700, and 800 ng/g, respectively, in wild mussels collected in the proximity of beaches with recreational activity (see Table 8.2). Filters 4-MBC and octocrylene (Fig. 8.10) are not readily biodegradable, and have been shown in laboratory to accumulate in marine mussels.[138] The environmental concern caused by these substances is based on their short-term toxicity, with LC_{50} for early life stages of marine organisms in the order of 100 µg/L, but mainly on their suspected endocrine disrupting properties. The filter 4-MBC and the similar molecule 3-BC showed remarkable estrogenic activity in vitro and increase uterine weight in rats.[139] OD-PABA has also been shown to have estrogenic effects in vitro. In addition, to the interference with sex hormones, 4-MBC, EHMC and OD-PABA have been shown in an insect to induce the expression of the ecdysone receptor gene.[140] Ecdysones are steroid hormones that play a central role in the molting of arthropods, including crustaceans.

Pharmaceuticals   The worldwide pharmaceutical market steadily grew in the 21st century from 390 billion US dollars revenue in 2001 to 1072 billion in 2015.[141] Pharmaceutically active products reach the natural environment through the urban effluents, and despite removal during wastewater treatment that commonly ranges between 30% and 99%,[142] they are increasingly detected in surface waters, including coastal ecosystems. Although most of these products are readily biodegradable, the continuous input through urban wastes makes the environmental situation comparable to that of persistent pollutants. Mussels and benthic fish have been reported to accumulate analgesics, antiinflammatories, antibiotics, psychiatric drugs,[143] contraceptive hormones,[144] and other active molecules such as caffeine or cocaine at concentrations in the order of ng/g DW.[145] When concentrations measured in the effluents are compared to acute aquatic toxicity data the highest risk seems to be posed by antibiotics.[146] However, the long-term ecological effects of these emerging pollutants remain to be assessed.

KEY IDEAS

- Synthetic OC chemicals with chlorine atoms bound to phenolic rings were massively produced for their use as pesticides because their environmental persistence granted them high efficacy as biocides. However, they caused unexpected deleterious effects on birds and other nontarget species and were phased out.

- Polychlorinated biphenyls (PCB), synthetized for their use in electric apparatus and many other industrial applications, also showed environmental persistence and high bioaccumulation potential, and were also phased out, although they are still present in electric equipments.

- Polybrominated diphenyl ethers (PBDE) are persistent, bioaccumulative synthetic chemical massively used as flame retardants in plastics, textiles, electronic components, furniture, etc. As a consequence, they showed increasing trends over the last decades of the 20th century, and were legally restricted from most applications during the beginning of the 21st century. They were mostly replaced by organophosphorus chlorinated flame retardants such as TCPP and TDCPP.

- Perfluorooctane sulfonate (PFOS) and related compounds are synthetic chemicals used in textiles that show high environmental persistence, bioaccumulation potential, and biomagnification across food chains. Unlike OCs, they are not lipophilic but proteophilic, and accumulate preferentially in the liver.

- Components of plastics, cleaning products, and cosmetics, such as bisphenol A, orthophthalates, nonylphenol, or certain benzophenones, have shown in vitro and in vivo potential to interfere with the function of natural hormones such as estrogens, androgens and thyroid hormones, and are thus termed endocrine disrupting chemicals (EDC).

- Household chemicals present in everyday life objects, including plastic additives, cleaning products, pharmaceuticals, and cosmetics, are emitted through the urban wastewaters into the coastal waterbodies where they can reach maximum concentrations of 1 μg/L. The continuous input may cause in receiving waters chronic effects similar to those of persistent pollutants.

Endnotes

1. Hellou J, Lebeuf M, Rudi M. Review on DDT and metabolites in birds and mammals of aquatic ecosystems. Environ Rev 2013;21:53−69.

2. Peakall DB. Pesticides and the reproduction of birds. Sci Am 1970;222(4):72−8.

3. Miller DS, Kinter WB, Peakall DB. Enzymatic basis for DDE-induced eggshell thinning in a sensitive bird. Nature 1976;259(15):122−124.

4. Prakash S, Tandon GS, Seth TD, et al. The role of reactive oxygen species in the degradation of lindane and DDT. Biochem Biophy Res Commun 1994;199(3):1284−1288.

5. Patil et al. Metabolic transformation of DDT, Dieldrin, Aldrin, and Endrin by marine micro-organisms. Environm Sci Technol 1972;6(7):629−632.

6. Gohil. PhD Dissertation, University of Florida; 2011.

7. Kwong RWM,Yu PKN, Lam PKS, et al. Uptake, elimination, and biotransformation of aqueous and dietary DDT in marine fish. Environ Toxicol Chem 2008;27(10):2053—2063.

8. Kwong RWM,Yu PKN, Lam PKS, et al. Biokinetics and biotransformation of DDTs in the marine green mussels *Perna viridis*. Aquat Toxicol 2009;93:196—20.

9. Bertgman A, Norstrom RJ, Haraguchi K, et al. PCB and DDE methyl sulfones in mammals from Canada and Sweden. Environ Toxicol Chem 1994;13:121—128.

10. Aguilar A. Relationship of DDE/ΣDDT in marine mammals to the chronology of DDT input into the ecosystem. Can J Fish Aquat Sci 1984;41:840-844.; Temporal trend of bis(4-chlorophenyl) sulfone, methylsulfonyl-DDE and -PCBs in Baltic guillemot (*Uria aalge*) egg 1971-2001 - A comparison to 4,4′-DDE and PCB trends. Jörundsdóttir H, Norström K, Olsson M, et al. Environ Pollut 2006;141:226—237.

11. Bellas J, Albentosa M, Vidal-Liñán L, et al. Combined use of chemical, biochemical and physiological variables in mussels for the assessment of marine pollution along the N-NW Spanish coast. Mar Environ Res 2014;96:105—117.

12. Ramu K, Kajiwara N, Sudaryanto A, et al. Asian Mussel Watch Program: Contamination Status of Polybrominated Diphenyl Ethers and Organochlorines in Coastal Waters of Asian Countries. Environ Sci Technol 2007;41:4580—4586.

13. Woodwell GM, Wurster CF, Issacson PA. DDT Residues in an East Coast Estuary: A Case of Biological Concentration of a Persistent Insecticide. Science 1967;156:821—823.

14. Lebeuf M. La contamination du béluga de l'estuaire du Saint-Laurent par les polluants organiques persistants en revue. Revue des Sciences de l'Eau 2009;22(2):199—233.

15. Lopez J, Boyd D, Ylitalo GN, et al. Persistent organic pollutants in the endangered Hawaiian monk seal (*Monachus schauinslandi*) from the main Hawaiian Islands. Mar Pollut Bull 2012; 64:2588—2598.

16. Frouin H, Lebeuf M, Hammill M, et al. Transfer of PBDEs and chlorinated POPs from mother to pup during lactation in harp seals *Phoca groenlandica*. Sci Total Environ 2012;417—418: 98—107.

17. Hellou et al. (2013) op. cit.

18. Bellas J, González-Quijano A, Baamonde A, et al. PCBs in wild mussels (*Mytilus galloprovincialis*) from the N—NW Spanish coast: Current levels and long-term trends during the period 1991—2009. Chemosphere 2011;85:533—541.

19. Frignani M, Bellucci LG, Carraro C, et al. Polychlorinated biphenyls in sediments of the Venice Lagoon. Chemosphere 2001;43:567-575.; Okay OS, Karacik B, Basak S, et al. PCB and PCDD/F in sediments and mussels of the Istanbul strait (Turkey). Chemosphere 2009;76:159—166.

20. Gaus C, Brunskill GJ, Weber R, et al. Historical PCDD inputs and their source implications from dated sediment cores in Queensland (Australia). Environ Sci Technol 2001;35(23): 4597—4603.

21. Sinkkonen S, Paasivirta J. Degradation half-life times of PCDDs, PCDFs and PCBs for environmental fate modeling. Chemosphere 2000;40:943—949.

22. Blanco SL, Sobrado C, Quintela C, et al. Dietary uptake of dioxins (PCDD/PCDFs) and dioxin-like PCBs in Spanish aquacultured turbot (*Psetta maxima*). Food Addit Contam 2007;24(4):421—8.

23. For biota see: Bayarri S, Baldassarra LT, Iacovella N et al. PCDD, PCDFs, PCBs and DDE in edible marine species from the Adriatic Sea. Chemosphere 2001;43:601-610. For sediments see: Tyler AO, Millward GE. Distribuition and partitioning of polychlorinated dibenzo-p-dioxins, polychlorinated dibenzofurans and polychlorinated biphenyls in the Humvber Estuary, UK. Mar Pollut Bull 1996;32(5):397—403.

24. Van den Berg M, Birnbaum L, Bosveld ATC, et al. Toxic Equivalency Factors (TEFs) for PCBs, PCDDs, PCDFs for Humans and Wildlife. Environmental Health Perspectives 1998;106(12): 775—792.

25. Tato T, Salgueiro-González N, León VM, et al. Ecotoxicological evaluation of the risk posed by bisphenol A, triclosan, and 4-nonylphenol in coastal waters using early life stages of marine organisms (*Isochrysis galbana, Mytilus galloprovincialis, Paracentrotus lividus*, and *Acartia clausi*). Environ Pollut 2018;232:173–182.

26. Careghini A, Mastorgio AF, Saponaro S, et al. Bisphenol A, nonylphenols, benzophenones, and benzotriazoles in soils, groundwater, surface water, sediments, and food: a review. Environ Sci Pollut Res 2015;22:5711–5741.

27. Isobe T, Takada H, Kanai M, et al. Distribution of Polycyclic Aromatic Hydrocarbons (PAHs) and phenolic endocrine disrupting chemicals in South and Southeast Asian mussels. Environ Monit Assess 2007;135:423–440.

28. Staniszewska M, Falkowska L, Grabowski P, et al. Bisphenol A, 4-tert-Octylphenol, and 4-Nonylphenol in The Gulf of Gdańsk (Southern Baltic). Arch Environ Contam Toxicol 2014;67:335–347.

29. Isobe T, Nishiyama H, Nakashima A, et al. Distribution and Behavior of Nonylphenol, Octylphenol, and Nonylphenol Monoethoxylate in Tokyo Metropolitan Area: Their Association with Aquatic Particles and Sedimentary Distributions. Environ Sci Technol 2001;35: 1041–1049.

30. Pintado-Herrera MG, González-Mazo E, Lara-Martín PA. Determining the distribution of triclosan and methyl triclosan in estuarine settings. Chemosphere 2014;95:478–485.

31. Fernandes D, Schnell S, Porte C. Can pharmaceuticals interfere with the synthesis of active androgens in male fish? An in vitro study. J Environ Monit 2011;13:801.

32. Pjana G, Gomiero A, Jonkers N, et al. Natural and synthetic endocrine disrupting compounds (EDCs) in water, sediment and biota of a coastal lagoon. Environ Int 2007;33: 929–936.

33. Dodder NG, Maruya K, Ferguson PL, et al. Occurrence of contaminants of emerging concern in mussels (Mytilus spp.) along the California coast and the influence of land use, storm water discharge, and treated wastewater effluent. Mar Pollut Bull 2014;81:340–346.

34. Vidal-Liñán L, Bellas J, Salgueiro-González N, et al. Bioaccumulation of 4-nonylphenol and effects on biomarkers, acetylcholinesterase, glutathione-S-transferase and glutathione peroxidase, in *Mytilus galloprovincialis* mussel gills. Environ Pollut 2015a;200:133–139.

35. Miller TR, Heidler J, Chillrud SN, et al. Fate of Triclosan and Evidence for Reductive Dechlorination of Triclocarban in Estuarine Sediments. Environ Sci Technol 2008;42: 4570–4576.

36. Kookana RS, Shareef A, Fernandes MB, et al. Bioconcentration of triclosan and methyltriclosan in marine mussels (*Mytilus galloprovincialis*) under laboratory conditions and in metropolitan waters of Gulf St Vincent, South Australia. Mar Pollut Bull 2013;74:66–72.

37. Salgueiro-González N, Turnes-Carou I, Viñas-Dieguez L, et al. Occurrence of endocrine disrupting compounds in five estuaries of the northwest coast of Spain: ecological and human health impact. Chemosphere 2015;131:241–247.

38. Jornet-Martínez N, Muñoz-Ortuño M, Moliner-Martínez Y, et al. On-line in-tube solid phase microextraction-capillary liquid chromatography method for monitoring degradation products of di-(2-ethylhexyl) phthalate in waters. J Chromatogr A 2014;1347:157–160.

39. Muñoz-Ortuño M, Argente-García A, Moliner-Martínez Y, et al. A cost-effective method for estimating di(2-ethylhexyl)phthalate in coastal sediments. J Chromatogr A 2014;1324: 57–62.

40. Muñoz-Ortuño M, Moliner-Martínez Y, Cogollos-Costa S, et al. A miniaturized method for estimating di(2-ethylhexyl) phthalate in bivalves as bioindicators. J Chromatogr A 2012; 1260:169–173.

41. Brown D, Thompson RS. Phthalates and the aquatic environment: Part II. The bioconcentration and depuration of di-2-ethylhexyl phthalate (DEHP) and di-isodecyl phthalate (DIDP) in mussels (*Mytilus edulis*). Chemosphere 1982;2(4):427–435.

42. Kim Y-E, Kim H-S, Choi H-G, et al. Contamination and Bioaccumulation of Polybrominated Diphenyl Ethers (PBDEs) in Gwangyang Bay, Korea. Toxicol Environ Health Sci 2012;4(1): 42−49.

43. Wurl O, Lam PKS, Obbard JP. Occurrence and distribution of polybrominated diphenyl ethers (PBDEs) in the dissolved and suspended phases of the sea-surface microlayer and seawater in Hong Kong, China. Chemosphere 2006;65:1660−1666.

44. Eljarrat E, De La Cal A, Larrazabal D, et al. Occurrence of polybrominated diphenylethers, polychlorinated dibenzo-*p*-dioxins, dibenzofurans and biphenyls in coastal sediments from Spain. Environ Pollut 2005;136:493−501.

45. Christensen JH, Platz J. Screening of polybrominated diphenyl ethers in blue mussels, marine and freshwater sediments in Denmark. J Environ Monit 2001;3:543−547.

46. Oros DR, Hoover D, Rodigari F, et al. Levels and Distribution of Polybrominated Diphenyl Ethers in Water, Surface Sediments, and Bivalves from the San Francisco Estuary. Environ Sci Technol 2005;39:33−41.

47. Allchin CR, Law RJ, Morris S. Polybrominated diphenylethers in sediments and biota downstream of potential sources in the UK. Environ Pollut 1999;105:197−207.

48. de Boer J, Wester PG, van der Horst A, et al. Polybrominated diphenyl ethers in influents, suspended particulate matter, sediments, sewage treatment plant and effluents and biota from the Netherlands. Environ Pollut 2003;122:63−74.

49. Gama AC, Sanatcumar P, Viana P, et al. The occurrence of polybrominated diphenyl ethers in river and coastal biota from Portugal. Chemosphere 2006;64:306−310.

50. Vidal-Liñán L, Bellas J, Fumega J, et al. Bioaccumulation of BDE-47 and effects on molecular biomarkers acetylcholinesterase, glutathione-S-transferase and glutathione peroxidase in *Mytilus galloprovincialis* mussels. Ecotoxicology 2015b;24:292−300.

51. Yamashita N, Kannan K, Taniyasu S, et al. A global survey of perfluorinated acids in oceans. Mar Pollut Bull 2005;51:658−668.

52. Nakata H, Kannan K, Nasu T, et al. Perfluorinated Contaminants in Sediments and Aquatic Organisms Collected from Shallow Water and Tidal Flat Areas of the Ariake Sea, Japan: Environmental Fate of Perfluorooctane Sulfonate in Aquatic Ecosystems. Environ Sci Technol 2006;40:4916−4921.

53. Jeon HK, Chung Y, Ryu JC. Simultaneous determination of benzophenone-type UV filters in water and soil by gas chromatography−mass spectrometry. J Chromatogr A 2006;1131: 192−202.

54. Sánchez-Quiles D, Tovar-Sánchez A. Are sunscreens a new environmental risk associated with coastal tourism? Environ Inter 2015;83:158−170.

55. Bachelot M, Li Z, Munaron D, et al. Organic UV filter concentrations in marine mussels from French coastal regions.

56. Johansson I, Héas-Moisan K, Guiot N, et al. Polybrominated diphenyl ethers (PBDEs) in mussels from selected French coastal sites: 1981−2003. Chemosphere 2006;64:296−305.

57. Barón E, Gago-Ferrero P, Gorga M, et al. Occurrence of hydrophobic organic pollutants (BFRs and UV-filters) in sediments from South America. Chemosphere 2013; 92:309−316.

58. Amine H, Gomez E, Halwani J, et al. UV filters, ethylhexyl methoxycinnamate, octocrylene and ethylhexyl dimethyl PABA from untreated wastewater in sediment from eastern Mediterranean river transition and coastal zones. Mar Pollut Bull 2012;64: 2435−2442.

59. Picot Groz M, Martinez Bueno MJ, Rosain D, et al. Detection of emerging contaminants (UV filters, UV stabilizers and musks) in marine mussels from Portuguese coast by QuEChERS extraction and GC−MS/MS. Sci Total Environ 2014;493:162−169.

60. Langford KH, Reid MJ, Fjeld E, et al. Environmental occurrence and risk of organic UV filters and stabilizers in multiple matrices in Norway. Environ Inter 2015;80:1−7.

61. Goksøyr A, Tollefsen KE, Grung M, et al. Balsa Raft Crossing the Pacific Finds Low Contaminant Levels. Environ Sci Technol 2009;43(13):4783−4790.

62. Giraldo A, Montes R, Rodil R, Ecotoxicological Evaluation of the UV Filters Ethylhexyl Dimethyl p-Aminobenzoic Acid and Octocrylene Using Marine Organisms *Isochrysis galbana*, *Mytilus galloprovincialis* and *Paracentrotus lividus*. Arch Environ Contam Toxicol 2017;72: 606−611.

63. Paredes E, Pérez S, Rodil R, et al. Ecotoxicological evaluation of four UV filters using marine organisms from different trophic levels *Isochrysis galbana, Mytilus galloprovincialis, Paracentrotus lividus,* and *Siriella armata*. Chemosphere 2014;104:44−50.

64. Negreira N, Rodríguez I, Rodil R, et al. Optimization of matrix solid-phase dispersion conditions for UV filters determination in biota samples. Intern J Environ Anal Chem 2013; 93(No. 11):1174−1188.

65. Naito W, Gamo Y, Yoshida K. Screening level risk assessment of di(2-ethylhexyl)phthalate for aquatic organisms using moinitoring data in Japan. Environ Monit Assess 2006;115: 451−471.

66. Bratkovics S, Sapozhnikova Y. Determination of seven commonly used organic UV filters in fresh and saline waters by liquid chromatography−tandem mass spectrometry. Anal. Methods 2011;3(12):2943−2950.

67. Alaee M, Arias P, Sjödin A, et al. An overview of commercially used brominated flame retardants, their applications, their use patterns in different countries/regions and possible modes of release. Environ Inter 2003;29:683−689.

68. EFSA. Scientific opinion on polybrominated diphenyl ethers (PBDEs) in food. European Food Safety Authority: Parma; 2011. 274 pp.

69. Vane CH, Ma Y-J, Chen S-J, et al. Increasing polybrominated diphenyl ether (PBDE) contamination in sediment cores from the inner Clyde Estuary, UK. Environ Geochem Health 2010; 32:13−21.

70. Hale RC, La Guardia MJ, Harvey EP, et al. Flame retardants. Persistent pollutants in land-applied sludges. Nature 2001;412(12):140−141.

71. Reviewed by Law RJ, Allchin CR, de Boer, J, et al. Levels and trends of brominated flame retardants in the European environment. Chemosphere 2006;64:187−208; see also Ramu et al. (2007) op. cit. for mussels; Gauthier LT, Hebert CE, Weseloh DVC, et al. Dramatic Changes in the Temporal Trends of Polybrominated Diphenyl Ethers(PBDEs) in Herring Gull Eggs From the Laurentian Great Lakes: 1982−2006. for seabirds.

72. Norén K, Meironyté D. Certain organochlorine and organobromine contaminants in Swedish human milk in perspective of past 20-30 years. Chemosphere 2000;40:1111−1123.

73. de Winter-Sorkina R, Bakker MI, Wolterink G, et al. Brominated flame retardants: occurrence, dietary intake and risk assessment. RIVM-report 320100002/2006; 2006. See also for mussels; Johansson et al. (2006) op. cit.

74. Law RJ, Herzke D, Harrad S, et al. Levels and trends of HBCD and BDEs in the European and Asian environments, with some information for other BFRs. Chemosphere 2008;73: 223−241.

75. See: EFSA. Scientific Opinion on Polybrominated Diphenyl Ethers (PBDEs) in Food1 EFSA Panel on Contaminants in the Food Chain (CONTAM). Sci Opin 2011:51.

76. Klamer HJC, Leonards PEG, Lamoree MH, et al. A chemical and toxicological profile of Dutch North Sea surface sediments. Chemosphere 2005;58:1579−1587.

77. Law et al. (2006) op. cit.

78. Tian S, Zhu L, Liu M. Bioaccumulation and distribution of polybrominated diphenyl ethers in marine species from Bohai Bay, China. Environ Toxicol Chem 2010;29(10):2278−2285; See also Allchin et al. (1999) op. cit.

79. Vidal-Liñán et al. (2015b) op. cit.

80. Ramu et al. (2007) op. cit.

81. EU. PolyBDEs EQS dossier 2011.

82. Isobe et al. (2001) op. cit.

83. Deanin RD. Additives in plastics. Environ Health Perspect 1975;11:35—39.

84. Kasprzyk-Hordern B, Dinsdale RM, Guwy AJ. The removal of pharmaceuticals, personal care products, endocrine disruptors and illicit drugs during wastewater treatment and its impact on the quality of receiving waters. Wat Res 2009;43:363—380.

85. Rodil R, Quintana JB, Concha-Graña E, et al. Emerging pollutants in sewage, surface and drinking water in Galicia (NW Spain). Chemosphere 2012;86:1040—1049.

86. EU. PFOS EQS Dossier 2011.

87. Perfluoroalkyl Contaminants in an Arctic Marine Food Web: Trophic Magnification and Wildlife Exposure. Environ Sci Technol 2009;43:4037—4043.

88. Kannan K, Corsolini S, Falandysz J, et al. Perfluorooctanesulfonate and Related Fluorinated Hydrocarbons in Marine Mammals, Fishes, and Birds from Coasts of the Baltic and the Mediterranean Seas.

89. Nakata et al. (2006) op. cit.

90. See EC. PFOS EQS Dossier 2011.

91. EU. European union risk assessment report. 4,4′-isopropylidenediphenol (Bisphenol-A). Luxembourg: European Communities; 2010.

92. EU (2010) op. cit.

93. Flint S, Markle T, Thompson, S, Wallace, E. Bisphenol A exposure, effects, and policy: a wildlife perspective. J Environ Manag 2012;104:19—34.

94. Yamaguchi A, Ishibashi H, Kohra S, et al. Short-term effects of endocrinedisrupting chemicals on the expression. Of estrogen-responsive genes in male medaka (Oryzias latipes). Aquat Toxicol 2005;72:239e49.

95. vom Saal FS, Nagel SC, Coe BL, Angle BM, Taylor JA. The estrogenic endocrine disrupting chemical bisphenol A (BPA) and obesity. Molec Cell Endocr 2012;354(1):74—84.

96. EU. 2011. Commission regulation no 10/2011, of 14 January 2011, on plastic materials and articles intended to come into contact with food.

97. EFSA. No consumer health risk from bisphenol A exposure. European Food Safety Authority. 2015. https://www.efsa.europa.eu/en/press/news/150121.

98. Tato et al. (2018) op. cit.

99. Gatidou G, Vassalou E, Thomaidis NS. Bioconcentration of selected endocrine disrupting compounds in the Mediterranean mussel, *Mytilus galloprovincialis*. Mar Pollut Bull 2010; 60(11):2111—2116.

100. Ying GG, Williams B, Kookana R. Environmental fate of alkylphenols and alkylphenol ethoxylates - a review. Environ Int 2002;28:215—226.

101. Loyo-Rosales JE, Rosales-Rivera GC, Lynch AM, et al. Migration of nonylphenol from plastic containers to water and a milk surrogate. J Agr Food Chem 2004;52:2016—2020.

102. Talmage SS. Environmental and Human Safety of Major Surfactants. Alcohol Ethoxylates and Alkylphenol Ethoxylates. A Report to the Soap and Detergent Association. Boca Raton: Lewis Publishers; 1994. p. 200.

103. David A, Fenet H, Gomez E. Alkylphenols in marine environments: distribution monitoring strategies and detection considerations. Mar Pollut Bull 2009;58(7):953—960. and citations therein.

104. Preuss TG, Gehrhardt J, Schirmer K, et al. Nonylphenol Isomers Differ in Estrogenic Activity. Environ Sci Technol 2006;40:5147−5153; Shioji H, Tsunoi S, Kobayashi Y, et al. Estrogenic activity of branched 4-nonylphenol isomers examined by yeast two-hybrid assay. J Health Sci 2006;52(2):132−141; Katase T, Okuda K, Kim YS, et al. Estrogen equivalent concentration of 13 branched para-nonylphenols in three technical mixtures by isomer-specific determination using their synthetic standards in SIM mode with GC-MS and two new diasteromeric isomers. Chemosphere 2008;70:1961−1972.

105. Shioji et al. (2006) op. cit.

106. National Library of Medicine: https://pubchem.ncbi.nlm.nih.gov/.

107. US-EPA. National recommended water quality criteria. Aquatic life criteria table; 2016. Available from: https://www.epa.gov/wqc/national-recommended-water-quality-criteria-aquatic-life-criteria-table.

108. US-EPA. Nonphyenol and nonylphenol ethoxylates action plan August. Washington D.C, USA: U.S. Environmental Protection Agency; 2010.

109. Hu JY, Jin F, Wan Y, et al. Trophodynamic behavior of 4-nonylphenol and nonylphenol polyethoxylate in a marine aquatic food web from Bohai Bay, North China: Comparison to DDTs. Environ Sci Technol 2005;39:4801−4807.

110. Karels AA, Manning S, Brouwer TH, et al. Reproductive effects of estrogenic and antiestrogenic chemicals on sheephead minnows (*Cyprinodon variegatus*). Environ Toxicol Chem 2003;22(4):855−865.

111. David RM. Proposed Mode of Action for In Utero Effects of Some Phthalate Esters on the Developing Male Reproductive Tract. Toxicologic Pathology, 2006;34:209−219.

112. Zhao B, Chu Y, Huang Y, et al. Structure-dependent inhibition of human and rat 11-hydroxysteroid dehydrogenase 2 activities by phthalates. Chemico-Biological Interactions 2010; 183:79−84.

113. California EPA, Toxicological profile for.di-(2-ethylhexyl) phthalate (dehp); September 2009.

114. Van der Veen I, de Boer J. Phosphorus flame retardants: Properties, production, environmental occurrence, toxicity and analysis. Chemosphere 2012;88:1119−1153.

115. Hallanger IG, Sagerup K, Evenset A, et al. Organophosphorous flame retardants in biota from Svalbard, Norway. Mar Pollut Bull 2015;101:442−447.

116. Evenset A, Leknes H, Christensen GN, et al. Screening of new contaminants in samples from the Norwegian Arctic. NIVA Rapport 1049/2009. Norwegian Pollution Control Authority; 2009.

117. Andresen JA, Muir D, Ueno D, et al. Emerging pollutants in the North Sea in comparison to Lake Ontario, Canada, data. Environ Toxicol Chem 2007;26(No. 6):1081−1089.

118. Sundkvist AM, Olofsson U, Haglund P. Organophosphorus flame retardants and plasticizers in marine and fresh water biota and in human milk. J Environ Monit 2010;12: 943−951

119. Hu M, Li J, Zhang B, et al. Regional distribution of halogenated organophosphate flame retardants in seawater samples from three coastal cities in China. Mar Pollut Bull 2014;86: 569−574.

120. Van der Veen and de Boer (2012) op. cit.

121. Singer H. Müller S, Tixier C, et al. Triclosan: Occurrence and Fate of a Widely Used Biocide in the Aquatic Environment: Field Measurements in Wastewater Treatment Plants, Surface Waters, and Lake Sediments. Environ Sci Technol 2002;36:4998−5004.

122. Dann AB, Hontela A. Triclosan: environmental exposure, toxicity and mechanisms of action. J Appl Toxicol 2011;31:285−311.

123. Reviewed by EC (2010) op. cit.; Dann and Hontela (2011) op. cit.; see also Pintado-Herrera et al. (2014) op. cit.

124. See Tato et al. (2018) op. cit.

125. Kookana et al. (2013) op. cit.

126. Schweizer HP. Triclosan: a widely used biocide and its link to antibiotics. FEMS Microbiology Letters 2001;202(1):1−7.

127. Latch DE, Packer JL, Arnold WA, et al. Photochemical conversion of triclosan to 2,8-dichlorodibenzo-*p*-dioxin in aqueous solution J Photochem Photobiol A: Chem 2003;158:63−66.

128. Health Canada. List of Prohibited and Restricted Cosmetic Ingredients. Canada Cosmetic Ingredient Hotlist; 2015.; Japan Ministry of Health. Standards for Cosmetics. Evaluation and Licensing; 2000. Division. Pharmaceutical and Food Safety Bureau. Notification No.331 of 2000.

129. EC. Preliminary Opinion on Triclosan-antimicrobial Resistance. Scientific Committee on Consumer Safety. European Commission, Brussels. SCCP/1251/09. 2010.

130. Lee DG, Chu K-H. Effects of growth substrate on triclosan biodegradation potential of oxygenase-expressing bacteria. Chemosphere 2013;93:1904−1911.

131. US-EPA. Office of Chemical Safety and Pollution Prevention 5/13/15 Letter in Response to Citizen Petition for a Ban on Triclosan. U.S. Enivronmental Protection Agency; 2015.

132. ECHA. Decision on substance evaluation pursuant to article 46(1) of regulation (EC) no 1907/2006; 2015.

133. Reviewed by Fent K, Kunz PY, Gomez E. UV Filters in the Aquatic Environment Induce Hormonal Effects and Affect Fertility and Reproduction in Fish. Chimia 2008;62(5):368−375.

134. Morohoshi K, Yamamoto H, Kamata R, et al. Estrogenic activity of 37 components of commercial sunscreen lotions evaluated by in vitro assays. Toxicol Vitro 2005;19:457−469.

135. Schlumpf M, Schmid P, Durrer S, et al. Endocrine activity and developmental toxicity of cosmetic UV filters—an update. Toxicology 2004;205:113−122.

136. Paredes et al. (2014) op. cit.

137. Ogawa Y, Kawamura Y, Wakui C, et al. Estrogenic activities of chemicals related to food contact plastics and rubbers tested by the yeast two-hybrid assay. Food Additives and Contaminants, April, 2006;(23)4:422−430.

138. Vidal-Liñán V, Villaverde-de Sáa E, Rodil R, et al. Bioaccumulation of UV filters in *Mytilus galloprovincialis* mussel. Chemosphere 2018;190:267−271.

139. Schlumpf M, Cotton B, Conscience M, et al. In Vitro and in Vivo Estrogenicity of UV Screens. Environ Health Perspect 2001;109(3):239−244.

140. Ozáez I, Martínez-Guitarte JL, Morcillo Y, Effects of in vivo exposure to UV filters (4-MBC, OMC, BP-3, 4-HB, OC, OD-PABA) on endocrine signaling genes in the insect *Chironomus riparius*. Sci Total Environ 2013;456−457:120−126.

141. https://www.statista.com/statistics/263102/pharmaceutical-market-worldwide-revenue-since-2001/.

142. Luo Y, Guo W, Ngo HH, et al. A review on the occurrence of micropollutants in the aquatic environment and their fate and removal during wastewater treatment. Sci Total Environ 2014;473−474:619−641.

143. Gomez E, Bachelot M, Boillot C, et al. Bioconcentration of two pharmaceuticals (benzodiazepines) and two personal care products (UV filters) in marine mussels (Mytilus galloprovincialis) under controlled laboratory conditions. Environ Sci Pollut Res 2012;19:2561−2569.

144. Caban M, Szaniawska A, Stepnowski P. Screening of 17B-ethynylestradiol and non-steroidal antiinflammatory pharmaceuticals accumulation in *Mytilus edulis trossulus* (Gould, 1890) collected from the Gulf of Gdansk. Int J Oceanograph Hydrobiol 2016;45(4):605−614.

145. Álvarez-Muñoz D, Rodríguez-Mozaz S, Maulvault AL, et al. Occurrence of pharmaceuticals and endocrine disrupting compounds in macroalgaes, bivalves, and fish from coastal areas in Europe. Environ Res 2015;143:56—64.

146. Verlicchi P, Al Aukidi M, Zambello E. Occurrence of pharmaceutical compounds in urban wastewater: Removal, mass load and environmental risk after a secondary treatment—A review. Sci Total Environ 2012;429:123—155.

Suggested Further Reading

- Álvarez-Muñoz D, Rodríguez-Mozaz S, Maulvault AL, Tediosi A, Fernández-Tejedor M, et al. Occurrence of pharmaceuticals and endocrine disrupting compounds in macroalgae, bivalves, and fish from coastal areas in Europe. Environ Res 2015;143B:56—64.

- Dann AB, Hontela A. Triclosan: environmental exposure, toxicity and mechanisms of action. J Appl Toxicol 2011;2011(31):285—311.

- Gu Y, Yu J, Hu X, Yin D. Characteristics of the alkylphenol and bisphenol A distributions in marine organisms and implications for human health: a case study of the East China Sea. Sci Total Environ 2016;539:460—9.

- Kelly BC, Ikonomou MG, Blair JD, et al. Perfluoroalkyl contaminants in an Arctic marine food web: trophic magnification and wildlife exposure. Environ Sci Technol 2009;43:4037—43.

- Laws EA. 10. Pesticides and persistent organic pollutants. In: Laws EA, editor. Aquatic pollution; an introductory text. 4th ed. Chichester: John Wiley & Sons Ltd.; 2018. p. 311—74.

- Oros DR, Hoover D, Rodigari F, et al. Levels and distribution of polybrominated diphenyl ethers in water, surface sediments, and bivalves from the San Francisco Estuary. Environ Sci Technol 2005;39:33—41.

- UK Environmental Agency. UV-filters in cosmetics — prioritisation for environmental assessment. 2008. Available on line, https://assets.publishing.service.gov.uk/government/uploads/system/uploads/attachment_data/file/291007/scho1008bpay-e-e.pdf.

- Yamaguchi A, Ishibashi H, Kohra S, et al. Short-term effects of endocrine-disrupting chemicals on the expression. Of estrogen-responsive genes in male medaka (*Oryzias latipes*). Aquat Toxicol 2005;72:239—49.

CHAPTER 9

Trace Metals and Organometallic Compounds

9.1 METALS OF MOST ENVIRONMENTAL CONCERN

Metals are chemical elements with low ionization energies that tend to form cations through electron loss, and are thus good conductors of electricity and heat. The properties of chemical elements are periodic functions of their atomic number. Taking a look at the periodic table of elements, metals are placed in the left and central parts of the table, while nonmetals are in the right hand. The electronegativity describes the tendency of an atom to attract electrons, and all metals show electronegativity values <2, while for nonmetals, the values range from 2 to 4. Nonmetals have high ionization energies and, except for the nonreactive group of the noble gases, tend to accept electrons and form anions in the environment. In the diagonal border between metals and nonmetals there are seven elements with mixed physical properties and hence difficult to classify. They are the metalloids: B, Si, Ge, As, Sb, Se, and Te.

Metals are elements that tend to yield electrons

The tendency of metals to form cations decreases as we move from left to right across a period and down a group. Thus, the elements from the first two columns of the periodic table, called alkali and alkaline-earth metals, have the lowest ionization energies and they are almost always present in the aquatic environment as cations (monovalent and divalent, respectively). In contrast, the **transition metals**, listed in the next columns, and particularly those on the far right next to the metalloids, show an increased tendency to form covalent bonds with organic molecules, which confers them special potential toxicity, due to either binding to macromolecules and impairing their metabolic roles, or forming organometallic molecules capable to diffuse across biological membranes.

Transition metals tend to bind to organic molecules

Another interesting property of the transition metals is that they form cations with different oxidation states ($+1$, $+2$, $+3$) because the energies of the

137

Marine Pollution. https://doi.org/10.1016/B978-0-12-813736-9.00009-X

successive outermost electrons are similar, which confers them potential to take part in electron transport chains and redox reactions.

Transition metals are termed heavy metals because of their high density or trace metals because of their low natural abundance

The term **heavy metals**, extensively used in ecotoxicology to refer to the transition metals of most environmental concern, stems from the fact that many of them have a density >5 g/cm^3. However, a relevant exception is Al, with a relative density of only 1.5 but toxic to freshwater fauna especially in acidic conditions. Turning our attention to their abundance in the Earth's crust, we can see that most minerals are mainly composed of Si, Mg, O, Fe, and Ca, all of them with even numbers of protons and neutrons that give remarkable stability to the nuclei. Other elements show abundances in mineral rocks ranging from 1 to 50 ppt, such as the metals Al, Mg, Ti, and Mn. Besides, the remaining metals, called **trace metals**, show abundances <1000 ppm, such as Cr, V, Ni, Zn, Cu, Co, Pb, Sn, or even below 1 ppm, such as Cd, Pt, or Hg.[1]

Their affinity for organic molecules confers them high potential toxicity

An alternative classification of metals take into account their affinity for either oxygen (class A metals: corresponding to alkali and alkaline-earth) or sulfur/ nitrogen (class B metals: Cu, Hg, Ag, Pb), and the borderline type (Cd, Zn, Fe, Cr, Co, Ni, As, V).[2] Class B metals show the highest acute toxicity to aquatic organisms (see EC$_{50}$ values in Table 9.1) because of their tendency to form covalent bonds with organic molecules. Therefore, a relationship can be identified between the acute toxicity of a metal and its affinity for the functional groups of organic macromolecules with vital roles in the metabolism, such as the amino, imino, and thiol groups. Divalent cations such as Cu^{2+} and Hg^{2+} bind to two adjacent $-SH^-$ groups, while monovalent cations such as CH$_3$Hg$^+$ bind to single $-SH^-$ group. This binding can alter the structure and function of the macromolecule affected, for example when it takes place in the active site of an enzyme that becomes inhibited.

Toxicity in aqueous environments is in fact vastly affected by chemical speciation (see Section 10.4), and by the solubility of the chemical species, more than by the total amount of each element. Concerning chronic toxicity, the potential for long-term bioaccumulation will be a decisive factor, and again the tendency to form covalent bonds with endogenous metabolites will enhance the bioaccumulation potential.

Environmental concern depends also on human enrichment

Another factor affecting the ecotoxicological relevance of a trace element is its mobilization from its mineral ores due to mining and other human activities, and thus its degree of anthropogenic enrichment in the biosphere. Because they are present in elemental state on the Earth's crust and can be used either individually or in alloys to build up hard and durable objects, many transition metals (Fe, Cu, Sn, Au, Ag, Pb) were traditionally exploited from their ores by humankind. Mining is still the main human activity that interferes with the natural biogeochemical cycle of many trace metals. Lead is used in storage batteries, pigments, and ammunition; copper in electrical equipment and pesticides; cadmium and zinc for galvanizing and surface treatments; silver in jewelry; and mercury in lights, batteries, amalgams, or industrial electrolysis.

Table 9.1 Main Anthropogenic Source, Main Global Input Into the Sea, Typical Concentrations of Trace Metals (in µg of Total Metal per L) in Different Liquid Environmental Matrices, Toxicity to ELS of Marine Organisms and Safety Margin, Estimated as Toxicity Thresholds Divided by Deep Seawater Concentrations. Toxicity Thresholds Were Obtained Applying a Factor of 0.1 to EC_{50} Values

	Main Source	River (ppb)[1]	Crude Oil (ppm)[1]	Main Input	Urban Sewage (ppm)[3]	Deep SW (ppb)	Main spp in SW	Vigo Harb SW (ppb)[4]	EC_{50} (ppb)[5]	Safety Margin
Hg	chlor-alkali ind.[6]	0.07	3.4	Riv[6]	n.d.	0.002[1]	$HgCl_4^{2-}$	n.m.	10	500
Cd	mining	0.02	0.01	Riv[2]	3-25 x10^{-3}	0.04[8]	$CdCl_2$	0.009–0.031	2219	5500
Pb	combustion[6]	0.3	0.3	Atm[2]	0.05–0.2	0.003[1]	$Pb(CO_3)_2^{2-}$	0.09–0.68	968	32,267
Cu	urban sewage[6]	2	0.7	Riv[2]	0.14–0.46	0.25[1]	$Cu\,CO_3$	0.6–2.9	24	9.6
Zn	mining	7	8.0	Atm[2]	0.19–1.6	0.6[1]	$ZnCl^+$	1.5–10.7	320	53
Cr	metallurgy[7]	1	0.12	Riv[2]	0.04–0.16	0.2[1]	$CrO_4^{2-[7]}$	n.m.	4564	2282

Atm, atmospheric; ELS, early life stages; n.d., no data; n.m., not measured; Riv, riverine; SW, seawater.
Sources: [1]Merlin[11]; [2]Chester and Jickells (1990); [3]Henry and Heinke (1996); [4]Beiras et al. [12]; [5]His et al. (2000); [6]Clark (2001); [7]Chiffoleau[13]; [8]Cossa and Lassus.[14]

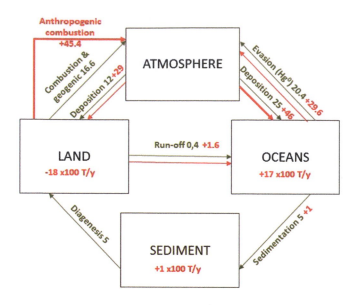

FIGURE 9.1

Current biogeochemical cycle of mercury indicating the natural fluxes (*green arrows*) and the anthropogenic impact (*red arrows*). Fluxes in ×100 tons/year. Notice that the natural fluxes between compartments, corresponding to a preindustrial age, are in equilibrium, whereas the current anthropogenic activities, mainly the combustion of fossil fuels, cause a depletion of the terrestrial stock of mercury and an enrichment of the marine compartment with 1700 tons/year. *Data from Selin (2009).*

Fig. 9.1 shows the current biogeochemical cycle of mercury where the natural fluxes (green arrows) are accelerated by anthropogenic activites (red arrows) that increased the amount of mercury cycling through the system by a factor of three.[3] The overall anthropogenic effect on the cycle depletes the underground mineral sources from land and eventually enriches the marine compartment, once again the main global sink for this contaminant.

Iron, manganese, copper, zinc, lead, titanium, and chromium are, in this order, the most consumed metals by industry.[4] If we divide human consumption by natural abundance then we can construct an index of **human enrichment** for the metal in the biosphere. Different estimates[5] point at mercury, lead, copper, cadmium, and zinc as the metals with the highest enrichment factors due to human activities at global scale.

9.2 ESSENTIAL VERSUS XENOBIOTIC TRACE ELEMENTS

Chemical elements can be essential or xenobiotic according to their interaction with organisms

The human body is mostly composed of O, C, H, N, P, and S, since water and the macronutrients: carbohydrates, lipids, proteins, and nucleic acids are composed of just these six chemical elements, which along with the Ca from the bones account for the 99% of the body weight. The major components of the remaining 1% are K, Na, Cl, and Mg, which are vital in many biochemical

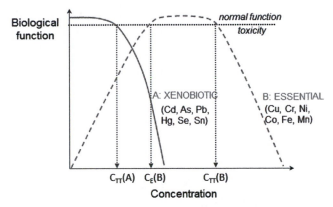

FIGURE 9.2

Contrasting patterns of response of a biological function of an organism (survival, reproduction, growth, etc.) to the external (or internalized) concentration of a chemical element for xenobiotic (A) and essential (B) elements.

processes, from osmotic regulation to the transmission of the nerve impulse. It may seem striking that the same 11 elements are the major components—with some exceptions like the Si in certain algae and sponges—of all living organisms. Besides, a short number of minority elements called micronutrients or trace nutrients are also essential for vital processes, such as those that act as coenzimes (Fe, Zn, Co, Cu, Mn, Se), or heteroatoms in essential organic molecules (Mo) or hormones (I). All these elements are called **essential** since they must be incorporated through the diet, and lacking any of them produces deficiencies that may give raise to illness. The biological role of other minority elements (Ni, B, Cr, F, V) is still controversial, and they do not seem to be universally required in the diet.

The remaining elements from the periodic table are considered as **xenobiotics**, meaning they do not play any biological role in any organism. These include Hg, Cd, Ag, Pb, Sn, or Al.

We can illustrate the different interaction of xenobiotic versus essential metals and metalloids with organisms by means of the hypothetical scheme depicted in Fig. 9.2. When a given biological function is represented as a function of the environmental (or internalized) concentration of the element, we can obtain either a type A curve for xenobiotics or a type B curve for essential elements.

Both curves differ in two features. First, generally speaking, the **toxicity thresholds** (C_{TT}) or concentrations above which the biological function is depressed are lower for xenobiotics than for essential elements. Indeed, many of the most toxic elements such as Hg, Cd, Ag, As, and Pb are xenobiotics. Second, essential elements need by definition a minimum level (C_E) within the organism to carry out normal biological functions. Below that level the **deficiency** causes a decrease in the biological function. For essential elements, the range between C_E and C_{TT} comprises the corresponding range of environmental concentrations needed for optimal biological functioning. The case of copper, the most toxic essential element for aquatic organisms, is particularly interesting since for this metal this range is especially narrow.

Due to their lower toxicity threshold, xenobiotic metals are particularly feared. In fact, the interest for metals as environmental pollutants stem from two episodes of chronic poisoning with human casualties happened in Japan by mid 20th century and caused by Hg (Minamata disease) and Cd (*Itai-Itai* disease) poisoning.

9.3   SOURCES AND DISTRIBUTION OF TRACE METALS IN THE SEA

Riverine versus atmospheric inputs

For most, but not all, trace elements, the main input in marine environment is the **riverine** pathway. Table 9.1 shows typical concentrations of trace metals in river and deep-sea waters, showing that except for Cd the earlier are about one order of magnitude higher than the latter. In the North Atlantic,[6] the estimated fluvial input of dissolved metals is 2−10 times larger than the atmospheric input for Ni, Co, Cr, or Cu, but both pathways are of similar importance for Cd, whereas for Zn and Pb the **atmospheric** input is estimated to be majoritarian due to combustion of fossil fuels. Comparing anthropogenic to natural sources of trace metals, Pb, Zn, and Hg are the most enriched in the atmosphere due to human activities.[7] Leaded petrol was traditionally by far the main anthropogenic source of Pb to the atmosphere, while Zn comes mainly from coal combustion. In contrast to Pb and Zn, the natural fluxes of Hg via atmosphere are quite large due to its high volatility. Dissolved metals are washed out from the atmosphere onto the sea surface by rainfall. Dry deposition is also significant, especially near the sources and for the particulate fraction of the atmospheric metals.

Costal sediments are enriched in metals

When they reach estuarine and coastal systems, riverine metal inputs suffer complex processes, briefly outlined in Section 1.6. The major amount of metals transported by river waters resides in the components of the crystal lattice of mineral silicates. A lesser but more biologically relevant amount is present as organic or inorganic complexes and adsorbed to clay particles. At salinities above 5 ppt the increasing Na^+, Mg^{2+}, and K^+ concentrations cause **desorption** of trace metal ions in the order Hg > Cu > Zn > Pb > Cr. Cd in contrast remains adsorbed to clay and mud and readily sediments to the bottom.[8] The solubilized metals are trapped in the bottom sediments by means of several mechanisms, including **complexation** with humic acids and flocculation of the complexes at salinities above 15 ppt due to **coprecipitation** with iron oxyhydroxides.[9] Particulate metal, in turn, tends to **settle** as current speed decreases in the mouth of the river due to estuarine circulation. Biological uptake and incorporation into Particulate Organic Matter (POM) that eventually settles as biodeposits on the seafloor is also a relevant mechanism.[10] As a consequence of all these processes trace metals are removed from the water column yielding seawater concentrations markedly lower than those in freshwater (see Table 9.1), whereas estuarine **sediments act as sinks** and get enriched by these pollutants (Table 9.2).

Tables 9.1 and 9.2 also show levels of metals in clean and polluted waters and sediments, the toxicity of dissolved metals quantified as their EC_{50} in sensitive marine organisms, and the sediment quality criteria ERL and ERM, which can be interpreted as estimates of their toxicity threshold. Notice that the safety margins, quantified as the ratio of toxic metal concentrations or toxicity thresholds to natural background levels, are narrower for sediments than for seawater. This means that ecologically deleterious effects of metals are more likely expected as a result of **sediment pollution** compared to water pollution. Since many natural (hydrodynamics, bioturbation) and anthropogenic processes (e.g., dredging) may resuspend the sediment into the water column, it is easy to see the role that coastal marine sediments play as repositories of conservative contaminants posing a risk to the coastal ecosystems.

Safety margins are lower for sediments than for the water column

Once in open sea, trace metals can present three different patterns of vertical distribution: **conservative distribution**, determined just by physicochemical variables and with a vertical maximum of abundance dependent on the salinity and temperature profiles (Rb, Cs, Mo), **surface-enriched distribution** (Pb, Sn), typical of elements with an important atmospheric input, and **distribution associated to nutrients** (Cu, Zn, Ni, Cr, Cd) typical of elements taken up by plankton (normally because they are essential trace nutrients but also typical of others like Cd that may mimic essential elements) and hence depleted from the surface and deposited into deep waters (Fig. 9.3).

Types of vertical distribution in the water column

In mussels and other filter feeders, trace metals are bioconcentrated up to ecotoxicologically relevant values, with bioconcentration factor (BCF) ranging from 100 to 10,000, and bivalves from metal polluted places may fail to meet the standards for human foodstuff (Table 9.3). In these organisms, **dissolved metal uptake via water** was frequently reported as quantitatively dominant compared to particulate metal.[25] Dissolved metals are taken up through the gills by a facilitated diffusion process, at a rate that is a linear function of the free metal ion concentration in the water (see Section 10.4). Free ions cannot diffuse across biological membranes but they pass through the ion channels of transmembrane proteins available for essential elements such as Ca^{2+}. These gateways are selective depending on the charge and dimension of the ion, though ions of similar size and charge such as Ca^{2+}, Cd^{2+} and Pb^{2+} may be taken up via the same gateways.[26] Experimental data report no saturation, indicating an excess of transport sites even at high trace metal concentrations. In filter feeders, temperature and thus the temperature-dependent pumping rate, does not affect uptake rate, indicating that the seawater passing over the exchange surfaces will lose only a minute amount of its metal load.[27] Metal uptake in contrast decreases as salinity increases,[28] as expected, by the decreased free metal ion activity caused by enhanced ionic strength in more saline media.[29]

Metals are taken up by marine biota through the gills, and reach particularly high levels in filter feeders

Table 9.2 Concentrations of Trace Metals (mg/Kg dry Weight) in the Ocean's Crust, Coastal Sediments From Clean (CA) and Polluted Sites (Vigo Harbor) and Background Values for the Galician Coast (Northwest Iberian Peninsula), Sediment Quality Criteria (ERL and ERM), and Safety Margin, Estimated as Toxicity Thresholds Divided by Background Concentrations. Toxicity Thresholds Were Obtained From ERL Values

	Ocean Crust (ppm)[1]	Backg (ppm)[2]	Ref (CA)[3]	Vigo harbor[3]	ERL	ERM	Safety Margin
Hg	0.02	0.02	0.16	0.35–2.6	0.15	0.71	36
Cd	0.13	0.1	0.13	0.46–1.79	1.2	9.6	96
Pb	0.89	51	73.6	63.0–496	47	218	4
Cu	81	21	33.2	70.3–367	34	270	13
Zn	78	110	146	216–1530	150	410	4
Cr	317	60	78.8	74.4–111	81	370	6

Sources: [1]Merian et al. (2004) op. cit. [2]Álvarez-Iglesias et al. [15] for Pb, Cu, Zn and Cr; Prego et al. [16] for Cd, Canário et al. [17] for Hg. [3]Beiras et al. (2012) op. cit.

Table 9.3 Metals Bioconcentration Factor (BCF), OSPAR Assessment Criteria (BAC), Concentrations of Trace Metals in Mussels From the US and a Polluted Site in the Iberian Coast, and EU Standards for Foodstuff

	Essential/ Xenobiotic	Bivalve BCF (L/Kg WW)[1]	Mussel BAC (mg/Kg DW)[2]	Mussel NOAA Medium (mg/Kg DW)[3]	Mussel NOAA High (mg/Kg DW)[3]	Vigo Harbor Wild Mussels (mg/Kg DW)[5]	EU Maximum in Food (mg/Kg WW)[4]
Hg	xenobiotic	12000[a]	0.09	0.18	0.36	0.099–0.255	0.5 (c. 2.5 mg/Kg DW)
Cd	xenobiotic	786	0.96	4	10	0.501–1.23	1 (c. 5 mg/Kg DW)
Pb	xenobiotic	688	1.3	4	7	3.65–15.02	1.5 (c. 7.5 mg/Kg DW)
Cu	essential	128[b]	6	17	40	5.92–109	not regulated
Zn	essential	314	63	140	321	257–411	not regulated

DW, dry weight; WW, wet weight.
[a]Value for mercuric chloride; methyl-Hg and phenyl-Hg accumulated four times more.[22]
[b]Oysters excluded; oysters show exceptionally high BCF for this metal.
Sources: [1]US EPA[18], [2]OSPAR[19], [3]Kimbrough et al.[20], [4]EC (2006)[21], [5]Beiras et al. (2012).

Assimilation efficiency of metals from phytoplankton or sediment particles in the digestive system of bivalves may range from 10% to 80%, and when **food** is significantly enriched in metal compared to water the contribution to total metal uptake by this via is very relevant.[30] Some particulate metal uptake can also take place by pinocytosis in the gills.[31]

The internal levels of essential metals may be regulated within a range of water concentrations

Fig. 9.4 illustrates the different behavior of nonessential and essential trace metals concerning uptake and bioaccumulation as a function of the metal concentration in the water. The internal concentrations of xenobiotic metals frequently reflect those of the surrounding environment. This is certainly a key assumption in the biological monitoring of water quality. In contrast, for essential elements, aquatic organisms are able to **regulate** internal concentrations by active uptake at below optimal concentrations and active outflux at supraoptimal levels. The regulatory capacity though may be exceeded above a given environmental threshold and excessive accumulation may take place. In both cases this accumulation may be innocuous to the organism since different mechanisms allow molecular sequestering (see metallothioneins in Section 12.6) or accumulation in nonbioavailable subcellular compartments (granules, vacuoles).

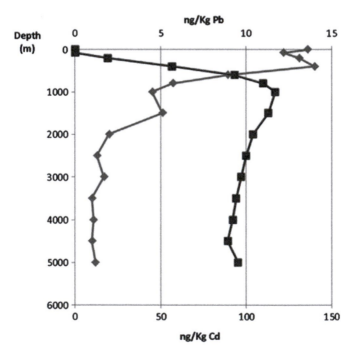

FIGURE 9.3

Vertical profile of dissolved Pb (*blue diamonds*) and Cd (*red squares*) in the North Pacific. Pb shows a surface-enriched distribution while Cd shown a nutrient-associated distribution with depleted values in the surface and maximum at subsurface waters, like nitrate and phosphate. *Data from Schaule BK, Patterson CC. Lead concentrations in the northeast Pacific: evidence for global anthropogenic perturbations Earth and Planetary Science Letters 1981;54:97—116 for Pb, and Bruland KW. Oceanographic distributions of cadmium, zinc, nickel, and copper in the North Pacific. Earth Planet Sci Lett 1980;47:176—198 for Cd.*

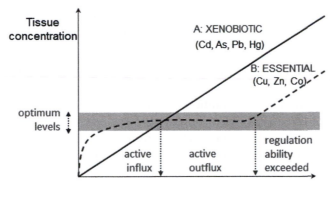

FIGURE 9.4

Schematic representation of the different relationship between water and tissular concentrations of a metal for the cases of a xenobiotic and an essential element. In the first case tissular concentrations reflect water levels multiplied by the respective BCF. In the second case the regulatory ability of the organism interferes with this linear relation and keeps constant optimum internal levels within a certain range of environmental concentrations.

9.4 COPPER AND ZINC

Copper and zinc are transition metals of correlative position in the periodic table (atomic numbers 29 and 30), and several other chemical similarities: atomic masses 63.5 and 65.4, densities 8.93 and 7.14 g/cm^3, atomic radius 128 and 134 p.m., ionic radius (divalent cation) 72 and 74 p.m.[32] Even though with different ores, both are relatively abundant and ubiquitous in the Earth's crust, and were traditionally used by many cultures as brass, an alloy of Cu and Zn, yet in European cultures the use of Cu alone or alloyed with Sn (bronze) is more ancient.

Cu and Zn share chemical and biological properties

Both are also **essential** trace metals because of their role as cofactors of multiple enzymes. Copper, though, is common in two oxidation states (+1 and +2), a property that confers it specific metabolic roles such oxygen transport and release in hemocyanin or electron transport in photosynthesis, not shared by Zn whose only common oxidation state is +2. In plants Cu and Zn deficiency is a more common problem than toxicity. Humans need a daily intake of Cu of 2 mg, and its deficiency may have severe effects especially in children. Clinical manifestations of zinc deficiency in humans have also been described.

Despite being essential, both elements can be **toxic**, and are used in different inorganic forms or organic complexes as biocides. Copper shows a specific mechanism of toxicity via production of reactive oxygen species, derived from its ability to donate electrons and change oxidation state. In aquatic animals, toxicity is very dependent on chemical speciation. Copper is, according to the Irving–Williams series the second (after Hg) metal that forms the most stable complexes with a series of ligands, including organic macromolecules. The presence of humic substances in the water greatly decreases dissolved copper toxicity by sequestering free ions and reducing their bioavailability. This will be discussed in depth in Section 10.4.

A direct input of Cu in the sea comes from antifouling paints

Specific uses of Cu include electric industrial applications (it is after Ag the best electricity conductor), antifouling paints, alguicides, agriculture pesticides, and pigments. The ban of tributyltin (TBT) in antifouling paints for boat hulls prompted a substitution by Cu in these formulations associated to an increase of the dissolved Cu concentrations in certain coastal environments[33] (see also Fig. 17.6 in Section 17.4). Specific uses of Zn include corrosion-resistant plating of iron, batteries, semiconductors, fluorescent lamps, pigments, wood preservatives, and pharmaceutical products.

Tables 9.1—9.3 show the background concentrations of Cu and Zn in different matrices. Zn is normally c. 3 times more abundant than Cu in natural waters, sediments, and bivalves. Although Zn has important atmospheric sources because it is, along with V and Mo, one of the most abundant trace elements in crude oil, Cu and Zn frequently covary in marine samples, pointing at common sources such as urban and industrial wastewaters. Both are the most abundant trace metals in urban sewage and run-off waters, with levels commonly ranging from 100 to 300 µg/L for Cu and up to 1600 µg/L for Zn.[34]

Mussels, but not oysters, can regulate internal levels of Cu

Since they are essential elements, the internal levels of Cu and Zn are subject to metabolic regulation.[35] Algae[36] and many aquatic animals such as crustacean decapods,[37] gastropods,[38] and benthic fish[39] have been reported capable to regulate the internal concentrations of Cu and Zn. Even the well-known "accumulator" *Mytilus edulis* regulates internal Cu levels and is suitable to indicate copper pollution only at very high environmental levels.[40] Relevant exceptions are **oysters** and polychaetes that can accumulate Cu in nonpolluted environments with high BCF values.

9.5   LEAD

Pb was formerly used in pipes, paints, and gasoline, but today mainly in electric batteries

Because of its abundance in the galena ore, softness, high density, and easiness to mold and alloy, Pb has been used by man since ancient times for multiple applications, from gunshot to water pipes. Thus, most of the Pb currently in the environment has been mobilized by human activities. Dated polar ice cores allowed identifying the historic events that caused a peak in the atmospheric levels of this metal, including the Roman Empire mining activities, the industrial revolution and its concomitant increase in coal combustion, and the spread of leaded petrol for automobiles (Fig. 9.5). The main source of anthropogenic Pb used to be the use of alkyl-lead additives, at concentrations up to 1 g Pb/L, as antiknock agents for car petrol. These additives were introduced during the 1930s but phased out in many countries during the 1990s.[41] Concurrently, in the Northern Hemisphere the levels of atmospheric Pb peaked in the 1970s and decreased afterwards,[42] but today more than 90% of the atmospheric Pb is still of anthropogenic origin,

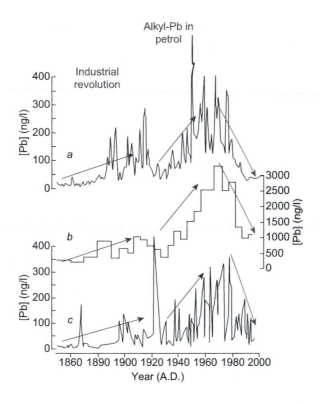

FIGURE 9.5

Concentration of lead in dated ice cores from Greenland (A), Alps (B), and North Canada (C), reflecting the changes in the atmospheric levels of Pb since the 19th century. Notice the steady increase during the industrial revolution, the sharp peak corresponding to the use of leaded petrol and the recovery after lead additives were phased out. *Modified from Osterberg E, Mayewski P, Kreutz K, et al. Ice core record of rising lead pollution in the North Pacific atmosphere. Geophy Res Lett 1994;35:L05810, doi:10.1029/2007GL032680*

according to its isotopic composition.[43] Lead occurs as a mixture of different stable isotopes, mainly 206, 207, and 208, with minor contents of 204 and the radioactive isotope 210. The isotopic ratios may help to identify the source of the metal in the environment.

Another important source of Pb in the environment is the lead oxide (minium), broadly used in corrosion-resistant paints for iron and steel structures, but also less common nowadays because of environmental concerns. Pb is also present in glassware, ceramics, cable sheadings, and some plastics. Around 85% of the use of Pb worldwide in 2012 was due to electric batteries, with pigments as the second most important use (5.5%).[44]

The main input of lead into the sea takes place via deposition from the atmosphere, to which anthropogenic Pb is emitted from mine smelting and **combustion** processes. As a result, unlike most other anthropogenically enriched trace elements, Pb is relatively abundant in central oceanic gyres, and its vertical distribution (see Fig. 9.3) shows surface maxima and concentrations decreasing with depth.[46]

The main input of Pb in the sea is atmospheric, and thus its levels are enriched in surface waters

Lead can be present with two different oxidation states, II and IV. In seawater Pb (II) carbonates and chlorides are the main species, with negligible or minor amounts of organic complexes and particulate Pb, while reduced environments promote its precipitation as insoluble PbS. Pb (IV) is only common in organic compounds.

Typical levels of Pb in water, sediments, and bivalves are shown in Tables 9.1−9.3. Preindustrial sediments show values < 20 μg/g DW. Widespread sediment quality criteria are ERL = 47 and ERM = 218 μg/g DW, though the earlier seems a too low value to be applied in European coasts, where an ecotoxicologically based alternative value of 58.7 has been proposed.[47]

Lead bioaccumulates in many organisms and their levels in foodstuff are regulated because of health issues

Lead **accumulates** in organisms with reported BCF values between 20 and 10,000, with higher values for crustaceans, intermediate for mollusks and lower for fish and algae.[48] Bivalves and crustaceans show soft tissue concentrations ranging between 0.1 and 10 μg/g DW, and preferential accumulation in the digestive gland and kidney. Fish show lower concentrations in muscle, but higher in nonedible parts such as liver and bones. Although methylated forms of Pb are present in the environment no evidence of biomagnification across trophic chains for inorganic or organic lead has been reported. Due to similar ionic radii, Pb^{2+} is believed to move across cellular membranes through Ca^{2+} channels. For the same reason Pb accumulates in the bones of vertebrates.

Acute toxicity of lead, including effects on the central nervous system, is well known. Lead poisoning, also called **plumbism**, is mainly a risk from occupational exposure of workers involved in activities such as battery recycling, paint production, etc. The activity of the enzyme **δ-aminolevulinic acid dehydratase (ALA-D)** in the erythrocytes is a sensitive biomarker of exposure to Pb, with a threshold value of detection of about 10 μg/dL. At Pb concentrations in blood above 70−80 μg/dL the activity of the enzyme ALA-D, involved in the synthesis of hemoglobin is almost completely inhibited.[49]

The early fabrication of **tetraethyl-Pb** for its use as additive in leaded gasoline caused in the factory workers many poisoning events with neurotoxic consequences, several of them fatal.[50] Lead shot ingestion has been recognized as a principal cause of waterfowl intoxication throughout the world, and also for terrestrial birds.[51]

The chronic toxicity of lead on humans is a debatable subject. Effects on kidney, circulatory, or reproductive system, and cognitive deficits in children have been reported, but these findings are controversial. The concentration of Pb in blood is generally accepted as an indicator of the individual's lead exposure. Adults should not have levels higher than 35−40 μg Pb in 1 dL blood.[52] Blood lead levels in children as low as 10 μg/dL were associated with decreased intelligence and impaired neurobehavioral development.[53] Lead is one of the three trace metals whose maximum contents in foodstuff are currently regulated in the

EU, with a maximum concentration in seafood ranging from 0.3 to 1.5 µg/g WW depending on the species.[54] The EU thus endorse the WHO recommendation of a tolerable weekly intake of 25 µg/kg body weight.

9.6 CADMIUM

Cadmium, named after the Greek name of the zinc ore calamine, occurs together with zinc, from which it must be separated, and its main anthropogenic source is zinc mining. Obtaining 1 ton of zinc originates about 3 kg of Cd as a by-product.[55] Continental waters bring most of this Cd into the sea. Unlike other transition metals used by mankind from ancient times, cadmium has only been massively consumed by industry for the past 60 years, mainly as **anticorrosive** coatings for metals, because electrodeposited Cd, even a layer of just a few µm thick, has excellent properties for protecting iron and steel against corrosion. Other uses include yellow, orange, and red pigments (Cd sulfides and sulfoselenides), and stabilizers for plastics, mainly PVC. Most of those uses are progressively being phased out, and currently, rechargeable nickel-cadmium **batteries** account for most of the Cd used by industry.[56] Urban solid waste and wastewaters can thus be considered as the second most important source of anthropogenic Cd. Atmospheric input of Cd into the sea is secondary compared to the fluvial pathway, since Cd occurs at c. 1 ppm levels in coal but at negligibly low concentrations in oil.

Cadmium is mainly used in anticorrosive coatings and rechargeable batteries

Background levels of Cd in water, sediments, and bivalves are shown in Tables 9.1–9.3. Oceanic seawater in the North Atlantic shows Cd concentrations between 10 and 40 ng/L, with a vertical profile associated to the nutrients (see Fig. 9.3) because of absorption by plankton and sedimentation as particulate matter.[57] Inputs of dissolved Cd in coastal environments may be caused by the natural upwelling of deep water masses.

Sediments from clean coastal areas show levels around 0.1–0.2 µg/g DW. Some internationally accepted sediment quality criteria are ERL = 1.2 µg/g DW and ERM = 9.6 µg/g DW (Table 9.2).

Cd shows moderate BCF values except for algae, where values up to 20,000 have been reported.[58] In mussels Cd has been reported to preferentially accumulate in the kidney.[59] Unlike other metals like Hg or Cu whose toxicity thresholds are more or less universal, Cd seems to be substantially more **toxic to crustaceans** than to other marine invertebrates, although data are very heterogeneous.[60]

In vertebrates Cd accumulates in kidney, and it is according to WHO a suspected carcinogenic agent. The cadmium ion (radius = 103 pm) is very close in size to the calcium ion (r = 106 pm), and this confers it a toxicologically relevant metabolic similarity to the major constituent of vertebrate bones and invertebrate exoskeletons. Chronic toxicity of Cd to humans has caused renal tubular dysfunction, and osteoporosis and other structural changes in

Cd interferes with Ca metabolism in crustaceans and vertebrates

bones, with severe symptoms in postmenopausal women. This was named in Japan *Itai-Itai* disease, and it was caused by ingestion of rice with high Cd contents coming from water polluted by wastes from Zn/Cd mines.

Maximum contents of Cd in seafood are regulated in Europe at levels from 0.05 to 1.0 µg/g WW depending on the species,[61] with the objective to set a maximum weekly intake of 2.5 µg/kg body weight.

9.7 MERCURY AND METHYLMERCURY

The main sources of Hg to aquatic ecosystems are chlor-alkali industries

Mercury, in Latin "liquid silver" (*hydrargyrum*), is a chemical element that is naturally present in the Earth's crust as cinnabar (mercury sulfide) or even in elemental metallic form that remains liquid at most temperatures occurring in the biosphere (melting point $-39.8°C$, boiling point $357°C$). It has been used by humankind throughout the history because of that and other properties, namely to dissolve other metals in amalgams, as a pesticide and antiseptic. Metallic mercury has a relatively high vapor pressure and lipid solubility, which confers it relevant toxicity well known at least since the Roman Empire. In addition, elemental mercury is easily oxidized into the highly soluble Hg^{2+}, which is present in seawater as inorganic complexes such as $HgCl_4{}^{2-}$ (see Table 9.1), but readily forms stable complexes with organic ligands through covalent bonds with N, S, and C atoms. The differences in solubility of the various inorganic salts of mercury are responsible for their corresponding broad differences in toxicity to aquatic organisms. Thus $HgCl_2$ has a water solubility of 69 g/L, Hg_2Cl_2 of 2 mg/L, and HgS of only 10 ng/L.[62]

In Europe and North America the quantitatively most important source of Hg in coastal ecosystems is the **chlor-alkali production**, that used to employ a Hg cathode for the electrolysis of NaCl. This method is currently being phased out due to environmental concerns, but caused chronic Hg pollution in many estuaries. In the Ria of Pontevedra (Northwest Iberian Peninsula) for example, Hg concentrations in sediment from the vicinity of a chlor-alkali plant reached 1.2 ppm (WW), corresponding to an enrichment factor of c. 60, with up to 25% methylmerculry (MeHg).[63] In other areas of the world with laxer environmental regulations plastic production and particularly artisanal gold mining are the most important sources of anthropogenic mercury. Other industrial uses of mercury include dental amalgams, measurement and control devices, fluorescent lighting, batteries, and other electric and electronic components.

Mercury mining is still active, especially in emerging countries

The biogeochemical cycle of mercury includes natural (volcanic activity, forest fires) and anthropogenic (coal and oil combustion) emissions of elemental mercury into the atmosphere. From the **atmosphere deposition** to the ocean surface as oxidized Hg^{2+} and further sedimentation associated to particles into the deep ocean, as well as evasion of elemental Hg produced by photochemical reduction from the hydrosphere back to the air are the major pathways.[64] According to recent estimations most of the current atmospheric mercury stems from anthropogenic sources.[65] The effect of man unbalanced the equilibrium among compartments by increasing emissions to the atmosphere ($+4540$ Mg/year), with subsequent deposition on the sea surface ($+4600$ Mg/year) and sedimentation into deeper waters ($+2000$ Mg/year), with a net enrichment of the oceans of 1700 Mg/year due to the influence of human activities. Eventually, marine sediments will be the final sink for this anthropogenic Hg, but the process is very slow and retention time of Hg in deep waters have been estimated in the order of many centuries.

Especially relevant for its ecotoxicological properties is the mercury present as organic compounds in the environment. Both aerobic (e.g., *Pseudomonas fluorescens*) and anaerobic (e.g., *Clostridium cochlearium*) bacteria in the sediments can catalyze the production of **methylmercury** (MeHg: CH_3Hg^+) from Hg^{2+}. The methyl group is provided by the coenzyme methylcobalamine, synthetized by the bacteria and excreted to the medium, similar in structure to vitamin B_{12}. Anaerobic sediment conditions and redox values within the range of $+150$ to -100 mv are ideal.[66] Mercury methylation can also be abiotic through chemical transmethylation, involving the exchange of the CH_3 group between metals, and photochemical methylation. CH_3Hg^+ can be further methylated to $(CH_3)_2Hg$, dimethyl mercury, which is volatile and evaporates into the atmosphere. Acid or neutral pH favors MeHg while dimethyl-Hg is favored at basic pH.

Methyl-Hg, produced by anaerobic bacteria, is more bioaccumulable and toxic than inorganic Hg

Organic forms of mercury rise more environmental concern because they are more **bioavailable** since they can readily diffuse across the lipid bilayer of biological membranes, including the blood—brain barrier and the placental barrier. MeHg is about 10 times more bioaccumulable and toxic than inorganic mercury, and was the responsible of the infamous Minamata disease, caused by the consumption of fish and shellfish from a chronically polluted area affected by MeHg discharges from the acetaldehyde plant of a plastic factory.[67] Alkylmercury antifungal seed dressing agents such as ethylmercury-p-toluene sulfonanilide led to many episodes of human poisoning in the 1950s.[68] Since those episodes the use of mercury products was progressively phased out, and global Hg consumption in western countries decreased by almost 10 times, though the pattern is not the same in other emerging economies.[69] In fact in recent years the global production of mercury by mining has risen again (Fig. 9.6).

Background levels of mercury in water, sediments, and bivalves are shown in Tables 9.1—9.3. For reliable measurement of Hg in natural water ultra-clean methods must be conducted, and data prior to the 1990s are vastly overestimated. Oceanic seawater shows Hg concentrations between 0.2 and 2 ng/L, with a subsurface maximum in the vicinity of the thermocline.[70] The **Mediterranean Sea** shows higher average values than the Atlantic Ocean, which may be due to natural mineralogical conditions.

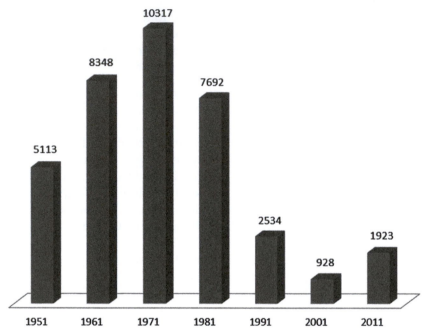

Global annual mercury mining production
(metric Tonnes)

FIGURE 9.6
Global mercury mining production. Notice that the decreasing trend from the 1970s due to the phasing-out of most Hg applications ended at the beginning of the century, and Hg mining is currently rallying. *Data from USGS.*

Clean sediments from pristine areas with no cinnabar show levels around 20 ng/g DW. Some internationally accepted sediment quality criteria are ERL = 0.15 μg/g DW and ERM = 0.71 μg/g DW. Standard analytical methods do not make the difference between organic and inorganic mercury, but speciation studies concluded that in seawater <2% is methylated, while in hypoxic sediments 3%–5% is methylated.

Hg accumulates in fish muscle and bird feathers

Hg accumulates in the aquatic organisms. Typical BCF values reported for marine bivalves from laboratory experiments are 12,000 for waterborne inorganic Hg and up to 55,000 for MeHg.[71] Bioaccumulation is even higher when uptake via **food** is considered. The percentage of methylated Hg in biota is much higher than in inorganic matrices, and it may raise up to 70%–90%. The age of the organism seems to play a role. In large fish the concentration of both inorganic and methylated Hg increases with body size, and thus with age.[72] The percentage of methylated Hg—but not the total Hg content—increased in cockles from 30% to 100% with age.[73] MeHg accumulates in fish **muscle** bound to the Cys rests of the muscle proteins. In birds, maximum

Hg concentrations are found in **feathers**. The levels of Hg in feathers from dated seabird specimens preserved in museum collections provided a mean to track historical changes in environmental mercury. Appelquist et al.[74] for example reported an increase in the Hg content of feathers of Baltic Guillemots (*Uria* sp.) along the 20th century with a peak up to c. 5 ppm around 1969 and a subsequent decrease associated to the ban of aryl-Hg pesticides.

Mercury is the only metal showing **biomagnification** across trophic webs, and the tissular levels of Hg in biota increase as the trophic position of the organism increases, from primary producers to top predators. Planktonic organisms typically show total Hg concentrations below 0.1 ppm (DW), filter feeding bivalves between 0.1 and 0.2 ppm, clupeid fish between 0.3 and 0.7 ppm, ray fish, sharks and thunnids between 1 and 10 ppm, and seabirds between 2 and 15 ppm. It is MeHg, and not the inorganic forms, the chemical species presumably responsible for the observed biomagnification.[75] In both mammals and fish MeHg shows much higher assimilation efficiency from food and a much longer metabolic half-life than inorganic Hg.[76] Although certain ability for methylation of mercury in fish gut has been reported, mediated by the intestinal flora, it is generally assumed that the high % MeHg in the tissues is mainly derived from MeHg ingestion instead of metabolic transformation.[77] In fact, the percentage of methylated Hg also increases with trophic position, being <50% in invertebrates,[78] between 50% and 70% in whole fish (though >90% in fish muscle), and above 97% in seabird blood.[79]

> **Hg levels increase c. 4× at each trophic step**

Biomagnification of Hg may be the result of at least two independent and overlapping phenomena. First, the bioaccumulation potential of Hg in aquatic organisms is higher than for any other metal due to the tendency of Hg species to covalent bonding to thiol groups of endogenous metabolites and the associated difficulty for its **excretion**. Second, Hg levels in fish (but not in bivalves[80]) have been reported to continuously increase with age at a linear rate in double logarithmic plot. Top predators show longer life spans and lower **turnover** rates than primary consumers. As a result, their biomass becomes enriched in conservative chemicals difficult to eliminate from the organism such as organic Hg species. Apex predators are usually large organisms, and Hg concentration was reported to positively correlate to body weight in different species across aquatic food webs.

> **This may be due to difficult excretion of organic Hg and higher life spans of predators**

Trophic Web Biomagnification Factors (**BMF$_{TW}$**; see Section 11.4) of 3.8 for total Hg and 6.5 for MeHg were found in an Arctic food web,[81] meaning that on average the concentrations magnified ×3.8 times and ×6.5 times respectively in each trophic transfer across the whole food web. However, these values are representative of the whole trophic web studied, and remarkable differences among organisms are expected. In fact, when biomagnification was quantified on single predator-prey trophic steps (see Section 11.4), values were quite heterogeneous and remarkably higher for seabirds compared to other organisms.

Fig. 9.7 illustrates the biomagnification of Hg in an estuarine food web from Korea.[82] Hg has been reported to biomagnify more markedly in pelagic than in benthic habitats. In fact, excluding benthic organisms from the Figure would result in higher BMF_{TW} values.

Toxicity thresholds for dissolved Hg may be as low as 1 ppb, but it can be accumulated in feathers at >10 ppm with no apparent deleterious effects

Toxicity of Hg is vastly affected by its chemical speciation. Dissolved Hg^{2+} shows acute toxicity to the most sensitive marine organisms at levels c. 10 μg/L,[83] and chronic toxicity at concentrations c. 1 μg/L.[84] Seabirds and marine mammals though can accumulate up to hundreds ppm in the plumage and liver with no apparent detrimental effect to their fitness. Selenium seems to play a role in Hg detoxification in these organisms.

Except for specific occupational exposure to Hg vapor, food provides the major pathway of Hg intake for human populations. The chemical speciation determines uptake from food as well. Gastrointestinal absorption of elemental mercury is negligible, inorganic salts are absorbed with about a 7% efficiency, while MeHg absorption is very high, c. 95%.[85] In the EU, maximum Hg contents in foodstuff are set to 0.5 μg/g WW, except for a list of predator fish species with allowance up to 1 μg/g. The objective is to achieve a maximum

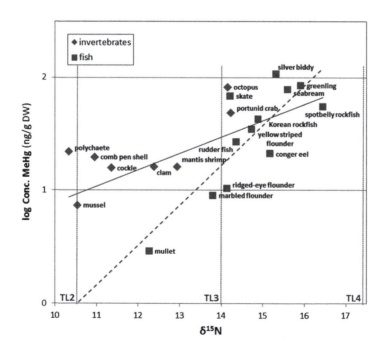

FIGURE 9.7

Increase in methylmercury (MeHg) concentration with trophic level (TL)—estimated from the $\delta^{15}N$—in a food web from a temperate estuary. The slope for all data (*solid line*) was 0.17. Notice that benthic invertebrates show similar MeHg levels disregarding $\delta^{15}N$ value. The slope of the regression for fish data only (*dotted line*) increases to 0.35. *Modified from Kim et al. (2012).*

weekly intake of 1.6 µg/kg body weight.[86] Surprisingly, the EU standard for environmental protection is more stringent than the food standards, with an EQS value for biota of 0.02 µg/g WW,[87] similar to the OSPAR BAC for mussels of 0.09 µg/g DW, which is approximately equivalent to 0.018 µg/g WW.

9.8 TRIBUTYLTIN (TBT)

Sources and environmental distribution

Organic compounds of tetravalent Sn such as TBT show physical, chemical, and biological properties useful for different industrial applications, such as stabilizers in PVC and other materials, catalysts, and biocides for a range of applications from wood preservatives to antifouling coatings.

In water, TBT is rapidly oxidized, which does not alter its toxicity, but dealkylation as a result of UV light and microbial degradation proceeds more slowly, with a half-life of several weeks. Dealkylation transforms TBT to dibutyltin (DBT) and then to monobutyltin (MBT), which are less toxic than the parental compound. Since only TBT is used in boat paints and it is slowly dealkylated in the sediments, the ratio (MBT + DBT)/TBT is directly related to the age of the TBT. Despite this, the direct input of TBT to the marine environment via antifouling paints for boat hulls caused concentrations in the seawater as high as 1 µg Sn/L in marinas and harbor areas.[88]

In the sediments, in absence of light and under low redox conditions, TBT can persist much longer. Sediment concentrations of TBT in harbor areas frequently reach concentrations in the order of 1–10 µg Sn/g DW.[89] Bivalves can accumulate TBT with BCF values >2000.[90] Oysters and other marine invertebrates from TBT-polluted areas, but not fish, reached concentrations up to 1 µg Sn/g DW.[91]

TBT affected oyster farming in Southwestern France

TBT is a toxic substance for mammals that induces apoptosis in vitro by interfering with ion regulation, and its oxide causes immunotoxicity in vivo.[92] But it was the use of TBT in **antifouling paints** for boat hulls that prompted one of the best studied marine ecotoxicological problems at global scale. In 1980 E. His and R. Robert reported for the first time the possible deleterious effects of the presence of TBT on oyster-farming areas.[93] As a result of that and subsequent studies, France banned the TBT-based antifouling paints in 1982, and the ban was extended for all Europe in 1989 (for small vessels[94]) and 2003,[95] and worldwide in 2008.[96]

In the 1970s boat paints containing TBT had become more popular than the traditional Cu-based paints because of their longer effectiveness. By the end of that decade the flourishing oyster industry from Soutwestern France was disturbed by several phenomena that caused its decline: poor growth, shell malformations characterized by considerable thickening of the valves that rendered the oyster unmarketable, and reduced settlement of field-collected spat. In the affected areas oyster culture coexisted with large numbers of pleasure boats and marinas. Several studies attributed to TBT from the boat paints the **shell-thickening**,[97] and failure at environmentally relevant concentrations of normal oyster larva development and settlement.[98] Juvenile mussel growth rate was reduced at TBTO concentrations above 0.1 µg/L,[99] and TBT resulted to rank among the most toxic substances to early life stages of marine invertebrates, with toxicity thresholds in the order of 10 ng/L.[100] In fact that is the reason for its efficacy as antifouling substance. After the TBT ban the oyster production in Southwestern France recovered the previous levels.

TBT causes gastropod's imposex   The second case of TBT toxicity is even better known. In 1981 B.S. Smith reported that TBT caused superimposed male sex organs (vas deferens, penis) in female snails of *Nassarius obsoletus*, a phenomenon first described by Blaber,[101] and called "**imposex**."[102] The same was found for *Nucella lapillus* and other marine snails. Imposex prevalence was particularly remarkable at locations in the vicinity of harbor areas. In extreme cases the overgrowth of vas deferens tissue can block the oviduct and prevent the release of egg capsules, rendering the female sterile.[103] This led to *N. lapillus* populations' decline in areas from Southwestern Great Britain severely contaminated by TBT.[104] Imposex could be induced in the laboratory following exposure to 20 ng/L of TBT, a concentration far below those recorded in polluted harbors. The quantitative relationship between the TBT bioaccumulation and the degree of anatomical alteration in female snails is so tight that imposex is currently used as biomarker of TBT pollution in monitoring programs (see also Chapters 14.3 and 16.5).

KEY IDEAS

- High atomic weight transition metals such as Hg, Cd, Cu, Pb, and Zn tend to form covalent bonds with organic molecules, which confers them special potential toxicity. In water these metals form divalent cations (e.g., Cu^{2+}) that bind to two adjacent $-SH^-$ groups of proteins. This binding can alter the structure or inhibit the function of the affected macromolecule.

- Aquatic humic matter can bind those cations and reduce their toxicity.

- Metals which do not play any role in the metabolism (e.g., Hg, Cd, Pb, Sn) are called xenobiotics, and their levels in the organism depend on those in the environment, while those needed at trace levels (e.g., Cu, Zn, Ni, Co) are called essential, and their internal levels may be regulated. Both can be toxic at concentrations above a threshold.

- Human activities have enriched the atmosphere in Hg and Pb, and they enter the oceans by deposition. The main inputs of Cu and Cd to the sea are the water courses. In both cases metals ultimately accumulate in the sediments.

- Background levels of trace metals in marine sediments depend on mineralogical factors, but are <0.1 mg/Kg dry weight (=ppm) for Hg, < 0.5 mg/Kg for Cd, and <50 mg/Kg for Pb and Cu.

- Mussels accumulate anthropogenic Hg, Cd, and Pb but are capable to regulate Cu until a certain limit is exceeded. Oysters accumulate Cu as well. Because of their bioaccumulation in many organisms and human health issues, levels of Hg, Cd, and Pb in foodstuff, including fishing products, are regulated.

- Anthropogenic Cu and Zn frequently show common sources and reach concentrations up to 1 mg/L in urban wastewaters. Cd and Pb are more frequently originated in industrial activities. The main source of Hg in the coastal environment is the chlor-alkali industry.

- Global Pb pollution decreased after leaded petrol was phased out, and this can be documented in dated ice cores.

- Mercury is deposited from the atmosphere in inorganic form, and it is present as Hg^{2+} in the water column, but anaerobic bacteria actively methylate it in the sediments.

- Mercury is the only metal showing biomagnification across trophic webs. This is mainly due to the high bioavailability of the methylmercury, which readily pass across biological membranes, and the difficult excretion of this Hg species that binds to the proteins.

- TBT from antifouling boat paints caused problems in the oyster culture in the French coast and imposex in gastropods worldwide.

Endnotes

1. Wedepohl KH. The composition of Earth's upper crust. Natural cycles of elements. Natural resources. Chapter I.1. In: Merian E, M. Anke M, Ihnat M, Stoeppler M. editors. Metals and their compounds in the environment. 2nd ed. Weinheim: Wiley VCH. 2004; pp. 3–16.
2. Nioboer and Richardson. Environ pollut 1980;1B:3–26.
3. Selin NE. Global Biogeochemical Cycling of Mercury: A Review. Annu Rev Environ Resourc 2009;34:43–63.
4. Wedepohl (2004) op. cit.
5. Förstner U, Müller G. Heavy metal accumulation in river sediments: a response to environmental pollution. Geoforum 1973;14:5361.
6. Chester R, Jickells T. Marine Geochemistry 3rd ed. Wiley-Blackwell; 2012.
7. Puxbaum H, Limbeck A. Metal compounds in the atmosphere. In: Merian E, M. Anke M, Ihnat M, Stoeppler M. editors. Metals and their compounds in the environment. 2nd ed. Weinheim: Wiley VCH. 2004; pp. 17–46.
8. EPA ambient water quality criterion for Cd.
9. Liber SM. An introduction to marine biogeochemistry. New York: J Wiley & Sons; 1992 Chapter 30.
10. Kennish MJ. Ecology of estuaries. Anthopogenic effects. Boca Raton:CRC Press; 1991. (p. 269).

11. Merian et al. (2004) op. cit.

12. Beiras R, Durán I, Parra S, et al. Linking chemical contamination to biological effects in coastal pollution monitoring. Ecotoxicology 2012;21:9−17.

13. Chiffoleau JF. Le chrome en milieu marin. Repères Océan N°8-1994. IFREMER; 1994.

14. Cossa D, Lassus P. Le cadmium en milieu marin. Biogéochimie et écotoxicologie. IFREMER 1989. Data for North Atlantic.

15. Álvarez-Iglesias P, Rubio, B, Pérez-Arlucea, M. Reliability of subtidal sediments as "geochemical recorders" of pollution input: San Simón Bay (Ría de Vigo, NW Spain) Estuarine, Coastal Shelf Sci 2006;70, 507−521.

16. Prego, R, Filgueiras, AV, Santos-Echeandía, J. Temporal and spatial changes of total and labile metal concentration in the surface sediments of the Vigo Ria (NW Iberian Peninsula): influence of anthropogenic sources. Mar Pollut Bull 2008;56:1031−1042.

17. Canario J., Prego R., Vale C. Branco V. Distribution of mercury and monomethylmercury. In Sediments of Vigo Ria, NW Iberian Peninsula Water Air Soil Pollut 2007;182:21−29.

18. US EPA. Screening level ecological risk assessment protocol for hazardous waste combustión facilities; August 1999.

19. OSPAR Commission. Levels and trends in marine contaminants and their biological effects − CEMP Assessment report 2012. 2013.

20. Kimbrough KL, Johnson WE, Lauenstein GG, Christensen JD, Apeti DA. An Assessment of two decades of contaminant monitoring in the Nation's Coastal Zone. Silver Spring, MD. NOAA Tech Memo NOS NCCOS 2008;74:1−105.

21. Commission Regulation (EC) No 1881/2006 of 19 December 2006 setting maximum levels for certain contaminants in foodstuffs. Commission regulation (EU) No 488/2014 amending Regulation (EC) No 1881/2006 as regards maximum levels of cadmium in foodstuffs.

22. Kopfler FC. The Accumulation of Organic and Inorganic Mercury Compounds by the Eastern Oyster (*Crassostrea virginica*). Bull Environm Contam Toxicol 1974;11(3):275−280.

23. Schaule BK, Patterson CC. Lead concentrations in the northeast Pacific: evidence for global anthropogenic perturbations Earth and Planetary Science Letters 1981;54:97−116.

24. Bruland KW. Oceanographic distributions of cadmium, zinc, nickel, and copper in the North Pacific. Earth Planet Sci Lett 1980;47:176−198.

25. Borchardt T. Influence of food quantity on the kinetics of cadmium uptake and loss via food and seawater In *Mytilus edulis*. Mar Biol 1983;76:67−76. Riisgård HU, Bjørnestad E, Møhlenberg F. Accumulation of cadmium in the mussel *Mvtilus edulis*: kinetics and importance of uptake via food and seawater. Mar Biol 1987;96:349−353.

26. Markich SJ, Jefree RA. Absorption of divalent trace metals as analogous of calcium by Australian freshwater bivalves: an explanation of how water hardness reduces metal toxicity. Aquatic Toxicol 1994;29:257−290.

27. Depledge MH, Rainbow PS. Models of regulation and accumulation of trace metals in marine invertebrates. Comp Biochem Physiol 1990;97C (No 1):1−7.

28. Wright DA, Zamuda CD. Copper accumulation by two bivalve molluscs: salinity effect is independent of cupric ion activity. Mar Environ Res 1987;23:1−14; Wang WX, Dei RCH. Factors affecting trace element uptake in the black mussel *Septifer virgatus*. Mar Ecol Prog Ser 1999;186:161−172; Chong K, Wang WX. Comparative studies on the biokinetiks of Cd, Cr, and Zn in the green mussel *Perna viridis* and the Manila clam *Ruditapes philippinarum*. Environ Pollut 2001;115:107−121; Blust R, Kockelbergh E, Bailleieul M. Effect of salinity on the uptake of cadmium by the brine shrimp *Artemia franciscana*. Mar Ecol Prog Ser 1992;84:245−254.

29. See Blust et al. (1992) op. cit. p. 252.

30. Ke C, Wang WX. Bioaccumulation of Cd, Se, and Zn in an estuarine oyster (*Crassostrea rivularis*) and a coastal oyster (*Saccostrea glomerata*). Aquatic Toxicol 2001;56:33–51; Chong and Wang (2001) op. cit.; Wang WX, Fisher NS. Assimilation efficiencies of chemical contaminants in aquatic invertebrates: a synthesis. Mar Ecol Prog Ser 1999;161:103–115.

31. George SF, Pirie BJS, Coombs TL. The kinetics of accumulation and excretion of ferric hydroxide in *Mytilus edulis* (L.) and its distribution in the tissues. J Exp Mar Biol Ecol 1976;23:71–84.

32. Kennish MJ. Practical handbook of marine science. 3rd ed. Boca Raton:CRC Press. 2001. (pp. 88–89).

33. O'Connor TP, Lauenstein GG. Status and trends of copper concentrations in mussels and oysters in the USA. Mar Chem 2005;97:49–59.

34. Henry JG. Chap. 12. Water pollution. In: Henry JG and Heinke GW, editors. Environmental Science and Engineering (2nd ed. New York: Prentice Hall; 1996.

35. Amiard JC, Amiard-Triquet C, Berthet B, et al. Comparative study of the patterns of bio-accumulation of essential (Cu, Zn) and non-essential (Cd, Pb) trace metals in various estuarine and coastal organisms. J Exp Mar Biol Ecol 1987;87(106):73–89.

36. McGeer JC, Brik KV, Skeaff JM, et al. Inverse relationship between bioconcentration factor and exposure concentration for metals: implications for hazard assessment of metals in the aquatic environment. Environ Toxicol Chem. 2003;22(5):1017–37.

37. Rainbow PS. The significance of trace metal concentrations in decapods. Symp. Zool Soc Lond 1988;59:291–313.

38. Bryan GW. 3. Heavy metal contamination in the sea. In: Johnston R. editor. Marine Pollution. London: Academic Press; 1976. pp. 185–302.

39. Bryan GW, Langston WJ, Hummerstone LG et al. A guide to the assessment of heavy metal contamination in estuaries using biological indicators. Mar Biol Assoc UK. Occasional Publication N° 4, 92 pp. 1985.

40. Bryan et al. (1985) op. cit.

41. Oudijk G. The Rise and Fall of Organometallic Additives in Automotive Gasoline. Environ Forensic 2010;11(1):17–49.

42. Shotyk W, Weiss D, Appleby PG, et al. History of Atmospheric Lead Deposition Since 12,370 14C yr BP from a Peat Bog, Jura Mountains, Switzerland. Science 1998:281:1635–1640; Boutron CF, Candelone J-P, Hong S. Past and recent changes in the large-scale tropospheric cycles of lead and other heavy metals as documented in Antarctic and Greenland snow and ice: A review. Geochimica et Cosmochimica Acta 1994;58(15):3217–3225.

43. Zheng J, Shotyk W, Krachler M. A 15,800-year record of atmospheric lead deposition on the Devon Island Ice Cap, Nunavut, Canada: Natural and anthropogenic enrichments, isotopic composition, and predominant sources. GLOBAL BIOGEOCHEMICAL CYCLES, 2007; VOL. 21, GB2027, doi:10.1029/2006GB002897.

44. http://www.ila-lead.org/lead-facts/lead-uses–statistics.

45. Osterberg E, Mayewski P, Kreutz K, et al. Ice core record of rising lead pollution in the North Pacific atmosphere. Geophy Res Lett 1994;35:L05810, doi:10.1029/2007GL032680.

46. Cossa D, Elbaz-Poulichet F, Gnassia-Barelli M, et al. Le plomb en milieu marin. Biogéochimie et écotoxicologue. IFREMER 1992.

47. Bellas J, Nieto O, Beiras R. Integrative assessment of coastal pollution: Development and evaluation of sediment quality criteria from chemical contamination and ecotoxicological data. Cont Shelf Res 31 2011:448–456.

48. Common Implementation Strategy for the Water Framework Directive. EQS datasheet priority substance No 20. Lead and its compounds; 2005.

49. Merian E. editor. Metals and their compounds in the environment. Weinheim:VCH Verlagsgesellschaft mbH; 1991. (p. 993)

50. Oudijk (2010) op. cit.

51. Fisher IJ, Pain DJ, Thomas VG. A review of lead poisoning from ammunition sources in terrestrial birds. Biological Conservation 2006;131:421–432.

52. Merian E. editor. Metals and their compounds in the environment. Weinheim:VCH Verlagsgesellschaft mbH; 1991. (p. 971)

53. Davis JM, Svendsgaard DJ, Lead and child development. Nature 1987;329:297–300.

54. Commission regulation (EC) N° 1881/2006.

55. Merian E. editor. Metals and their compounds in the environment. Weinheim:VCH Verlagsgesellschaft mbH; 1991. (p. 807)

56. Morrow, H. Markets and applications for cadmium. In: Eighth International Cadmium Conference 10–13 Nov 2011. Kuming, China; 2011.

57. Cossa & Lassus (1989) op. cit.

58. Common Implementation Strategy for the Water Framework Directive. EQS datasheet priority substance No 6. Cadmium and its compounds. 2005.

59. George SG, Coombs TL. The effects of chelating agents on the uptake and accumulation of cadmium by Mytilus edulis. Mar Biol 1977;39:261–268.

60. Cossa & Lassus (1989) op. cit.

61. Commission Regulation (EU) No 488/2014.

62. Von Burg R, Greenwood MR. Mercury. In: Merian E. editor. Metals and their compounds in the environment. Weinheim:VCH Verlagsgesellschaft mbH; 1991. pp. 1045–1088.

63. Cela R, Lorenzo RA, Rubi E, et al. Mercury speciation in raw sediments of the Pontevedra estuary (Galicia-Spain). Environ Technol 1992;13:11–22.

64. Selin (2009) op. cit.

65. Sunderland EM, Mason MP. Human impacts on open ocean mercury concentrations. Global biogeochemical cycles, vol. 21, GB4022; 2007. doi:10.1029/2006GB002876.

66. Cossa D, Thibaud Y, Romeo M, et al. mercure en milieu marin. Biogéochimie et écotoxicologie. Rapports scientifiques et techniques de l'IFREMER, No 19. IFREMER; 1990.

67. Ekino S, Susa M, Ninomiya T, et al. Minamata disease revisited: an update on the acute and chronic manifestations of methyl mercury poisoning. J. Neurol. Sci. 2007;262:131–144.

68. Bakir F, Damluji S, Amin-Zaki L, et al. Methylmercury poisoning in Iraq. An interuniversity report. Science 1973;181:230–241.

69. UNEP Global Mercury Assessment; 2013.

70. Cossa et al. (1990) op. cit.

71. US EPA. Screening level ecological risk assessment protocol for hazardous waste combustión facilities. August 1999. See also McGeer et al. 2003 and common implementation strategy for the water framework directive. EQS datasheet priority substance No 21. Mercury and its compounds; 2005.

72. Cossa et al. (1990) op. cit.

73. Møhlenberg F, Riisgård HU. Partitioning of inorganic and organic mercury in cockles Cardium edule (L.) and C. glaucum (Bruguiére) from a chronically polluted area: influence of size and age. Env Pollut 1988;55:137–148.

74. Appelquist H, Drabaek I, Asbirk S. Variation in Mercury Content of Guillemot Feathers over 150 Years. Mar Pollut Bull 1985;16(6):244–248.

75. Watras CJ, Back RC, Halvorsen S, et al. Accumulation of mercury in pelagic freshwater food webs. Sci Total Environ 1998;219:183–208.

76. Wang R, Wong M-H, Wang W-X. Mercury exposure in the freshwater tilapia Oreochromis niloticus. Environ Pollut 2010;158:2694–2701 for fish; Merian (1991) op. cit. p. 1059 & 1061–1062 for mammals.

77. Wang R, Feng X-B, Wang W-X. In Vivo Mercury Methylation and Demethylation in Freshwater Tilapia Quantified by Mercury Stable Isotopes. Environ Sci Technol 2013;47: 7949–7957.

78. Kim E, Kim H, Shin K, et al. Biomagnification of mercury through the benthic food webs of a temperate estuary: Masan Bay, Korea. Environ Toxicol Chem 2012;31(6):1254–1263.

79. Lavoie RA, Hebert CE, Rail J-F, et al. Trophic structure and mercury distribution in a Gulf of St. Lawrence (Canada) food web using stable isotope analysis. Sci Total Environ 2010;408: 5529–5539.

80. Bryan GW, Langston WJ, Hummerstone LG. The use of biological indicators of heavy metal contamination in estuaries with special reference to an assessment of the biological availability of metals in estuarine sediments from SW Britain. Mar Biol Assoc UK Occasional publications N°1, 73 pp. 1980.

81. Lavoie et al. (2010) op. cit.

82. Kim et al. (2012) op. cit.

83. His E, Beiras R, Seaman MNL. The assessment of marine pollution- Bioassays with bivalve larvae. Advances in Marine Biology 2000; 37:1–178. San Diego:Academic Press.

84. USEPA. Mercury; 1980. Ambient water quality criteria document.

85. Merian E. editor. Metals and their compounds in the environment. Weinheim:VCH Verlagsgesellschaft mbH; 1991. (p. 1059).

86. See Endnote 54.

87. Directive 2013/39/EU.

88. Alzieu C. L'étain et les organoétains en milieu marin biogéochimie et ecotoxicologie. IFREMER; 1989.

89. Díez S, Jover E, Albaigés J, et al. Occurrence and degradation of butyltins and wastewater marker compounds in sediments from Barcelona harbor, Spain. Environ Inter 2006;32: 858–865; Rodríguez JG, Solaun O, Larreta J, et al. Baseline of butyltin pollution in coastal sediments within the Basque Country (northern Spain), in 2007–2008. Mar Pollut Bull 2010;60:139–151; Alzieu (1989) op. cit.

90. Li Q, Osada M, Takahashi K, et al. Accumulation and depuration of tributyltin oxide and its effect on the fertilization and embryonic development in the Pacific oyster *Crassostrea gigas*. Bull Environ Contam Toxicol 1997;58:489–496; Laughlin RB, French W. Concentration dependence of bis(tributyl)tin oxide accumulation in the mussel, *Mytilus edulis*. Environ Toxicol Chem 1988;7:1021–1026.

91. Morcillo Y, Borghi V, Porte C. Survey of Organotin Compounds in the Western Mediterranean Using Molluscs and Fish as Sentinel Organisms. Arch Environ Contam Toxicol 1997; 32:198–203.

92. Ballantyne B, Marrs T, Syversen T, editors. General and applied toxicology. 2nd ed. london: Macmillan Reference Ltd; 1999. vol. 1. p. 188–189 (in vitro toxicity to cell cutures); Ballantyne et al. (1999) op. cit. vol. 2. p. 1008 (in vivo toxicity to rats).

93. His E, Robert R. Action d'un sel órgano-métallique, l'acétate de tributyle-étain sur les œufs et les larves D de *Crassostrea gigas* (Thunberg). Conseil International pour l'Exploration de la Mer. Comité de la Mariculure. C.M. 1980/F:27; 1980. 10 pp.

94. Directive 89/677/CEE.

95. Regulation (EC) No 782/2003.

96. IMO. International convention on the control of harmful anti-fouling systems on ships. I. M. organization; 2001. The Convention only entered to force in 2008.

97. Alzieu C, Héral M, Thibaud Y, Dardignac MJ, Feuillet M. Influence des peintures antisalissures à base d'organostanniques sur la calcification de la coquille de l'huitre Crassostrea gigas. Rev. Trav. Inst. PêchesMarit 1982;45:100–116. Waldock MJ, Thain J. Shell thickening in *Crassostrea gigas*: organotin antifouling or sediment induced? Mar Pollut Bull 1983;14(11):411–415.

98. His E, Robert R. Développement des véligères de *Crassostrea gigas* dans le Bassin d'Arcachon. Etudes sur les mortalités larvaires. Rev Trav Inst Pêchesmarit 1985;47:63–88. His E. Embryogenesis and larval development in *Crassostrea gigas*: experimental data and field observations on the effect of tributyltin compounds. In: Chapman M, Seligman PF, editors. Organotin. London: Chapman & Hall; 1995. pp 239–258.

99. Strømgren T, Bongard T. The effect of Tributyltin oxide on growth of Mytilus edulis. Mar Pollut Bull 1987;18(1):30–31.

100. His E, Maurer D, Robert R. Estimation de la teneur en acétate de tributylétain dans l'eau de mer par une méthode biologique. J Moll Stud 1983;Suppt 12A:60–68; Lapota D, Rosenberg DE, Platter-Rieger MF, et al. Growth and survival of *Mytilus edulis* larvae exposed to low levels of dibutyltin and tributyltin. Mar Biol 1993;115:413–419.

101. Smith BS. Tributyltin Compounds Induce Male Characteristics on Female Mud Snails *Nassarius obsoletus = Ilyanassa obsolete*. Jornal of Applied Toxicology, 1981;1(3):141–144.

102. Blaber SJM. The occurrence of a penis-like outgrowth behind the right tentacle in spent females of *Nucella lapillus* (L.). Proceedings of the Malacological Society of London 1970;39: 231–233.

103. Gibbs PE, Bryan GW. Reproductive failure in populations of the dog-whelk, Nucella lapillus, caused by imposex induced by tributyltin from antifouling paints. J mar boil Ass UK 1986; 66:767–777.

104. Bryan, GW, Gibbs, PE, Hummerstone, LG, Burt, GR. The decline of the gas-tropod Nucella lapillus around the south-West England: evidence for the effectoftributyltin from antifouling paints. J Mar Biol Assoc UK 1986;66:611–640; Gibbs, PE, Bryan, GW. TBT-induced imposex in neogastropod snails: mas-culinization to mass extinction tributyltin: case study of an environmental contaminant. Cambridge University Press; 1996.

Suggested Further Reading

- Merian E, Anke M, Ihnat M, Stoeppler M, editors. Elements and their compounds in the environment: occurrence, analysis and biological relevance. 2nd ed. Weinheim: Wiley-VCH; 2004.

- Wang W-X, Fisher NS. Modeling the influence of body size on trace element accumulation in the mussel *Mytilus edulis*. Mar Ecol Prog Ser 1997;161:103–15.

- His E. Embryogenesis and larval development in *Crassostrea gigas*: experimental data and field observations on the effect of tributyltin compounds. In: Champ MA, Seligman PF, editors. Organotin. Dordrecht: Springer; 1996.

- Kim E, Kim H, Shin K, et al. Biomagnification of mercury through the benthic food webs of a temperate estuary: Masan Bay, Korea. Environ Toxicol Chem 2012;31(6):1254–63.

Part II: Marine Ecotoxicology

CHAPTER 10

Distribution of Pollutants in the Marine Environment

10.1 MODELING THE DISTRIBUTION OF POLLUTANTS IN MARINE ECOSYSTEMS

Given a specific input of contamination, we may wish to predict the distribution and eventual concentration of that particular contaminant in the different compartments of the marine environment to assess its potential biological effects on the organisms inhabiting those compartments. To do that we first need to divide the marine environment into discrete compartments. Guided by the parsimony principle, modeling imposes some simplification of the actually highly heterogeneous environment, and definition of a reduced number of homogeneous compartments or phases where a chemical in theory is considered to behave in a uniform manner and be distributed uniformly within them. General environmental models often define three large abiotic compartments: **air** (atmosphere), **surface waters** (hydrosphere) and **land**, and a biotic compartment, the organisms, or **biota**. For convenience, the aquatic compartment can be further divided into **surface layer**, **water column**, and bottom **sediment**, with the second operationally divided into dissolved and particulate matter, and the third into pore water and sediment particles (Fig. 10.1).

Atmospheric pollutants with low vapor pressure such as polycyclic aromatic hydrocarbons (PAHs) and polychlorinated biphenyls (PCBs) tend to sorb to aerosol particles submitted to dry deposition by gravity or to much faster wet deposition associated to rainfall. The surface layer of the water is the interface for contaminants coming from the atmosphere but also where oils and microplastics of terrestrial origin, with lower density than water, accumulate. When an oil slick is present on the water surface it offers a hydrophobic phase for the accumulation of nonpolar organics such as organohalogenated compounds and other hydrophobic organics. The surface layer can be operationally defined in function of the sampling devices used (plates, screens, rotating drums) as the first few tenths to hundreds of microns of the water column (in this case surface microlayer), although other devices such as neuston nets collect the particles present in the first few centimeters of the water column. The surface microlayer can be enriched up to 500 times in pollutants compared to the underlying bulk water.[1]

The marine environment can be divided in three abiotic compartments: surface layer, water column, and sediment

Marine Pollution. https://doi.org/10.1016/B978-0-12-813736-9.00010-6

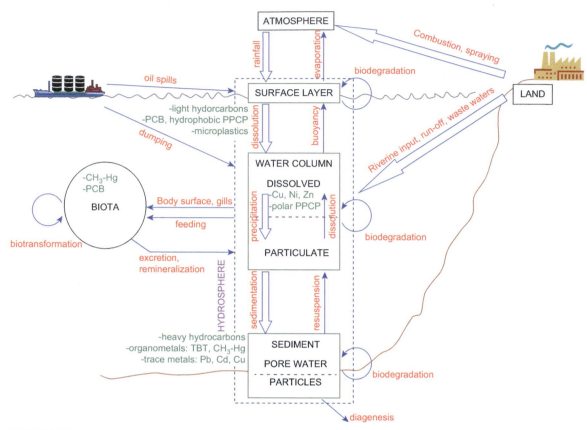

FIGURE 10.1

Environmental fate of contaminants in the biotic (circle) and abiotic (boxes) compartments. Transport processes are indicated by straight arrows and degradation processes as coiled arrows. Note that marine sediments act as final sink for the quantitatively most important routes of transport (thick arrows). For some contaminants (green), preferred compartments are indicated.

In oceanographic research it is a convention to quantify in the water column the suspended particulate matter (total particulate matter, TPM, dry weight) by filtration, typically through a 0.45 μm filter. Thus, particulate and dissolved metals and other contaminants are separately assessed in the water column. The organic content of the particulate fraction (particulate organic matter, POM, organic weight) is normally assessed through loss of weight after combustion (loss on ignition). Occasionally, the carbon content rather than the total organic content is measured. This is indicated as dissolved organic carbon (DOC) and total organic carbon (TOC). Carbon typically represents about a half of the total organic matter.

Concerning sediments, some sampling techniques are available for in situ or ex situ extraction of the interstitial water, which in estuarine sediments may have chemical properties very different to those of the water column. In contrast when the bulk sediment is analyzed concentrations are generally expressed on a dry weight basis, and liophilization is preferred as drying method to mini-mize loss of volatile contaminants. Organic content of sediments may be assessed by loss on ignition or by the more accurate measurement of TOC. Since organic pollutants tend to sorb to organic matter, concentrations of these pollutants in sediments are frequently expressed after normalization for a stan-dard organic content, e.g., 2.5% TOC.[2]

It is worth noting already that the quantitatively most important routes of transport of contaminants (thick arrows in Fig 10.1), disregarding their origin, converge in the marine sediments, which act as a sink where pollutants accu-mulate at concentration orders of magnitude above those in the water. Due to their large surface:volume ratio, **fine sediments are particularly relevant as sink of pollutants**. In 1922 C.K. Wentworth developed a scale still in use based on the $\log_2$ diameter (in mm) of the sediment particles that classifies fine sediments as those below 63 µm (2^{-4} mm). Fine-grain sediments, in addi-tion, frequently reach anoxic values with depth, and biodegradation rates under those conditions slow down, retaining the contaminants for much larger periods of time.

The main routes of transport of contaminants converge in the sediments

In Section 1.6 the three main pathways that allow entrance of contaminants into the marine compartments were already discussed: (1) **atmospheric input** of gases and aerosols originated from combustion, evaporation and spraying application of biocides, and washed into the sea surface by the rainfall; (2) **riverine input**, that includes also run-off waters over urban and rural land and point source liquid effluents; and (3) **direct input** of contaminants into the sea due to accidental or deliberate dumping and spillages.

In open-system kinetic models the contaminant can flow in and out of each compartment but at steady state total inputs equal total outputs

Once the environmental compartments have been defined, we can consider them open systems with the possibility for a given contaminant to flow into and out of the system. The distribution of the contaminant among the so defined environmental compartments will be determined by **state variables**, the amount (mass) of contaminant present in a compartment of a certain volume, and **flux variables**, expressed in mass·time^{-1} units, determined by transportation and elimination processes. The straight arrows connecting compartments in Fig. 10.1 represent **transportation** processes, while the coiled arrows within each compartment represent **degradation** processes, both expressed as functions of time. The first take the contaminants unchanged from one compartment to another (evaporation, rain-washing, sedimentation, uptake, and excretion by biota, etc.), while the latter eliminate the contaminant from each compartment by changing its molecular structure and properties (biotransformation, microbiological, and chemical degradation, etc.).

Once all compartments are connected with the respective fluxes the system can now be modeled just by considering mass balance equations and stating that for each compartment, the summation of contaminant influxes equals the summa-tion of outfluxes (including degradation rates if relevant). Differential equations in time must thus be introduced for each compartment and the system solved through integration. The increased level of algebraic complexity is currently treatable, resorting to a number of computer software and web-based programs available to allow this type of predictive environmental modeling.

In closed-system thermodynamic models the contaminant partitions between compartments according to its chemical properties and a picture of the final equilibrium is obtained

A second kind of models very common in environmental applications, despite their stringent assumptions, is the **thermodynamic models**. The purely thermodynamic models consider the environmental compartments as closed systems, on the sense that physical transportation processes are neglected. A static environment is assumed and thermodynamic equilibrium, or a mathematically undistinguishable approximation termed steady state, where for each compartment influxes equal to outfluxes is assumed to rule the distribution of chemicals among environmental compartments. As in kinetic models, compartments are viewed as homogeneous chemical phases. However, the variables connecting compartments are no longer time-dependent fluxes but functions of the physicochemical properties of the substance that determine its equilibrium partition between the phases involved. For example, evaporation of a nonpolar solute from the water surface into the atmosphere is determined by Henry's law that can be stated as follows:

$$P = H \cdot R \cdot T \cdot C_w \qquad \text{(Eq. 10.1)}$$

where P is the partial pressure (in International System pressure units, pascals, Pa) that drives the evaporation of that particular chemical, H is the dimensionless Henry's constant typical of each chemical[a], R is the ideal gas constant $(L \cdot Pa \cdot mol^{-1}/K)$, T is the temperature (K degrees), and C_w the water concentration $(mol \cdot L^{-1})$. In other words, the flux of the chemical from the water to the atmosphere is a direct function of its volatility (determined by H), temperature, and concentration in the original compartment. For a constant temperature (conventionally 298 K) then $H \cdot R \cdot T = H_D$ is the Henry's dimensional $(Pa \cdot L \cdot mol^{-1})$ constant, and evaporation will be a simple linear function of concentration in water according to the expression

$$P = H_D \cdot C_w \qquad \text{(Eq. 10.2)}$$

10.2 PARTITION EQUILIBRIUM OR FUGACITY MODELS

Fugacity is the escaping tendency of a chemical from a compartment

The approach presented earlier in the chapter for modeling environmental distribution of contaminants is based on the thermodynamic equilibriums achieved at steady state in the partition of the chemical between adjacent environmental phases with different affinities for that particular chemical. This approach broadly developed by Mackay[3] was very successful and exploits the

---

[a] Care must be taken when consulting H values in tables, since the original investigations concerned the solubility of gases, i.e., the fugacity of a gas from air into the water, rather than its volatility into the air, and thus the values may be the reciprocal, 1/H.

preexistent concept of **fugacity** (f_i) of a chemical from a given phase, i, defined as the escaping tendency of the chemical from that phase, and quantified as:

$$f_i = C_i/Z_i \qquad \text{(Eq. 10.3)}$$

where C_i is the concentration and Z_i the fugacity capacity constant or affinity constant, that quantifies the affinity of the chemical for that phase (Fig. 10.2).

For two adjacent phases 1 and 2 in equilibrium $f_1 = f_2$; i.e., $C_1/Z_1 = C_2/Z_2 = K_{12}$, where K_{12} is the partition coefficient between phases 1 and 2.

We have just introduced the case of two adjacent environmental phases: water and air, where the force that drives the flux of a neutral substance from water to air, quantified by the partial pressure (P), can now be reinterpreted as the fugacity from the water phase, $f_w = P = H \cdot R \cdot T \cdot C_w$ (Eq. 10.1), whereas f_a will be, according to Eq. (10.3), $f_a = C_a/Z_a$. Hence, in steady state, $f_w = f_a$; i.e., $H \cdot R \cdot T \cdot C_w = C_a/Z_a$

Taking $Z_a = 1/RT$ and rearranging:

$$C_a/C_w = H \qquad \text{(Eq. 10.4)}$$

Fugacity (f): tendency of a chemical contaminant to escape from an environmental compartment

$$f = c / Z$$

C: concentration of the contaminant in the compartment
Z: fugacity constant, or affinity between the chemical and the compartment

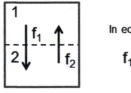

In equilibrium:

$$f_1 = f_2$$

$$K_{12} = C_1/C_2 = Z_1/Z_2$$

K_{12}: partition coefficient of the comtaminat between the two compartments

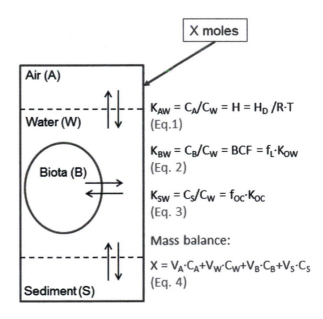

$K_{AW} = C_A/C_W = H = H_D /R\cdot T$
(Eq.1)

$K_{BW} = C_B/C_W = BCF = f_L\cdot K_{OW}$
(Eq. 2)

$K_{SW} = C_S/C_W = f_{OC}\cdot K_{OC}$
(Eq. 3)

Mass balance:

$X = V_A\cdot C_A + V_W\cdot C_W + V_B\cdot C_B + V_S\cdot C_S$
(Eq. 4)

FIGURE 10.2
A particularly successful type of environmental models is the fugacity or equilibrium partition models. These models predict the equilibrium partition of a chemical among two neighbor compartments on the basis of the affinities of each compartment to the chemical. The affinities are equated as functions of basic chemical properties of the substance (volatility, water solubility, octanol-water partition coefficient, etc.), and the environment (temperature, organic carbon, lipid content of biota, etc.).

This expression describes the partition of the substance between the water and air phases at thermodynamic equilibrium.

Provided we can relate the fugacities that drive the fluxes from one phase to the next one to physic-chemical properties of the substances we can obtain expressions similar to Eq. (10.4) for each pair of contiguous phases. For example for the sediment:water partition: $C_S/C_W = K_{SW} = f_{OC} \cdot K_{OC}$, where f_{OC} is the fraction of organic carbon of the sediment and K_{OC} is the organic carbon:water partition coefficient, available from the literature for many organics.

Similarly, the organism:water partition can be modeled as: $C_t/C_W = BCF$ where C_t is the concentration of the substance in the tissues of the organism on wet weight basis[b], and BCF is the bioconcentration factor. For many lipophilic chemicals BCF can be estimated from K_{OW} (see Section 11.3), and thus $C_t/C_W = f_L \cdot K_{OW}$, were f_L is the fraction of lipids in the total wet weight.

In summary, considering a system composed by air, water, sediment and aquatic biota (subscripts A, W, S and B) in thermodynamic equilibrium, a given chemical introduced in the system at a known amount x (mol) will partition according to the following partition coefficients:

$$K_{AW} = C_A/C_W = H = H_D/R \cdot T \tag{Eq. 10.5}$$

$$K_{SW} = C_S/C_W = f_{OC} \cdot K_{OC} \tag{Eq. 10.6}$$

$$K_{BW} = C_B/C_W = BCF = f_L \cdot K_{OW} \tag{Eq. 10.7}$$

and H_D for a given chemical can be calculated from its vapor pressure (P_A), water solubility (S_W, $g \cdot m^3$) and molar mass (M, $g \cdot mol^{-1}$) as

$$H_D = P_A/(S_W/M)$$

Now a simple mass balance equation,

$$x = V_A \cdot C_A + V_W \cdot C_W + V_S \cdot C_S + V_B \cdot C_B \tag{Eq. 10.8}$$

and Eqs. (10.5)−(10.7) provide a system of equations that allow prediction of the concentrations of the chemical in each compartment.

Fugacity models are used in a priori ERA of chemicals

Therefore, if we know the total mass of contaminant released to the environment, the volume of each compartment and basic characteristics of the substance (vapor pressure, solubility, K_{OW}, etc.) and the environment (T, organic fraction of sediments, lipid content of biota, etc.) we can predict, assuming a static environment where transportation processes were negligible, all the environmental concentrations at thermodynamic equilibrium. Despite their theoretical restrictions, these kinds of models are broadly used with regulatory

---

[b] Notice that chemists tend to report concentrations of pollutants in organisms on a dry weight basis in order to enhance the precision of the measurement, since water content of the biological tissues depends on temperature and other sampling and storage conditions. However, the physiologically relevant concentration in terms of bioaccumulation is the in vivo concentration, i.e., g of pollutant per Kg of tissue wet weight (ww). Thus, BCF units result to be: (g/Kg ww)/(g/L), i.e., L/Kg ww.

purposes for environmental risk assessment of new chemicals. In fact, these models were expanded to incorporate the possibility of transportation (e.g., advection) and reaction (e.g., degradation) processes and thus introducing time-dependent $(mol \cdot h^{-1})$ rates. The European Union developed EUSES, a model intended to enable public and private users to carry out rapid assessments of the general risks posed by chemical substances.[4]

10.3 ENVIRONMENTAL DEGRADATION VERSUS PERSISTENCE

A key aspect for assessing the environmental relevance of a pollutant and trying to predict its level under a certain scenario is its **environmental persistence**. Concerns prompted by persistent chemicals were paramount in the rise of the environmentalist movement in the 1960s. Persistence is the first trait usually evaluated for a chemical to be included in a priority list. We should bear in mind, though, that continuous input of nonpersistent pollutants used in everyday activities yield a similar scenario in receiving waters than discrete inputs of persistent chemicals.

Environmental persistence in a given compartment is assessed by the environmental half-life

The environmental persistence is commonly assessed by the **half-life ($t_{1/2}$)**, or time needed for a decay of 50%[c]. However, unlike the $t_{1/2}$ of radioactive isotopes, whose decay rate is independent of environmental conditions, the $t_{1/2}$ of a chemical is strongly dependent on temperature, light, oxygen, pH, or the amount and composition of the community of biodegrading microorganism. Therefore, degradation rates are quite different in the different abiotic environmental compartments depicted in Fig. 10.1. This is because both the abiotic and biotic degradation reactions are strongly dependent on environmental factors such as temperature, light, dissolved oxygen, redox potential (in sediments), nutrients for the decomposers, composition, and abundance of autochthonous microbiota, etc. Thus, the $t_{1/2}$ value is always associated to a given environmental compartment, for example the water column or the sediment, and standard methods are available to set fixed values to the main factors affecting degradation rates.

To understand this variability, we must pay attention first to the more relevant chemical transformations that an organic chemical may undergo in the environment. Degradation processes can be first classified in abiotic and biotic, the latter mediated by microorganisms and termed **biodegradation**. Abiotic degradation processes can be further classified in **photodegradation**, **hydrolysis** reactions, and redox, normally **oxidation** reactions.

Abiotic degradation of organics may be caused by exposure to light, or chemical reactions such as hydrolysis and redox

---

[c] Considering that environmental degradation frequently follows negative exponential kinetics, a substantial amount of the chemical remains after two half-lifes (25% in case of simple first order kinetics). Within the field of the study of pesticides, the term environmental persistence is sometimes defined as the time needed for the biological activity of a chemical to decrease by 95%. This term is not commonly used in aquatic studies.

10.3.1   Photodegradation

Aromatic rings and other molecular structures may be sensitive to light, particularly to the most energetic wave-length lower end of the solar spectrum, and are readily oxidized upon light exposure. The $t_{1/2}$ of the TCDD in surface water, for example, is only 6 days. Light may be absorbed by the pollutant itself (direct photolysis), or by a second chemical that subsequently reacts with the pollutant (sensitized photolysis); for example, excitation of oxygen to yield the highly reactive oxygen singlet.[5]

Due to the low penetration of UV light in the water column this mechanism is only relevant in very shallow waters, under very low turbidity and high irradiance conditions. Laboratory experiments using germicide lamps with wavelength emissions that are mostly filtered by the atmosphere are not suitable for the study of environmental scenarios.

10.3.2   Hydrolysis

The reaction of water with a substrate resulting in the cleavage of a covalent bond in the substrate and formation of a new covalent bond with oxygen is termed hydrolysis. Hydrolizable functional groups include halogenated (X) aliphatic (R_1-CH_2-X → R_1-CH_2-OH + XH), esters (R_1-O-R_2 → R_1-O^- + R_2-O^-), amides (R_1-CO-NH-R_2 → R_1-COO^- + R_2-NH_2), carbamates, ureas, and epoxides. Hydrolysis rate is strongly affected by the pH. Each hydrolysable functional group has a particular range of pH values within which it is more susceptible to nucleophilic attack by OH^- or H_2O.[6]

10.3.3   Redox Reactions

The oxidation-reduction or redox reactions involve the transfer of electrons between an electron donor (that becomes oxidized) and an electron acceptor (that becomes reduced). Oxidizing agents possess a strong affinity for electrons while reducing agents readily give them up.

Thus, a redox reaction may be decomposed in two half reactions, one for the oxidation of the electron donor and one for the reduction of the electron acceptor. The oxidation of a substance follows a second-order kinetics dependent not only on the substrate concentration but also on the oxidizing agent concentration. The same holds for the reductions with regard to the reducing agent concentrations. The difficulty in predicting redox reaction rates in natural environments is the identification of the naturally occurring oxidizing (or reducing) agent responsible for the oxidation (or reduction) of the chemical of interest.

Although oxygen is the most abundant oxidizing agent in the water column, metal ions, and mineral oxides containing Fe(III), Mn(III), and Mn(IV) play a more significant role as oxidation catalysts in aquatic environments. Less is known about abiotic reducing agents, although surface bound and structural (but not dissolved) $Fe(II)$ is assumed to be a relevant chemical reductant. Iron-bearing mineral oxides have been shown to effect the reductive dechlorination of halogenated aliphatics.

10.3.4 Biological Degradation

An ensemble of **heterotrophic** microorganisms (mainly **bacteria and fungi**) use the energy contained in the covalent bonds of reduced organic molecules as energy source. Unlike biotransformation, this process causes fundamental changes in the structure of the molecule and may lead to the assimilation of the intermediate metabolite by the microbiota or to the ultimate oxidation of the organic chemical to CO_2, according to the following general steps: alkane $\rightarrow$ alcohol $\rightarrow$ aldehyde $\rightarrow$ fatty acid.... (β-oxidation)...$\rightarrow CO_2$.

The molecular structure determines the biodegradation rate. Linear alkanes are readily biodegraded, with half-lives increasing as molecular weight increases. Chain length, branches, and rings increasingly rend the molecule less biodegradable. Aromatic rings are thus the most recalcitrant hydrocarbons in the environment. Halogen, nitro, and sulfonate substituents further inhibit biodegradation. In halogenated hydrocarbons increasing number of halogen substituents decreases aerobic biodegradation rates.

The most active biodegrading microorganisms are aerobic and their proliferation is limited by their high O_2 requirements and the slow diffusion of O_2 in water (for oil biodegradation see Section 7.4). Typically, degradation rates are orders of magnitude slower in anoxic, fine-grained sediments than in the water column. Substrate levels also influence biodegradation rates. In the event of an organic spillage, the population of decomposers undergoes a lag phase before exponential growth is achieved. If the chemical in question is difficult to degrade autochthonous bacteria may not be suitable for decomposing it, and active bioremediation may be necessary. Below a certain threshold concentration of substrate rates of biodegradation are extremely low, perhaps also because the enzymes involved are not induced to activity.

Degradative aerobic pathways are different in eukaryotes and prokaryotes (Fig. 10.3). In both cases degradation is initiated with an oxidation that demands molecular oxygen. This first step limits the oxidative biodegradation rate. However, while in fungi and other eukaryotes this oxidation is catalyzed by a **monooxygenase** that incorporates a single atom of the oxygen molecule into the hydrocarbon to form an epoxide, readily hydrolyzed to two hydroxyl groups, in bacteria it is catalyzed by a **dioxygenase** that incorporates both oxygen atoms to directly produce the two hydroxyl radicals. The resulting dihydroxylated molecule is termed dihydrodiol, and the isomeric configuration is different for monooxygenases (trans-dihydrodiol) than for dioxygenases (cis-dihydrodiol). Also, while in procaryotes the degradative pathway continues with ring fission and assimilation or complete mineralization of the degradation products, in algae and fungi the hydroxylations lead to

Rates of microbial degradation depend on oxygen levels and the structure of the molecule

Biodegradation in aerobic environments is mediated by dioxygenases in prokaryota and monooxygenases in eukaryota

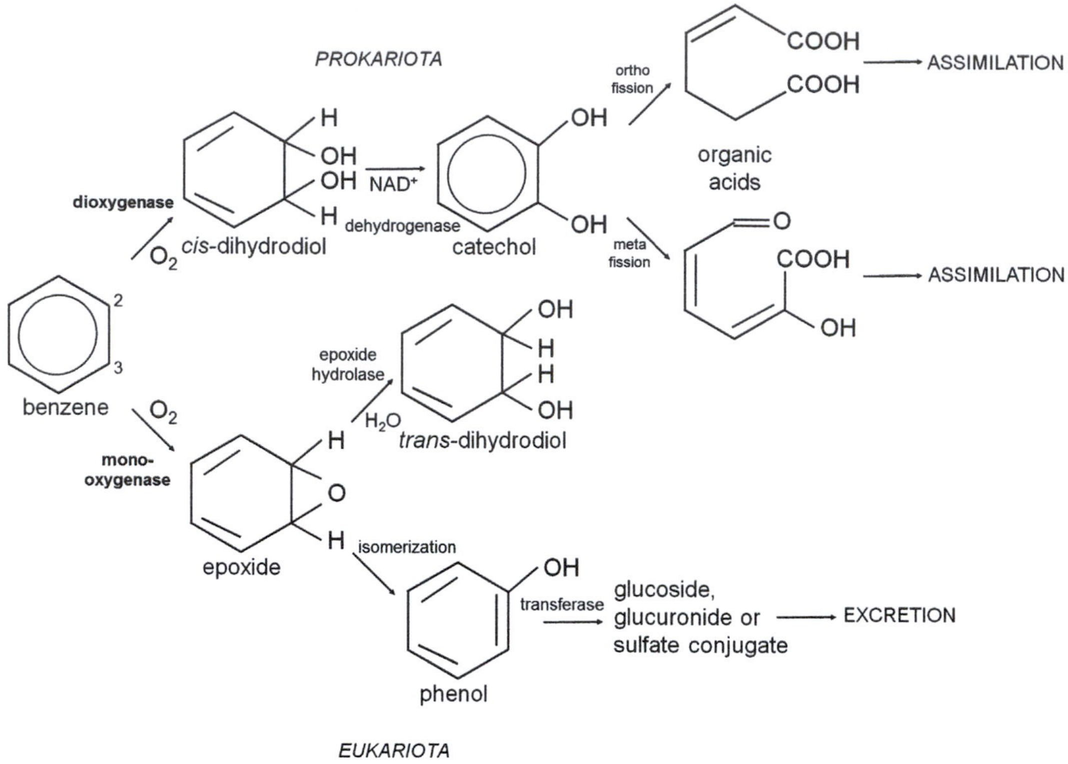

FIGURE 10.3

Biodegradation or aromatic hydrocarbons mediated by aerobic microorganisms differ in prokariota, where it is catalyzed by dioxygenases, and eukariota such as fungi, where the same role is played by the cytocrhome P450 monooxygenases. In the first case the biodegradation leads to ring cleavage and assimilation of the resulting organic acids by the bacteria, since in the second case the biodegradation forms phenolic intermediates that are conjugated to polar metabolites and eventually excreted. *Modified from Cerniglia CE, Heitkamp MA. Chapter 2. Microbial degradation of polycyclic aromatic hydrocarbons (PAH) in the aquatic environment. In: Varanasi U, editors. Metabolism of polycyclic aromatic hydrocarbons in the aquatic environment. Boca Raton: CRC Press; 2000. pp. 41–68.*

conjugation and excretion of the resulting metabolites (See Section 12.2), with no cleavage of the aromatic ring.

Particularly recalcitrant is the covalent bond between a C atom from a benzene ring and a halogen. Thus, number of halogen atoms in both aliphatics and aromatics increases persistence. Positions of the substitutions also affect biodegradability. For example, PCBs are oxidatively degraded by the dioxygenases of certain aerobic bacteria that mediate the 2-3- and 3-4-dioxygenation and the eventual cleavage of the ring in ortho or metha positions, respectively, provided there were no chlorine atoms in those positions.[8] This route is described in Fig. 10.4.

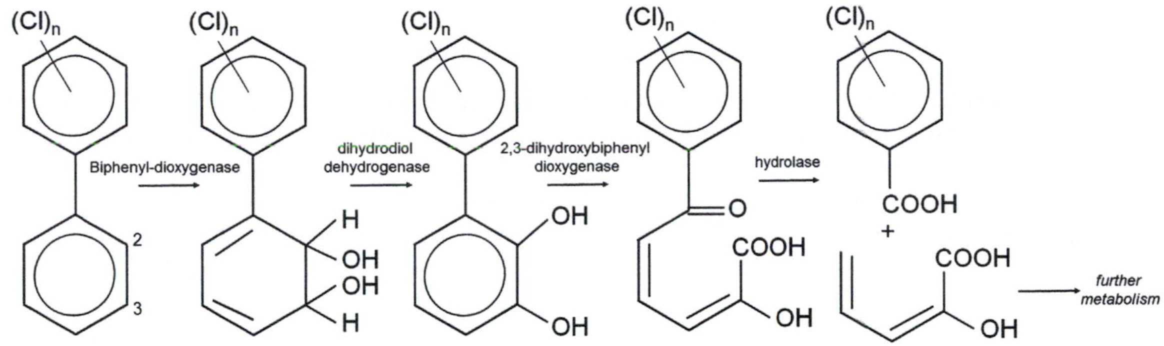

FIGURE 10.4

PCB degradation by aerobic heterotrophic bacteria. The biphenyl dioxygenase catalyzes the initial oxidation in the presence of molecular oxygen to insert an oxygen atom in carbons 2 and 3 of the less chlorinated ring and yield a 2,3-dihydrodiol, next dehydrogenated by the dihydrodiol dehydrogenase. The ring is cleaved by the 2,3-dihydroxybiphenyl dioxygenase (meta-cleavage) and the resulting metabolite is hydrolyzed to yield a chlorobenzoic acid as end-product and an organic acid that may be further metabolized. *Modified from Furukawa (2000).*

Anaerobic degradation routes, typically occurring in fine-grain sediments, are much slower than the aerobic metabolism. However, these routes may lead to complete mineralization of chlorinated aliphatics of environmental concern, such as trichloroethylene, cometabolized by facultative or strict anaerobes, such as methanogenic bacteria.[9] An instance of anaerobic degradation, the reductive dechlorination of halogenated hydrocarbons, plays a key role in the elimination of PCB and organochlorine insecticides from the environment. In PCBs, dechlorination preferentially removes chlorines from meta- and para-positions, but it is not effective in ortho position.[10] This mechanism explains the dominance of mono and dichlorinated PCB congeners in anaerobic sediments.[11] The process bears very relevant ecotoxicological implications since PCBs with lower levels chlorination are less bioaccumulative and, on the other hand, the resulting enrichment in ortho congeners reduces the proportion of dioxin-like (nonortho) PCBs in the mixtures.[12]

> Reductive dechlorination plays a key role in the biodegradation of organochlorines and reduces the toxicity of PCB in anoxic sediments

10.4 CHEMICAL SPECIATION AND BIOAVAILABILITY

Atoms and molecules may show different oxidation states and electric charges, and may bind by means of interactions of very variable strength to other molecules and particles of very different size and charge. Each of these forms, termed chemical species, may show a distinct interaction with living organisms in terms of uptake, accumulation, and toxicity.

> Chemical speciation of pollutants affects interactions with organisms

In aquatic studies, standard procedures make the difference between dissolved and particulate fractions, although colloids challenge this arbitrary convention. In aqueous media a metal atom may be free as a cation, bound to inorganic (chlorides, hydroxides, carbonates …) or organic (aminoacids, organic acids, humic substances …) ligands, adsorbed to the surface of mineral or organic particles, or covalently bound in an organometallic molecule.

Bioavailability may be defined as the ratio between the dose at the site of action and the exposure dose

Environmental toxicology uses the term bioavailable to refer to the fraction of the substance present in chemical species with biological activity, either suitable for passing through a biological membrane or capable of any other biochemical interaction leading to a relevant effect for the organism. The concept has been defined in several different ways. At the light of the basic principles of toxicology (Section 13.1) the bioavailable fraction can be very strictly defined as the ratio between the dose at site of action divided by the exposure dose. However, quantification of that fraction is not straightforward because bioavailability is a relative term applicable only within the context of a given chemical substance present in a defined stable medium, and related to a specific biological interaction (uptake, distribution, toxicity). The same chemical species may not be bioavailable for a certain biochemical pathway but bioavailable for another. For example, certain particles may not pass through biological membranes but be available upon ingestion through the digestive system, and structural features unique for certain taxa, such as cell walls in diatoms, may impose specific limitations to the uptake of some metal species not relevant for naked cells. Also, changes in the environmental conditions (salinity, pH, redox potential, dissolved organic matter, etc.) may change the bioavailable fraction of a substance.

Finally, living organisms possess mechanisms to sequester or compartmentalize potentially toxic chemicals, and the internalized dose may depart largely from the dose at the site of action. Bioavailability is thus a quantitative property dependent on both environmental and biological conditions.

Dissolved metal bioavailability can be predicted by the Free Ion Activity Model

Bioavailability has been particularly well studied for trace metals. It follows that the total amount of a metal present in a given medium does not inform us on its biological effects. In addition, we need to know its chemical speciation, which will be dependent on environmental factors.

According to the Debye–Hückel theory, the chemical reactivity of a metal cation, or any other electrolyte dissolved in an aqueous medium, depends on its activity, i.e., its concentration corrected for the shielding effect of anions electrostatically attracted to the metal, which increases with the ionic strength of the medium. Morel[13] extended this thermodynamic model to explain the biological activity of metal ions as an exclusive function of the chemical activity of the free ion present, disregarding the other chemical species. This is termed the Free Ion Activity Model (FIAM). The fundamental basis of FIAM is that the biological effect takes place through the interaction of the metal with a cellular membrane receptor to which only the free ion can bind, and thus the amount of effect is proportional to the free ion only, disregarding the remaining chemical species. The underlying assumptions, reviewed in detail by Campbell,[14] are that the system is in equilibrium, diffusion of the ion from the bulk solution to the site of action (i.e., the biological membrane) and the kinetics of the reaction

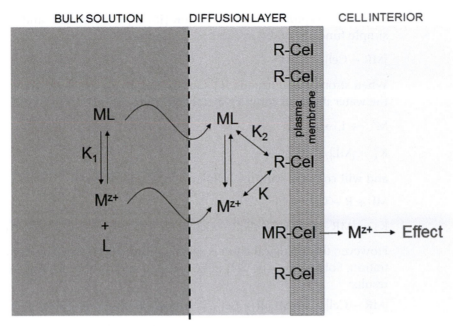

FIGURE 10.5

Schematic representation of the Free Ion Activity Model (FIAM) stating that the biological activity of a metal in solution is a simple function of the free ion concentration, disregarding the presence of inorganic and organic ligands (L) that compete with the receptor to bind the metal. The effect (toxicity, bioaccumulation, etc.) is mediated by the binding of the free ion (M^{Z+}) to a cellular receptor (R-Cel). Diffusion from the bulk solution to the cell membrane and number of cell receptors are assumed to be not limiting. *Modified from Campbell (1995) op. cit.*

between ion and ligand are not limiting, and binding sites for the ion in the biological membrane are in excess.

The FIAM was extended afterwards following the work by Playle[15] and Di Toro et al.[16] into the Biotic Ligand Model (BLM), which explicitly includes the competition of metals for the cellular receptor—now called biotic ligand—with protons and other cations. This is especially useful in freshwaters, where pH and water hardness may experience sharp variations influencing bioavailability.

As illustrated in Fig 10.5, a given biological effect of a dissolved metal (M^{Z+}) on an organism will be a consequence of the binding of M^{Z+} with a receptor in the cellular membrane R-Cel determined by the equation[18]:

$$M^{Z+} + R - Cel \rightarrow MR - Cel$$

with a stability constant:

$$K = [MR - Cel]/[M^{Z+}][R - Cel]$$

from which the effect can be quantified as a function of the complex formed:

$$[MR - Cel] = K[M^{Z+}][R - Cel].$$

Since the receptor is in excess, then K [R-Cel] is a constant, and the effect is a simple function of the free ion concentration:

$$[MR - Cel] = a \, [M^{Z+}] \tag{Eq. 10.9}$$

When strong metal ligands (L) such as dissolved organic matter are present in the water they will complex metal according to

$$M^{Z+} + L \rightarrow ML$$

$$K_1 = [ML]/[M^{Z+}][L] \tag{Eq. 10.10}$$

and will compete with the cellular receptor according to

$$ML + R - Cel \rightarrow MR - Cel + L$$

$$K_2 = [MR - Cel][L] \, /[ML][R - Cel] \tag{Eq. 10.11}$$

However, the biological effect is again a simple function of the free ion concentration. Solving for [MR-Cel] in Eq. (10.11) and combining with Eq. (10.10) it results:

$$[MR - Cel] = K_2[ML][R - Cel]/[L] = K_2K_1[R - Cel][M^{Z+}]$$

i.e., analogously to Eq. (10.9) again the biological effect is a simple function of the free ion concentration

$$[MR - Cel] = a'[M^{Z+}] \tag{Eq. 10.12}$$

Since standard analytical methods measure total metal (M_T) but the biological effect is expected to be a function of the free ion only, we may be interested in knowing, given a total amount of metal, the fraction corresponding to M^{Z+}, which will decrease as the ligand concentration increases. The simplest scenario may consider a single type of ligand, L, with a reaction stoichiometry of 1:1. Under this conditions it can be demonstrated[19] that

$$[M^{Z+}] = \left(- a + \left(a^2 + 4M_T/K_1\right)^{1/2}\right)\Big/2 \tag{Eq. 10.13}$$

where a= $(-[M_T] + [L] + 1/K_1)$

Using Cu as model metal, and humic acids (HA) with a given Cu-complexing capacity (N, μmol Cu/g HA) as model ligand, and considering the relation [L] = [HA]N, then Eq. (10.13) predicts a decrease in the percentage of free ion, Cu^{2+}, as a function of the dissolved HA concentration (mg/L) of the form depicted in Fig. 10.6.

This model was validated by electrochemical measurements of labile copper (Cu′), since in a stable, chemically defined, well buffered medium as seawater Cu^{2+} is a constant fraction of Cu′.

In real environmental scenarios the situation is much more complex, but computer models such as MINEQL+[21] or Visual MINTEQ[22] are available to predict, on the basis of the thermodynamic equilibrium constants, the free ion activity

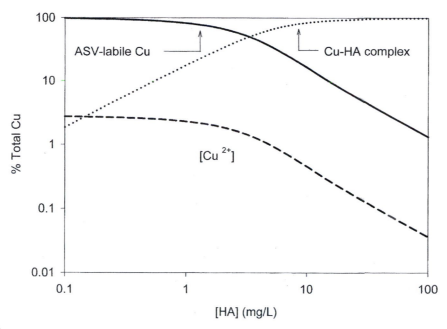

FIGURE 10.6

Theoretical complexation model for 1 mM Cu and humic acids (HA) representing the decrease in anodic stripping voltammetry (ASV)—labile Cu (Cu', solid line) and increase in Cu-AH complexes (dotted line) as HA concentration increases). Theoretical free-Cu concentrations (dashed line) are also shown. *From Lorenzo et al. (2005) op. cit.*

as a function of the concentration of the different ligands present and the relevant environmental parameters, such as salinity and pH.

The ecological relevance of chemical speciation of metals can be illustrated by the experimental results of J. I. Lorenzo and coworkers who exposed the sensitive early life stages of sea urchins to different combinations of Cu and humic substances (in this case fulvic acids). As shown in Fig. 10.7A, larval growth at total Cu concentrations around 1 µM varies from normal to virtually null depending on the amount of complexing organic matter present. Total Cu is measured after elimination by strong mineralization of the sample's organic matter, including the humic substances, and does not take into account chemical speciation. In contrast, when the biological response is plotted against the labile copper (Cu'), i.e., the fraction of metal not bound to strong ligands (Fig. 10.7B), the observed biological response fits well to a conventional logistic function of metal concentration. Cu' can be measured by electrochemical techniques and include the free ion (Cu^{2+}) plus inorganic complexes with weak ligands.

Electrochemically measured labile Cu in seawater predicts Cu toxicity

The issues on chemical speciation for the water column are applicable to the interstitial or pore water surrounding sediment particles. Thus sediment toxicity can be better predicted from the study of the pore water concentration of chemicals, after bioavailability issues are taken into account, than from bulk

Metal bioavailability in sediments is mainly determined by sulfides

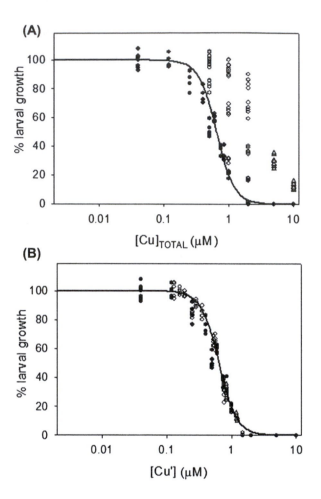

FIGURE 10.7

Effect of fulvic acid (FA) additions on Cu toxicity to sea urchin larvae. Black dots represent treatments with Cu only, and open dots represent treatments with various amounts of FA. Triangles correspond to treatments with extremely high Cu concentrations (5 and 10 μM) in the presence of high FA content. Larval growth is depicted versus total (A) and labile (B) Cu concentration. The solid line represents the logistic toxicity curve for the Cu-alone experiments. *From Lorenzo et al. (2006) op. cit.*

sediment analysis.[24] However, in contrast with water, sediments are chemically complex matrices where some of the major variables affecting speciation, such as pH and redox potential, show strong variations with depth within the sediment, and also with manipulation and storage of samples, which greatly complicates their study.

In fine-grain sediments with low redox values trace metal bioavailability is affected by the presence of sulfide ions (S^{2-}) that readily form insoluble sulfides removing the trace metals from the sediment interstitial water, and thus markedly reducing their bioavailability. These S^{2-} ions originated from decaying organic matter may be in the form of labile sulfides such as FeS and MnS. The speciation will thus depend on metal displacement reactions determined by the differential affinity of the metals for the S^{2-} according to the following order: $Mn^{2+} < Fe^{2+} << Ni^{2+} < Zn^{2+} << Cd^{2+} < Pb^{2+} << Cu^{2+} << Hg^{2+}$. Unfortunately, Hg, the metal with the strongest affinity for S^{2-}, is rarely analyzed in sediment speciation studies.

The labile Mn^{2+} and Fe^{2+} sulfides are termed acid-volatile sulfides (AVS) because they can be measured by cold acid extraction. In this situation the SEM/AVS stoichiometric proportion, where SEM is the concentration of simultaneously extracted metals using the same acid extraction than for AVS measurement, should provide a more accurate estimation of potential metal bioaccumulation and toxicity than the total metal concentration, i.e., the proportion of SEM moles exceeding AVS moles should estimate the amount of bioavailable metal. The model has received support from laboratory experiments spiking sediments with Cd and Ni ,[25] but in more complex situations toxicity predicted by the SEM/AVS ratio failed to outperform toxicity predictions conducted at the light of the classical sediment quality guidelines based on total dry weight normalized metal concentrations.[26]

KEY IDEAS

- Heterotrophic bacteria and fungi are capable of degrading most organic pollutants, but limited by oxygen availability and presence of the adequate inoculum. Branches, rings, and halogen substituents slow down biodegradation rates.

- Microbial degradation of hydrocarbons follow different pathways in prokaryota, where it is catalyzed by **dioxygenases** using both atoms of the O_2 molecule, than in eukaryota, where this role is played by **monoxygenases**. For aromatics, only dioxygenases lead to ring fission and assimilation or complete mineralization of the molecule.

- Organochlorine pesticides and PCBs may be biodegraded through anaerobic bacteria involving **reductive dechlorination**. In the case of PCBs these bacteria use tri-, tetra-, and penta-chloro congeners and yield less bioaccumulative and less toxic mono and di-chloro congeners. These less chlorinated PCBs may be further biodegraded in aerobic environments by bacterial dioxygenases.

- Chemical pollutants are present in different **chemical forms** (species) with very different ability to pass across biological membranes, and thus chemical **speciation** determines **bioavailability**.

- **Bioavailability** may be defined as the ratio between the dose at the site of action and the exposure dose. In practice, bioavailability is a relative term that depends on both environmental and biological conditions, applicable only within the context of a defined medium, and related to a specific biological interaction (uptake, distribution, toxicity).

■ For metals in aqueous environments of known chemical properties (pH, salinity, dissolved organic matter), bioaccumulation and toxicity are a function of the **free metal ion** concentration. This concentration may be estimated from electrochemical measurements.

■ Metal bioavailability in sediments is mainly determined by sulfides.

Endnotes

1. For chemical pollutants see: Wurl O, Obbard JP. A review of pollutants in the sea-surface microlayer (SML): a unique habitat for marine organisms. Mar Pollut Bull 2004;48: 1016−1030. For microplastics see: Song YK, Hong SH, Jang M, et al. Large accumulation of micro-sized synthetic polymer particles in the sea surface microlayer. Environ Sci Technol 2014;48:9014−9021.

2. OSPAR. Agreement on CEMP assessment criteria for the QSR 2010; 2009.

3. Mackay D. Multimedia environmental models. 2nd ed. Boca Raton: CRC Press; 2001.

4. https://ec.europa.eu/jrc/en/scientific-tool/european-union-system-evaluation-substances.

5. Schüürmann G, Markert B. editors.Ecotoxicology. Ecological fundamentals, chemical exposure, and biological effects. New York: Wiley; 1998. (p. 306).

6. Schüürmann & Markert (eds.) op. cit. (p. 262).

7. Cerniglia CE, Heitkamp MA. Chapter 2. Microbial degradation of polycyclic aromatic hydrocarbons (PAH) in the aquatic environment. In: Varanasi U, editors. Metabolism of polycyclic aromatic hydrocarbons in the aquatic environment. Boca Raton: CRC Press; 2000. pp. 41−68.

8. Furukawa K, Fujihara H. Microbial degradation of polychlorinated biphenyls: biochemical and molecular features. J Biosci Bioeng 2008;105(5):433−449.

9. Chaudhry GR, Chapalamadugu S. Biodegradation of halogenated organic compounds. Microbiol Rev 1991;55(1):59−79.

10. Rhee GT, Cho YC, Ostvofiky EB. Microbial degradation of PCBs in contaminated sediments. Remediation J. 1999;10(1):69−82.

11. Furukawa & Fujihara (2008) op. cit.

12. Abramowicz DA. Aerobic and anaerobic PCB biodegradation in the environment. Environ Health Perspect 1995;103(Suppl 5):97−99.

13. Morel FMM. Principles of aquatic chemistry. New York: Wiley−Interscience; 1983. p. 301.

14. Campbell PGC. Interactions between trace metals and aquatic organisms, a critique of the free-ion activity model. In: Tessier A, Turner, DR, Editors. Metal speciation and bioavailability in aquatic systems. IUPAC series on analytical and physical chemistry of environmental systems. Chichester: J Wiley & Sons; 1995. pp. 45−102.

15. Playle RC. Modeling metal interactions at fish gills. Sci Total Environ 1998;219:147−163.

16. Di Toro D, Allen HE, Bergman HL, Meyer JS, Paquin R, Santore RC. Biotic ligand model of the acute toxicity of metals. 1. Technical basis. Environ Toxicol Chem 2001;20:2383−2396.

17. Campbell (1995) op. cit.

18. Campbell (1995) op. cit.

19. Lorenzo JI, Nieto O, Beiras R. Effect of humic acids on speciation and toxicity of copper to *Paracentrotus lividus* larvae in seawater. Aquat Toxicol 2002;58:27–41.

20. Lorenzo JI, Beiras R, Mubiana VK. et al. Copper uptake by *Mytilus edulis* in the presence of humic acids. Environ Toxicol Chem 2005;24(4):973–980.

21. http://www.mineql.com/.

22. http://vminteq.lwr.kth.se/.

23. Lorenzo JI, Nieto O, Beiras R. Anodic stripping voltammetry measures copper bioavailability for sea urchin larvae in the presence of fulvic acids. Environm Toxicol Chem 2006;25(1): 36–44.

24. See: Ankley GT, Di Toro DM, Hansen DJ. et al. Assessing the ecological risk of metals in sediments. Environ Toxicol Chem 1996;15(12):2053–2055; and cites therein.

25. DiToro DM, Mahony JD, Hansen DJ. et al. Toxicity of Cd in sediments: the role of acid-volatile sulfide. Environ Toxicol Chem 1990;9:1487–1502. DiToro DM, Mahony JD, Hansen DJ. et al. Acid volatile sulfide predicts the acute toxicity of cadmium and nickel in sediments. Environ Science Technol 1992;26:96–101.

26. Long ER, MacDonald DD, Cubbage JC. et al. Predicting the toxicity of sediment-associated trace metals with simultaneously extracted trace metal: acid-volatile sulfide concentrations and dry-weight normalized concentrations: a critical comparison. Environ Toxicol Chem 1998;17(5):972–974. Data for not spiked sediments from DiToro et al. (1992) op. cit.

Suggested Further Reading

- Mackay D. Multimedia environmental models. 2nd ed. Boca Raton: CRC Press; 2001.
- Fiedler H, Lau C. 10. Transformation of chlorinated xenobiotics in the environment. In: Schüürmann G, Markert B, editors. Ecotoxicology. Ecological fundamentals, chemical exposure, and biological effects. New York: Wiley; 1998. p. 283–316.
- Metal speciation and bioavailability in aquatic systems. In: Tessier A, Turner DR, editors. IUPAC Series on Analytical and Physical Chemistry of Environmental Systems, vol. 3. Chichester: John Wiley & Sons; 1995.

CHAPTER 11

Bioaccumulation

11.1 UPTAKE AND FATE OF TOXIC POLLUTANTS IN THE ORGANISMS: TOXICOKINETICS

Aquatic organisms are exposed to chemical pollutants by two different pathways: pollutants dissolved in the water (water column or interstitial water) can be taken up through the gills and body surface, while pollutants associated to particles or live food can be ingested and assimilated in the digestive system. In laboratory experiments, when only waterborne uptake is considered, the ratio between the concentration in the organism tissues (C_t) and that in the surrounding water (C_w) at steady state is termed **bioconcentration factor (BCF)**,

$$BCF = C_t/C_w \qquad \text{(Eq. 11.1)}$$

When additional uptake via food[1] or sediment particles[2] is considered, the same ratio is termed **bioaccumulation factor (BAF)**. The difference between dissolved (or waterborne) and particulate (potentially ingested) fractions may be subtle, and for ecological relevance and comparative value with field studies both pathways should be considered. Many dissolved chemicals tend to adsorb to organic or inorganic particles, living or inert, moving from the dissolved to the particulate fraction through complex equilibriums dependent on environmental factors. In common pollution monitoring studies conducted in the field with no focus on the route of uptake both terms (BCF and BAF) are indistinctly used.

Finally, we may be interested in assessing the specific weight of the uptake of a given chemical via food. This may be relevant, for example, in regulations concerning secondary poisoning intended to protect natural predators and human health. In this case we can quantify the **trophic transfer** of the chemical from the food (or prey) to the consumer (or predator). The ratio between the concentrations in the predator tissues and that in the pray is sometimes termed biomagnification factor,[3] but a more specific term, such as **trophic transfer factor (TTF)**,[4] is more advisable in order to make the difference with the actual concept of biomagnification, which is a food web phenomenon that should not be applied to single predator-prey interactions[5] (See Sections 9.7 and 11.4).

Aquatic organisms take chemical pollutants from both the water and the food

Marine Pollution. https://doi.org/10.1016/B978-0-12-813736-9.00011-8

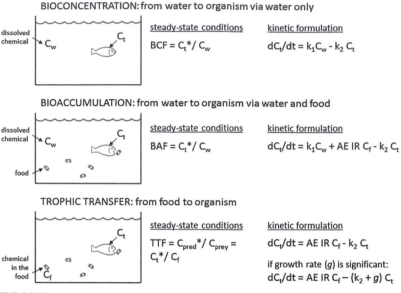

FIGURE 11.1

Experimental approaches to study the tendency of a chemical to accumulate in the tissues of a fish. The uptake may occur via water only, via water plus food, or via food only. The corresponding parameters bioconcentration factor (BCF), bioaccumulation factor (BAF), and trophic transfer factor (TTF) are sometimes confused in the technical literature. The asterisk in C_t^* denotes that steady-state must have been reached. For initials, see text.

Knowing the accumulation potential of chemicals in aquatic organisms is a key aspect in the identification of priority pollutants and the implementation of protective water quality standards. The different experimental approaches used, and the corresponding concepts of bioconcentration, bioaccumulation, and trophic transfer, are explained in Fig. 11.1.

The data obtained from those experiments can be fitted to either kinetic models, where accumulation is expressed as a function of time (See Section 11.2), or partition equilibrium models (See Section 11.3) that assume thermodynamic equilibrium or at least steady-state conditions have been reached.

Dissolved pollutants are taken through the gills and particulate pollutants via digestive system

We can now pay attention to the different processes that affect the uptake and distribution of the chemical among the fish tissues. Those processes, together with the metabolic transformation and elimination mechanisms are termed toxicokinetics. Dissolved and particulate pollutants are taken up following different pathways, through the gills the first and through the digestive system the latter (Fig. 11.2). For waterborne pollutants, only the bioavailable fraction is taken up and distributed by the organism. Regarding ingested particulate

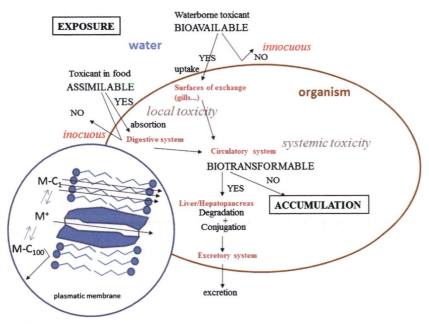

FIGURE 11.2

Pathways of exposure and fate of contaminants in an aquatic organism leading to their potential bio-accumulation. Insert: the bioavailability of a molecule (M) is greatly affected by its chemical speciation. For explanation, see text.

matter, upon digestion only the assimilable fraction is absorbed in the intestine, taken up by the enterocytes, the epithelial cells of the gut's lumen. The rest is egested as feces and hence innocuous for the organism.

Disregarding the pathway, gills body surface, or gut epithelium, to be taken up the substance must move across a biological membrane (see Fig 11.2, insert) that acts as a barrier to the diffusion of water-soluble molecules. The ability of a molecule to cross plasmatic membranes is mainly determined by its lipophilicity, size, and ionization. Low molecular weight nonpolar molecules (M-C_1 in the figure) can readily diffuse across the lipid bilayer. Although only nonpolar chemicals can move straight across the lipid bilayer, organisms have evolved to take up essential ions through a variety of mechanisms that can also be used by xenobiotics. Metal binding proteins such as carriers and ionophores are capable of surrounding ions and carrying them across membranes either by diffusion in the lipid bilayer or by interaction with membrane receptors, formation of vesicles, and endocytosis.[6] Thus, charged molecules and ions (M^+ in the figure) can pass through transmembrane protein channels with (active transport) or without (facilitated transport) energy expenditure.

To enter the body chemicals must move across biological membranes

Therefore, as discussed in Section 10.4, the chemical speciation of a substance will greatly affect its degree of bioavailability. For example, organometallic species of Hg such as MeHg can diffuse across the plasmatic membrane, while the divalent cation can only penetrate through ion exchange channels, and the Hg complexed with humic substances cannot use either of those pathways. MeHg and mercury cations are termed bioavailable species, while the later are regarded as not bioavailable.

Although gill and gut cell biological membranes are similar, due to the different physic-chemical conditions in sea water and gastrointestinal tract, the assimilable fraction is likely to be different than the bioavailable fraction. For example, P. Sánchez-Marín and coworkers[7] showed that humic-bound metal unavailable for gill uptake plays a role in metal bioaccumulation by filter feeders due to potential uptake via digestive system.

Bioconcentration will depend on the balance between uptake and elimination

After absorption either through the gills, body surface, or digestive tract, the chemicals are diluted in the blood and distributed to the whole organism through the circulatory system. In vertebrates the portal vein supplies blood into the liver, where most biotransformation processes take place. In invertebrates the equivalent role is played by the hepatopancreas or digestive gland.

The plasma concentration of the chemical is very important since it often directly relates to the concentration at the site of action, and according to the basic principles of toxicology (see Section 13.1) the biological effects of a chemical are directly related to that dose at the site of action. The relationship between the exposure concentration (normally the environmental concentration) and the concentration at the site of action is determined by the toxicokinetics processes; including uptake, distribution, and elimination (see Fig. 13.1). The balance between uptake and elimination, once a steady state is reached, will determine the BCF.

11.2   MODELING BIOACCUMULATION AND CALCULATING BIOCONCENTRATION FACTORS, I. KINETIC MODELS

Uptake and elimination processes frequently follow first-order kinetics

Most uptake and elimination processes follow first-order kinetics, meaning that uptake (or elimination) rate is directly related to the concentration (C) of the substance in the source compartment: $dC/dt = kC$, where k is the uptake (or elimination) rate constant, with t^{-1} units. First-order kinetics is typical of passive diffusion or protein binding processes, for example. In contrast, enzymatic activities at saturated substrate concentrations or active transport at limiting energy expenditure are examples of zero order kinetics, where $dC/dt = k$, and uptake or elimination rates, are independent of the concentration.

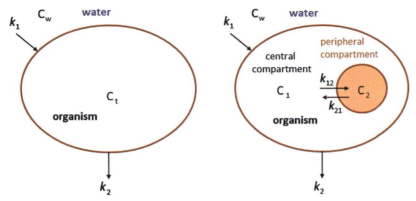

FIGURE 11.3
State variables and fluxes for one-compartment (left) and two-compartment (right) models of bioaccumulation applied to aquatic organisms exposed to waterborne contaminants.

Assuming first-order kinetics and considering first for simplicity the whole body as a single compartment (Fig. 11.3, left), the bioaccumulation of a water-borne substance in an aquatic organism can be modeled according to the expression:

$$dC_t/dt = k_1 C_w - k_2 C_t \qquad \text{(Eq. 11.2)}$$

where: C_t = tissular concentration, C_w = water concentration, k_1 = uptake rate constant, k_2 = excretion rate constant.

The units of the different terms in Eq. (11.2) deserve some attention; C_t: mass·mass^{-1}; C_w: mass·volume^{-1}; k_1: volume·mass^{-1}·time^{-1}, and k_2: time^{-1}.

Assuming C_w was constant, then we can integrate Eq. (11.2) to:

$$C_t = C_w k_1/k_2 \left(1 - e^{-k_2 t}\right) \qquad \text{(Eq. 11.3)}$$

For a substance with an initial tissular concentration of zero, Eqs. (11.2) and (11.3) describe a quick and almost linear accumulation ($k_1 C_w \gg k_2 C_t$) that decelerates as C_t increases and reaches a plateau when $k_1 C_w \approx k_2 C_t$, i.e., when C_t reaches such an amount that causes excretion to offset uptake. In mathematical terms, the asymptotic value of C_t corresponding to $t = \infty$ in Eq. (11.3) predicts a tissular concentration in equilibrium (C_t^*) that will be,

$$C_t^* = C_w(k_1/k_2) \qquad \text{(Eq. 11.4)}$$

Such a simple model describes surprisingly well the kinetics of accumulation of metals and other polar contaminants in aquatic organisms. Fig 11.4 shows the lead concentration in the amphipod *Hyalella azteca* when exposed to 480 nM of dissolved lead (uptake phase) and after depuration in clean water (elimination phase). Note the good fit of actual concentrations to a first-order kinetics model of bioaccumulation (A) and depuration (B).

Accumulation according to first-order kinetics predicts a quick initial increase in tissular concentration followed by a plateau

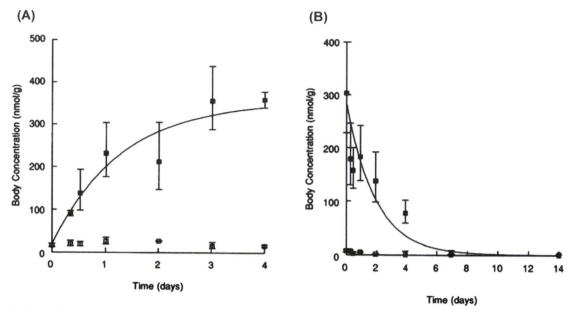

(A)   **(B)**

FIGURE 11.4

Concentration of lead in the tissues of the amphipod *Hyalella azteca* exposed to 480 nM of dissolved Pb for 4 days and further incubated in clean water for depuration during 14 days. The curves represent first-order kinetic models of uptake (Eq. 11.3) and elimination (Eq. 11.7). Empty symbols represent control animals. *Source: MacLean RS, Borgmann U, Dixon, DG. Bioaccumulation kinetics and toxicity of lead in Hyalella azteca (Crustacea, Amphipoda). Can J Fish Aquat Sci 1996;53:2212–2220.*

Short-term laboratory experiments with bivalves frequently reported metal accumulation rates that are a simple linear function of dissolved metal concentration,[9] as predicted by the first-order kinetic model. However, in most laboratory experiments using bivalves, the tissue concentration did not reach saturation even with exposure periods up to two months, preventing the calculation of BCF values.[10]

BCF is independent of water concentration, and thus we can monitor water pollution using bioaccumulator organisms

An interesting feature of Eq. (11.4) is that it predicts a bioconcentration factor independent of the concentration of the substance in the water. In effect, the BCF was defined in Section 11.1 as the ratio between the concentration of the pollutant inside the organism at steady state divided by the environmental concentration: $BCF = C_t^*/C_w$, and substituting C_t^* by its value from Eq. (11.4),

$$BCF = k_1/k_2 \tag{Eq. 11.5}$$

Thus, under the assumptions applied to model bioconcentration according to first-order kinetics, BCF is a function of the uptake and depuration rate constants only. These constants, k_1 and k_2, are characteristic of each organism and chemical, and will depend on physical factors that control the ability of the chemical to get across biological membranes (lipid solubility, molecular size, degree of ionization) and biological factors related to the ability of that particular organism to distribute and eliminate that particular chemical. Since for bioaccumulative chemicals the k_1/k_2 ratio has a positive value > 1, disregarding those chemical and biological factors a linear relation between C_t^* and C_w will hold. Indeed, combining Eqs. 11.4 and 11.5,

$$C_t^* = BCF \cdot C_w \qquad\qquad (Eq.\ 11.6)$$

Therefore, in the equilibrium the tissular concentration of the pollutant under study is k_1/k_2 times higher than its concentration in the water, disregarding the level of pollution. This is the rationale supporting the use of bioaccumulator organisms to monitor water quality. If BCF is independent of the environmental concentration, as predicted by the first-order kinetic model, then C_w can be estimated from the bioaccumulated concentration as $C_w = C_t^*/BCF$. Since for bioaccumulative pollutants C_t^* is orders of magnitude higher than C_w, and much more stable in time and space, it is easy to see the advantages of this approach for pollution monitoring.

For regulatory purposes and within the context of a priori risk-assessment studies, the BCF of a chemical can be experimentally determined under standardized laboratory conditions using model test species. For example, OECD (2012)[11] guidelines for the measurement of bioconcentration in fish recommend flow-through or semistatic exposure to a concentration below 1 percent of the acute LC_{50} for 28 days followed by depuration of 95% of the accumulated chemical, and admit a maximum variation in actual C_w of 20% and a maximum mortality of 10%. Table 11.1 shows fish BCF values compiled by the US Environmental Protection Agency (US-EPA). Notice that BCF is a ratio of concentrations, and thus dimensionless, but for simplicity tissue concentrations are commonly expressed as mg/Kg while water concentrations are expressed as mg/L, originating BCF values expressed in L/Kg. Care must also be taken about weight units that should be expressed on a wet weight basis (see Section 10.2).

The BCF is also used in regulations for chemical pollutants

Table 11.1 Bioconcentration Factors (BCF) of Organic Chemicals in Fish and Their Respective Octanol-Water Partition Coefficent (K_{ow}). Notice BCF Values are Expressed on a Wet Weight (WW) Basis

Chemical	BCF (L/Kg WW)	Log K_{ow}
Vinyl chloride	1.17	1.17
Chloroform	3.75	1.97
Benzene	5.2	2.12
Trichloroethylene	10.6	2.38
Aldrin	28	5.30
Fluorene	1300	4.20
Trichlorobenzene	2800	4.30
Chlordane	14,000	3.32
DDT	54,000	6.19
PCBs	100,000	6.04

Data from U.S. EPA. Superfund Public Health Evaluation Manual. Office of Emergency and Remedial response. Exhitit A-1. Washington DC; 1986.

These values allow estimating the pollutant concentrations reached in the organism or target organ under simulated exposure conditions by multiplying the BCF value (Table 11.1) by the water concentration predicted by environmental distribution models such as those discussed in the previous chapter.

Applications of BCF in environmental monitoring and risk assessment assume a linear relationship between environmental and tissular concentrations

The linearity between the water and tissue concentrations expressed in Eq. (11.6) has been experimentally proved in laboratory experiments (e.g., Zaroogian[13]), although important limitations must be taken into account. The tissue concentrations of essential trace elements, for example, can be regulated by many aquatic organisms (see Chapter 16). The BCF values calculated from field scenarios where water concentrations are generally much lower than those used in laboratory testing provide much higher values than those from laboratory exposures. Even when the analysis is restricted to the range of concentrations potentially toxic, a significant inverse relationship between BCF and C_w is still found.[14] Tissular compartmentalization, active biotransformation, and homeostatic control of bioaccumulation, particularly in the case of essential metals, may account for this lack of linearity. Also, the limitations derived from the assumptions made by the kinetic models (constant water concentration, equilibrium between water and organism, first-order kinetics) must also be taken into account when interpreting field monitoring data.

Metabolic half-life

Depuration experiments such as those depicted in Fig. 11.4B allow the calculation of another parameter of great interest in toxicology, the **metabolic half-life**, typical of each chemical, which should not be confounded with the environmental half-life previously discussed (Section 10.3). Given a certain concentration accumulated in an organism, the metabolic half-life ($t_{1/2}$) is the time required for the tissular concentration to drop to a half of the initial value.

Assuming again first-order kinetics, in clean water ($C_w = 0$) Eq. (11.2) simplifies to: $dC_t/dt = -k_2C_t$. This can be integrated to:

$$C_t = C_0e^{-k_2t} \qquad \text{(Eq. 11.7)}$$

Taking logarithms in Eq. (11.7):

$$\ln C_t = \ln C_0 - k_2t$$

At the time $t = t_{1/2}$ when $C_t = C_0/2$, then,

$$\text{Ln}(C_0/2) = \ln C_0 - \ln 2 = \ln C_0 - k_2t_{1/2}$$

and thus,

$$t_{1/2} = \ln 2/k_2 = 0.693/k_2$$

As expected, the metabolic half-life (in time units) is an inverse function of the excretion rate constant, and independent of C_t.

Any inducible biotransformation process activated above a certain threshold of C_t, or the mobilization of the chemical to a different biological compartment, would invalidate the assumed linearity between body burden and excretion rate, and thus could not be treated by this simple model.

Distribution of a chemical inside an organism is rarely homogeneous. In mussels, for example, metals are preferentially accumulated in the digestive gland, compared to the mantle.[15] More than one compartment must be considered when there is a widely different distribution between high perfusion (e.g., gills) and low perfusion (e.g., fat) tissues.[16] Therefore, a more realistic and only slightly more complex approach is obtained when the organism is divided into two interconnected compartments (Fig. 11.3 right): a **central** compartment (C_1) that exchanges the chemical with the environment, and a **peripheral** compartment (C_2) not connected with the outside with a higher affinity for the chemical (liver, fat tissue…). Being k_{12} and k_{21} the rate constants for the influx and outflux from the peripheral compartment respectively, the remaining assumptions of the first-order kinetics model can otherwise be applied.

Two-compartment models

Under these assumptions, and provided distribution ($k_{12} + k_{21}$) was not much faster than uptake (k_1), then during the accumulation phase C_1 will not reach a plateau but will eventually drop due to transport of the chemical to the peripheral compartment.[17]

In addition, a single compartment model predicts during the elimination phase a rapid and complete depuration determined by the k_2 value (Fig. 11.4B). In contrast, depuration kinetics in two-compartment models will be much slower than predicted by one-compartment models, due to the affinity of the peripheral compartment for the chemical. Laboratory experiments describing metal depuration in mussels sometimes conform to one-compartment models,[18] but more often describe depuration rates markedly lower than expected.[19] For example, F. Boisson et al.[20] showed that Pb depuration in mussels adjusted well to a two-compartment model with a rapid loss of Pb from a central compartment ($t_{1/2} = 1.4$ days) but a slow depuration from a peripheral compartment, were 62% of the Pb accumulated, with a $t_{1/2} = 2.5$ months.

The common kinetic models illustrated in Fig. 11.3 take into account uptake of waterborne substances only. However, **uptake via food** can be easily incorporated if the concentration of the substance in the food (C_f), the mass-specific ingestion rate (IR, mass of food consumed per mass of organism per time), and the food assimilation efficiency in the digestive system (AE, mass absorbed per mass ingested) are known:

Incorporating uptake via food …

$$dC_t/dt = C_w k_1 + AE \cdot IR \cdot C_f - k_2 C_t \qquad \text{(Eq. 11.8)}$$

If we can consider C_w, C_f and the feeding parameters as constant, then:

$$C_t = (C_w k_1 + AE \cdot IR \cdot C_f)/k_2 \left(1 - e^{-k_2 t}\right) \qquad \text{(Eq. 11.9)}$$

and in the equilibrium:

$$C_t^* = (C_w k_1 + AE \cdot IR \cdot C_f)/k_2 \qquad\qquad\qquad \text{(Eq. 11.10)}$$

Measuring IR and AE requires quantification and analysis of ingested food and excreted feces. For substances difficult to excrete (low k_2) and highly assimilable by the consumer (AE approximating to 1) this model predicts a relevant contribution of the food pathway and a concentration of the substance at steady state higher in the consumer compared to its food. This issue will be further discussed in Section 11.4.

... growth, and other physiological factors

The kinetic models can also be modified to incorporate the dilution in tissular concentrations derived from **growth** of the organism by replacing k_2 in Eqs. (11.8)–(11.10) by $(k_2 + g)$, where g is the mass-specific growth rate (in units of $mass \cdot mass^{-1} \cdot time^{-1}$).

More complex models including the incorporation of multiple sources, many elimination components (due for example to metabolic transformations), or nonconstant water concentrations were proposed.[21,22] These modifications may contribute to a more realistic prediction of actual bioaccumulation and hazard assessment under realistic environmental scenarios, but they also increase the data requirements to feed and test those models.

11.3   MODELING BIOACCUMULATION AND CALCULATING BIOCONCENTRATION FACTORS, II. PARTITION EQUILIBRIUM MODELS

K_{ow} predicts the bioaccumulation of nonpolar organics

Kinetic models based on the balance between elimination and uptake were first developed by pharmacologists interested in the time course of the effects of drugs administered at discontinuous pulses, and they are very useful in biomedical research. If we turn our attention to environmental pollutants, polar compounds such as metal ions are readily eliminated from the organism, and the correspondingly high k_2 values (see Eq. 11.3) caused steady-state conditions to be achieved in the order of a few days to weeks. However, for highly lipophilic compounds, such as organohalogenated hydrocarbons, the elimination, if any, is so slow that it renders the kinetic approach unpractical. In those cases, a useful alternative for the estimation of the BCF values is the **partition equilibrium models**. These are time-independent models, which assume thermodynamic equilibrium between the organisms and their environment. The equilibrium partition of a chemical between two immiscible phases, A and B, with different affinities for the chemical, is described by the partition coefficient, K_{AB}, calculated as $K_{AB} = C^*_A/C^*_B$, where C^* denotes the concentrations of solute reached in both phases at equilibrium.

The partition coefficient most commonly used in ecotoxicology is the **octanol-water partition coefficient (K_{ow})** that measures the degree of hydrophobicity of a substance by comparing its solubility in an organic solvent, octanol, to its aqueous solubility (for methods, see OCDE (2006), Test guideline 123). As reflected in Table 11.1 for fish, and observed also in mussels,[23] the tendency to bioaccumulate of a chemical is often directly related to its K_{ow} value. In other words, octanol seems to be a medium representative of the average affinity of the animal tissues that accumulate organic chemicals. This fact is commonly exploited in environmental fate and risk assessment modeling, since the measurement of the K_{ow} is much more rapid and, in principle, simpler than experimental measurement of BCF in animal testing.

The more general approach that advocates modeling the biological activity (bioaccumulation, toxicity, physiological rate …) of chemicals as a function of chemical properties (partition coefficients, water solubility, molecular weight …) dependent from the structure of the molecule is termed **quantitative structure-activity relationships (QSAR)**.[24]

If we view an organism as consisting of a number of phases among which fat shows by far the highest affinity for hydrophobic chemicals, then the BCF can be reinterpreted as a lipid–water partition coefficient, and BCF will be a one-constant function of K_{ow}. Taking this approach, D. Mackay[25] reviewed the values of both parameters for aquatic organisms and chemicals with K_{ow} values ranging from 2 to 6, and found that

Bioaccumulation of nonpolar organics takes place in the body fat

$$logBCF = logK_{ow} - 1.32 \qquad \text{(Eq. 11.11)}$$

i.e.,

$$BCF = 0.048K_{ow}; \left(r^2 = 0.95\right) \qquad \text{(Eq. 11.12)}$$

These expressions allowed estimation of BCF with a factor of error of 1.8, remarkably precise considering the double logarithmic relationship. The fitting parameter in Eqs. (11.12), 0.048, would be related to the average proportion of body fat of the organisms and the ratio between the solubility of the chemical in body fat and octanol. Other studies found slopes in the double logarithmic relationship between BCF and K_{ow} slightly lower than 1, but always >0.8, and intercept values ranging from −0.7 to −1.3.[26]

The prediction can be even more accurate if the actual value for the lipid content of the organism is available. Taking for example the organochlorine insecticide lindane, H.J. Geyer et al.[28] found that the experimentally obtained BCF showed a coefficient of variation of 37.8 percent when expressed on a total wet weight basis ($BCF_w = 450 \pm 170$ L/Kg WW), but only 2.3 percent when expressed on a lipid weight basis ($BCF_L = 11,000 \pm 250$ L/Kg LW).

The success of these thermodynamic models must not distract us from their underlying assumptions; they are only expected to apply precisely to solutes that comply with two requirements: (1) unionized compounds that readily diffuse through hydrophobic phases, and (2) nonmetabolizable compounds whose biotransformation rates were negligible compared to the diffusive fluxes. Also, while K_{ow} itself is a simple parameter, its measurement is not so

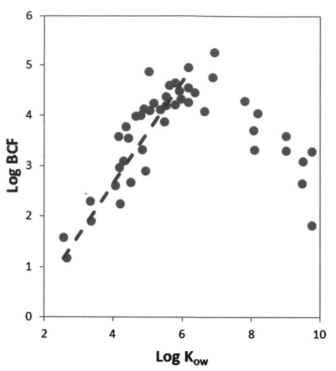

FIGURE 11.5

Relationship between the BCF in fish and the K_{ow} for organic chemicals. Within the range of log K_{ow} values between 2 and 6 BCF is a linear function of K_{ow} in a double logarithmic scale. The dashed line corresponds to Eq. (11.11). Notice that further increases in K_{ow} though cause a decrease in BCF. For those extremely hydrophobic chemicals octanol is no longer a good surrogate for cell membrane lipids. *Modified from Connel DW, Hawker DW. Use of polynomial expressions to describe the bioconcentration of hydrophobic chemicals by Fish. Ecotoxicol Environ Safety 1988;16:242–257.*

straightforward, and large differences among records from different sources for the same chemical are commonplace. K_{ow} values for highly hydrophobic chemicals obtained by shaking both phases are frequently underestimated since small droplets of octanol tend to remain in suspension in the aqueous phase.[29]

It must be borne in mind that the linear relationship stated in Eq. (11.12) only holds within a certain range of K_{ow} values (Fig. 11.5). Extremely hydrophobic substances do not bioaccumulate as much as those with intermediate log K_{ow} values. For example, the brominated diphenylether congener deca-BDE, with a log $K_{ow} > 10$ bioaccumulates less than penta-BDE, with log $K_{ow} = 6.57$ (see also Section 8.2).

In addition, the use of lipophilicity to predict bioaccumulation has been challenged by the emergence of nonlipophilic bioaccumulative chemicals such as perfluorinated compounds consisting of a 7–8 C chain and a polar group (e.g., PFOS, PFOA). These chemicals (see Section 8.3) are proteinophilic and their bioaccumulation potential cannot be predicted from K_{ow} or any other parameter measuring affinity for lipid phases.[30]

For benthic organisms, and particularly for deposit-feeders, it seems reasonable to hypothesize that sediment particles will be the major source of uptake of contaminants. In this case sediment-biota partitioning equilibrium models can be used. These models are applicable to nonpolar organics, and assume that the sediment organic matter concentrates all except for a negligible amount of the chemical in the environment, and that the body fat makes the same inside the organism. Thus when the steady state is reached the BAF, here termed biota-sediment accumulation factor (BSAF), can be estimated as:

Bioaccumulation from sediments depend on their organic content

$$BSAF = (C_t/f_L)/(Cs/f_{OC}) \qquad \text{(Eq. 11.13)}$$

Where: C_t = concentration in the organism (μg/g WW), f_L = lipid content of the organism (lipid weight/wet weight), Cs = concentration in the sediment (μg/g DW), f_{OC} = fraction of organic carbon in the sediment (g OC/g sediment).[31] The normalization of the tissue concentrations by lipids and the sediment concentrations by organic carbon reduces the variability of the BAF values among sediments and organisms.[32]

11.4 TROPHIC TRANSFER OF POLLUTANTS; BIOMAGNIFICATION

We may be interested to assess the relevance of the accumulation of pollutants via food, and thus the potential **trophic transfer** of a pollutant, for example, to implement environmental quality criteria taking into account secondary poisoning. For chemicals assimilable by the gut but difficult to eliminate by the organism, bioaccumulation kinetic models predict a relevant contribution of the food pathway and a concentration of the substance at a steady state higher in the consumer compared to its food. If this phenomenon is repeated through the different links of a trophic chain, top predators may be endangered even when primary consumers showed low bioaccumulation. This phenomenon called **biomagnification** will be formally defined here as an increase in pollutant body burden directly related to the trophic level occupied by the organisms. Fig. 9.7 illustrated the trend toward an increase in methylmercury content as trophic position increased in consumers from an estuarine food web. While remarkable bioaccumulation, disregarding the uptake route, has been demonstrated for several trace metals and many hydrophobic organic pollutants, biomagnification was consistently reported for Hg and organohalogen hydrocarbons only.

Trophic transfer takes place when food is a relevant via of uptake and BAF/ BCF > 1

For a given chemical, the most simple approach to study its trophic transfer potential would be to compare its concentration in the organism (or predator, C_{pred}) and in the food (or prey, C_{prey}) at steady state, and to calculate the **TTF**, according to the expression shown in Fig 11.1:

$$TTF = C_{pred}/C_{prey} \qquad \text{(Eq. 11.14)}$$

For lipophilic pollutants, such as organochlorines, both C_{pred} and C_{prey} must be expressed on a lipid weight basis (mg/Kg LW). Laboratory studies with small fish[33] found TTF values for organochlorine compounds ranging from <0.1 to 1, while field studies[34] found values only slightly higher (from 0.7 to 10), which do not support trophic transfer as the mechanism underlying biomagnification for these chemicals. TTF as defined assumes no growth, steady state of the chemical in both predator and prey, and consumption of a single prey. These requirements are hardly applicable to field situations. Furthermore, notice that in aquatic animals sampled from the field, higher concentrations of a chemical in a predator compared to its prey can result simply from a higher BCF of the predator, with no trophic transfer implied. An alternative experimental approach would be to compare BCF and BAF calculated as depicted in Fig 11.1. If **BAF/BCF>1** then food will be a relevant uptake route.

Biomagnification takes place when tissular concentration of a chemical is directly related to the trophic position of the organism

A more ecologically relevant concept is that of the **Biomagnification Factor in a trophic web** (BMF_{TW}). This can be defined as the overall degree of biomagnification of a chemical from producers (or another suitable baseline) to apex consumers in an entire food web, and it can be calculated from the simple linear regression:

$$Log\ C = a + b \cdot TL \qquad (Eq.\ 11.15)$$

where TL is the trophic level of an organism and C is the corresponding concentration of pollutant in its tissues, and

$$BMF_{TW} = 10^b \qquad (Eq.\ 11.16)$$

The difficulty here is to quantify the trophic position of each member of the food web. Traditional studies relied on the gut contents of consumers to establish trophic links. An important step forward was introduced with the analysis of the stable isotopes of C and especially N.

As explained in Section 3.1, the ratio between the heavier and the common stable isotopes of N, termed $\delta^{15}N$, changes as nitrogen circulates through certain metabolic routes because the light isotope may be mobilized faster, a process termed isotopic fractionation. This is the case of the transfer of organic matter to the consumer from the food. D.M. Post[35] found an average increase in $\delta^{15}N$ of 3.4‰ associated to each trophic transfer, which allows relating trophic position with isotopic fractionation according to the expression:

$$TL_a = 1 + (\delta^{15}N_a - \delta^{15}N_{baseline\ pp})/3.4 \qquad (Eq.\ 11.17)$$

where TL_a is the trophic level of a given organism (a) in the food web and $\delta^{15}N_a$- $\delta^{15}N_{baseline\ pp}$ is its ^{15}N fractionation compared to a representative primary producer from that ecosystem, taken as baseline. It is easy to see that if the organism *a* has the same $\delta^{15}N$ that the baseline then its TL will be 1, if it is enriched in ^{15}N by the 3.4‰ characteristic of one trophic step then its TL will be $1 + (3.4/3.4) = 2$, etc.

Finding an adequate baseline may be critical, and abundant primary consumers such as mussels or limpets have been proposed. In this case, since the baseline organism occupies trophic level 2 (primary consumer, [pc]) Eq. (11.17) is modified to:

$$TL_a = 2 + (\delta^{15}N_a - \delta^{15}N_{baseline\ pc})/3.4 \qquad \text{(Eq. 11.18)}$$

As biomagnification achieved the status of paradigm, confounding factors that may explain increased concentrations in top predators due to reasons not related to trophic transfer became somewhat overlooked. It is common, for example, to measure the pollutant in the whole body whereas only certain especially **accumulative organs** (liver, feathers, blubber) are used in the predators, because of their larger size. Concentrations of xenobiotics in organs specialized in its storage, such as liver or blubber, will always be orders of magnitude higher than total body burdens, disregarding trophic position. **Size** is another factor that can explain by itself higher concentrations in predators, usually larger, since excretion rates per unit of mass decrease as size (and volume to gill area ratios) increases. Another confounding factor that must be taken into account when measuring the concentrations of lipophilic chemicals such as organochlorines is the **lipid content** of the animal. For example, typical percentage of lipids in marine phytoplankton, invertebrates, and fish are 0.5%, 1.8%, and 5.4%, respectively.[36] In addition, lipid content may increase with age, and the life span of predators is usually longer than that of their preys. Since higher trophic levels organisms tend to have a higher lipid concentration, normalization of data to unit lipid concentration should be common practice, but many data sets do not use such procedures and naively interpret results as evidence for biomagnification.[37]

> **Nontrophic processes may also cause higher concentrations of pollutants in predators compared to low trophic-level organisms**

Finally physiological differences may also affect feeding habits and thus the main mechanism of uptake for certain chemicals. Marine mammals and seabirds are top predators but also **air-breeding animals** which cannot take up dissolved chemicals through the gills, as invertebrates and fish do. This is a very relevant factor. For example, PBDEs are generally considered to show biomagnification, but close inspection of data shows that there is no clear evidence of biomagnification from invertebrates to fish, and increased levels are only remarkable in air-breeding marine vertebrates.[38]

KEY IDEAS

- Chemical pollutants taken from the water via gills or from the food via digestive system accumulate in aquatic organisms. The bioaccumulation can be studied using kinetic models, where the concentration in the animal is expressed as a function of time, or partition equilibrium models that assume thermodynamic equilibrium and predict bioconcentration as a function of chemical properties of the pollutant, such as its hydrophobicity.

- One-compartment kinetic models predict a bioconcentration factor independent of the water concentration, i.e., the concentration of a chemical reached in the organisms is a linear function of its concentration in the water, providing the basis for the use of accumulator organisms in water quality monitoring.

- The BCF (L/Kg WW) for chemicals with K_{ow} values ranging from 2 to 6 can be accurately predicted from the expression: BCF = 0.05 K_{ow}, where 0.05 approximates the average proportion of fat to wet weight in the studied organisms. The prediction may be even more accurate if bioconcentration is expressed on a lipid weight basis.

- Trophic transfer takes place when food is a relevant via of uptake, and when this phenomenon is repeated in the different links of a food chain leads to highly elevated concentrations of the chemical in the top predators. This is termed biomagnification, and it is exclusive of mercury and some organohalogenated compounds.

- As the organic matter passes through the links of a food chain it becomes slightly enriched in the heavier stable isotopes of the N, the ^{15}N, by a fixed amount. Measuring the ^{15}N of a given organism and comparing it to that of a baseline organism of known trophic position allows obtaining a quantitative estimate of the trophic level occupied by that organism.

- Biomagnification in aquatic food webs has been demonstrated for methylmercury, organochlorine pesticides, PCB, PBDE, and PFOS. However, in some instances nontrophic factors such as lipid content, size (gill surface to body size ratio) or respiratory metabolism may account for this phenomenon.

Endnotes

1. Walker CH, Sibly RM, Hopkin SP, Peakall DB. Principles of ecotoxicology. 4th ed. CRC Press; 2012. Wright DA, Welbourn P. Environmental toxicology. Cambridge: Cambridge University Press; 2002. (p. 268).

2. Landrum PF, Lydy MJ, Lee H. Toxicokinetics in aquatic systems: Model comparisons and use in hazard assessment. Environ Toxicol Chem 1992;11:1709–1725. Newman MC. Fundamentals of ecotoxicology. 4th ed. Boca Raton: CRC Press; 2015.

3. Connell DW, Lam P, Richardson B. et al. Introduction to ecotoxicology. Wiley-Blackwell; 1999. (p. 39).

4. Mathews T, Fisher NS. Evaluating the trophic transfer of cadmium, polonium, and methylmercury in an estuarine food chain; Environ Toxicol Chem 2008;27(5):1093–1101. Guo F, Yang Y, Wang WX. Metal bioavailability from different natural prey to a marine predator *Nassarius siquijorensis*, Aquat Toxicol 2012;126:266–273.

5. Wright & Welbourn (2002) op. cit. (p. 268).

6. Taylor S. Transport of metals across membranes. In: Tessier A, Turner DR, editors. Metal speciation and bioavailability in aquatic systems. Chichester: J Wiley & Sons; 1995. pp 1–44.

7. Sánchez-Marín et al. Aquat Toxicol. Sep 2016;178:165–170.

8. MacLean RS, Borgmann U, Dixon, DG. Bioaccumulation kinetics and toxicity of lead in *Hyalella azteca* (Crustacea, Amphipoda). Can J Fish Aquat Sci 1996;53:2212−2220.

9. e.g. Wang W-X, Fisher NS, Luoma SN, Kinetic determinations of trace element bioaccumulation in the mussel *Mytilus edulis*. Mar Ecol Prog Ser 1996;140:91−113. Lee B-G, Wallace WG, Luoma SN. Uptake and loss kinetics of Cd, Cr and Zn in the bivalves *Potamocorbula amurensis* and *Macoma balthica*: effects of size and salinity. Mar Ecol Prog Ser 1998;175: 177−189. Lorenzo JI, Beiras R, Mubiana VK. et al. Copper uptake by *Mytilus edulis* in the presence of humic acids. Environ Toxicol Chem 2005;24(4):973−980. Sánchez-Marín P, Lorenzo JI, Mubiana VK, et al. Copper uptake by the marine mussel, *Mytilus edulis*, in the presence of fulvic acids. Environ Toxicol Chem 2012;31(8):1807−1813.

10. e.g. Borchardt T. Influence of food quantity on the kinetics of cadmium uptake and loss via food and seawater In *Mytilus edulis*. Mar Biol 1983;76:67−76. Riisgård HU, Bjørnestad E, Møhlenberg F. Accumulation of cadmium in the mussel *Mvtilus edulis*: kinetics and importance of uptake via food and seawater. Mar Biol 1987;96:349−353.

11. OECD. Guidelines for testing of chemicals. 305; 2012. Bioaccumulation in fish, aqueous and dietary exposure.

12. U.S. EPA. Superfund Public Health Evaluation Manual. Office of Emergency and Remedial response. Exhitit A-1. Washington DC; 1986.

13. GE Zaroogian. Crassotrea virginica as an indicator of cadmium pollution. Mar Biol 1980;58: 275−284.

14. McGeer JC, Brix KV, Skeaff JM et al. Inverse relationship between bioconcentration factor and exposure concentration for metals: implications for hazard assessment of metals in the aquatic environment. Environ Toxicol Chem 2003;22(5):1017−1037.

15. Walsh, O'Halloran. The Accumulation of chromium by mussels *Mytilus edulis* (L.) as a function of valency, solubility and ligation Mar Environ Res 1997;43(1/2):41−53.

16. Barron MG, Stehly GR, Hayton WL. Pharmacokinetic modeling in aquatic animals. I. Models and concepts. Aquat Toxicol 1990;18:61−86.

17. See Fig. 3.36, p. 136 in: O'Flaherty EJ. Toxicants and drugs: kinetics and dynamics. New York: J Wiley & Sons; 1981. 398 pp.

18. e.g. Fig 1 in Wang and Fisher (1997) op. cit. for Co and Zn.

19. For Cd and Se see: Wang and Fisher (1997) op. cit. For Cu see Fig. 4 in: Lorenzo JI, Aierbe E, Mubiana VK, et al. Indications of regulation on copper accumulation in the blue mussel *Mytilus edulis*. In: Villalba A, Reguera B, Romalde JL, Beiras R, editors. Molluscan shellfish safety. 2003; pp. 533−544. For Pb see Fig. 3 in: Sánchez-Marín P, Bellas J, Mubiana VK, et al. Pb uptake by the marine mussel *Mytilus* sp. Interactions with dissolved organic matter. Aquat Toxicol 2011;102:48−57.

20. Boisson F, Cotret O, Fowler SW, Bioaccumulation and retention of lead in the mussel *Mytilus galloprovincialis* following uptake from seawater. Sci Tot Environ. 1998;222:55−61.

21. Landrum et al. (1992) op. cit.

22. Newman MC. Fundamentals of ecotoxicology. 4th ed. Boca Raton: CRC Press; 2015. (pp. 117−119).

23. Geyer H, Sheehan P, Kotzias D, et al. Prediction of ecotoxicological behavior of chemicals: relationship between physico-chemical properties and bioaccumulation of organic chemicals in the mussel *Mytilus edulis*. Chemosphere 1982;11(11):1121−1134.

24. For a review see Donkin and Widdows (1990).

25. MacKay D. Correlation of bioconcentration factors. Environ Sci Technol 1982;16:274−278.

26. Veith GD, DeFoe DL, Bergstedt DV. Measuring and estimating the bioconcentration factor of chemicals in fish. J Fish Res Board Can 1979;36:1040−1048; Geyer et al. (1982) op. cit.; Hawker DW, Connell DW. Bioconcentration of lipophilic compounds by some aquatic organisms. Ecotoxicol Environ Safety 1986;11:184−197.

27. Connel DW, Hawker DW. Use of polynomial expressions to describe the bioconcentration of hydrophobic chemicals by Fish. Ecotoxicol Environ Safety 1988;16:242−257.

28. Geyer HJ, Scheunert I, Brüggermannn R, et al. Half-lives and bioconcentration of lindane (γ-HCH) in different fish species and relationship with their lipid content. Chemosphere 1997;35(1/2):343−351.

29. Hermens JLM, de Bruijn JHM, Brooke DN. The octanol-water partition coefficient: strengths and limitations. Environ Toxicol Chem 2013;32(4):732−733.

30. Conder JM, Hoke RA, De Wolf W, et al. Are PFCAs bioaccumulative? A critical review and comparison with regulatory criteria and persistent lipophilic compounds. Environ Sci Technol 2008;42(4):995−1003.

31. Ankley GT, PM Cook, AR Carlson et al. Bioaccumulation of PCBs from sediments by oligochaetes and fishes: Comparison of laboratory and field studies. Can J Fish Aquat Sci 1992; 49:2080−2085.

32. Landrum et al. (1992) op. cit.

33. Leblanc GA. Trophic level differences in the bioconcentration of chemicals: implications in assessing environmental biomagnification. Environ Sci Technol 1995;29:154−160.

34. Thommann RV. Bioaccumulation model of organic chemical distribution in aquatic food chains. Environ Sci Technol 1989;23:699−707.

35. Post DM. Using stable isotopes to estimate trophic position: models methods and assumptions. Ecology 2002;83(3):703−718.

36. Leblanc GA (1995) op. cit.

37. Gray JS. Biomagnification in marine systems: the perspective of an ecologist. Mar Pollut Bull 2002;45:46−52.

38. Boon JP, Lewis WE, Tjoen-A-Choy MR et al. Levels of Polybrominated Diphenyl Ether (PBDE) flame retardants in animals representing different trophic levels of the North Sea food web. Environ Sci Technol 2002;36:4025−4032.

Suggested Further Reading

- Depledge MH, Rainbow P. Models of regulation and accumulation of trace metals in marine invertebrates. Comp Biochem Physiol 1990;97C(1):1−7.

- Hermens JLM, de Bruijn JHM, Brooke DN. The octanol-water partition coefficient: strengths and limitations. Environ Toxicol Chem 2013;32(4):732−3.

- O'Flaherty EJ. 2. Absorption, distribution, and elimination of toxic agents. In: Williams PL, James RC, Roberts SM, editors. Principles of toxicology. Environmental and industrial applications. 2nd ed. New York: John Wiley & sons, Inc; 2000. p. 35−55.

- Vercauteren K, Blust R. Bioavailability of dissolved zinc to the common mussel *Mytilus edulis* in complexing environments. Mar Ecol Prog Ser 1996;137:123−32.

CHAPTER 12

Biotransformation

12.1 METABOLIC TRANSFORMATION AND ELIMINATION OF POLLUTANTS

Living organisms possess different means of getting rid of foreign chemicals, termed **xenobiotics**, bearing potential toxicity. This set of biochemical pathways for the elimination of xenobiotics are frequently referred to as **biotransformation** processes. The strategy may consist of immobilization and neutralization of the potential toxicity of the chemical (sequestration), removal of the toxic molecules or radicals (scavenging), or enzymatic transformation to facilitate excretion. A more formal definition for **biotransformation** is the in vivo metabolic pathways intended to prevent toxic effects of xenobiotic substances and eventually eliminate them from the organism.

Most biotransformation processes take place in the **liver** of vertebrates, and in the hepatopancreas or digestive gland in mollusks and crustaceans (or the anterior intestine in echinoderms), while the excretion of the biotransformed metabolites may take place in the kidney via urine (or alternative excretory system of invertebrates) or through the bile.

When a xenobiotic chemical is taken up by an organism it may be readily excreted if its chemical properties allow so, i.e., if its polar nature allows it to be eliminated through the excretory system. Alternatively, it can be bound to endogenous molecules or cellular structures (granules, membranes) where it is retained and cannot exert toxicity, a process termed **sequestration**. Extracellular insoluble **granules** rich in otherwise toxic trace metals have been described for instance in bivalves.[1] Exposure to heavy metals also induce the synthesis of low molecular weight proteins called **metallothioneins** (see Section 12.6), rich in cysteine residues whose thiol groups bind divalent cations (Cd^{2+}, Hg^{2+}, Cu^{2+}, etc.) preventing toxicity by inappropriate binding of the metal ion to enzymatic or structural proteins.

Some toxic molecules can be removed just by nonenzymatic reaction with endogenous chemicals present in high concentrations, which is termed

Biotransformation: the metabolism of foreign chemicals

Biotransformation may be achieved by sequestration, scavenging, or reaction catalyzed by detoxification enzymes

205

scavenging. This is the case of the tripeptide glutathione (GSH), which contains cysteine and acts as scavenger of otherwise highly toxic **reactive oxygen species** (ROS). Oxidative stress can also be fought, though, via enzymatic pathways such as those involving catalase and peroxidases (see Section 12.5).

Detoxification enzymes are induced in the presence of xenobiotics and may thus be used as biomarkers of pollution

Highly lipophilic xenobiotics tend to bioaccumulate in the body fat. However, they can also be metabolized through biotransformation routes catalyzed by **detoxification enzymes** that normally proceed in a stepwise manner as described in the next section. The synthesis of the enzymes and other molecules (stress proteins, cofactors) involved in the detoxification metabolism is induced by the presence in the environment of chemical pollutants. This has been used as a biological tool for monitoring chemical pollution, and the inducible molecules involved can be used as **biomarkers**. This issue will be treated in depth in Chapter 16.

12.2 PHASES OF THE ENZYMATIC BIOTRANSFORMATION

Biotransformation of xenobiotics proceeds through two steps: oxidation plus conjugation

Enzymatically mediated biotransformation of organic xenobiotics frequently proceeds through the steps summarized in Fig. 12.1. In **Phase I** reactions the molecule is normally oxidized to increase polarity and to provide more reactive groups for further transformation. If the substance is highly lipophilic Phase I can involve several oxidative steps. This is the case of the degradation of polycyclic aromatic hydrocarbons (PAHs) (see Fig. 7.2, Section 7.3). The first oxidation of organohalogenated compounds is frequently the limiting reaction for

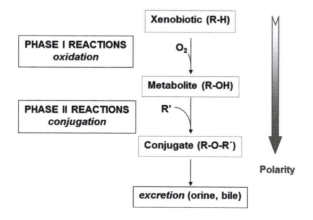

BIOTRANSFORMATION OF ORGANIC XENOBIOTICS

FIGURE 12.1

Schematic representation of a typical biotransformation process of a lipophilic organic xenobiotic. In Phase I the chemical is oxidized by a cytochrome p450-dependent monooxygenase (CYP). In Phase II the resulting metabolite, frequently more toxic than the parental compound, is conjugated to an endogenous nontoxic polar molecule. The resulting conjugate is suitable for elimination by the excretory system.

the overall elimination process. In **Phase II** reactions, the oxidized metabolite is conjugated with an endogenous nontoxic metabolite to form a water-soluble higher molecular weight product that can be excreted through the bile, the kidney, or equivalent excretory system.

Biotransformation affects toxicity. Oxidized derivatives of xenobiotics can be highly reactive and even more toxic than the parent compound. This is, for example, the case of paracetamol, which is normally conjugated with nontoxic endogenous metabolites but can also be oxidized by a Phase I reaction into a toxic quinone that must be conjugated to suppress its toxicity to the organism. Therefore, Phase 2 reactions, to increase the polarity of the molecule, have also the function of buffering the potential toxicity of the degradation product formed after Phase I.

The chemical nature of the xenobiotic determines the biotransformation pathway. If it is a very hydrosoluble substance it can be readily excreted in the urine or directly undergo Phase II by conjugation through one polar group. The excretion rate is related to the polarity of the molecule. This is again illustrated by the case of paracetamol, whose conjugates are 10–20 times more rapidly excreted than the parental compound.[2]

12.3 PHASE I: CYTOCHROME P450 DEPENDENT OXIDATIONS

Phase I reactions (Table 12.1) include different kinds of oxidation (hydroxylation, epoxidation, dealkylation, desulphuration) but can also consist of hydrolysis or in some cases (for instance DDT) even reduction. These reactions are catalyzed by enzymes called **cytochrome P450 monooxygenases (CYP monooxygenase** or simply **CYP**), also termed in the past **mixed function oxidases (MFOs)** (see Section 7.3). These enzymes are located in the **smooth endoplasmic reticulum** (SER) of hepatocytes and some other cellular types. Induced activity of detoxification enzymes is used as biomarker of chemical

Most Phase I oxidations take place in the smooth endoplasmic reticulum (SER) of hepatocytes...

Table 12.1 Main Types of Phase I Biotransformation Reactions, Enzymes Involved, and Examples of Substrates

Phase I Reaction	Enzymes	Examples of Substrates
OXIDATIONS	CYP monooxygenases	
Hydroxylation		Benzene, aliphatic hydrocarbons, 4-MBC,[3]
Epoxidation		BaP and other PAHs, PCB
Desulphuration		Parathion
Dealkylation		Ethoxyresorufin
HYDROLYSIS	Esterases	Carbamates
REDUCTION	Reductases	DDT, PCB

pollution. Invertebrates show in general much lower CYP activities than aquatic vertebrates, and this contributes to explain higher BCF values for many xenobiotics. For biochemical assay, the SER of hepatocytes or other cells is isolated by differential ultracentrifugation in the so-called **microsomal fraction**, and so the associated enzymes like CYP are called microsomal, in contrast to the fewer oxidation enzymes present in the cytosol.[a]

...because they involve several membrane-bound electron transport proteins

The overall enzymatic complex is termed cytochrome P450 monooxygenase system

Indeed, the apparently simple Phase I scheme, from the lipophilic xenobiotic, R—H to the oxidized metabolite R—OH in the presence of oxygen (see Fig. 12.1), is in fact a quite complex process, and CYP mediated reactions take place in the SER because they involve a number of membrane-bound proteins responsible for electron transport. Those include cytochrome p450, a hemeprotein with an iron protoporphyrin group that forms a complex with the substrate and the oxygen, an electron donner -the NADPH-, a NADPH cytochrome P450 reductase, a second electron donor -the NADH-, a NADH cytochrome b_5 reductase, and its coenzyme, the cytochrome b_5. The whole system (for details, see[4]) is loosely termed the cytochrome P450 monooxygenase system, and the global reaction catalyzed by this system is:

$$R-H + NADPH + H^+ + O_2 \rightarrow R-OH + NADP^+ + H_2O. \qquad \text{(Eq. 12.1)}$$

Different CYP families catalyze the biotransformation of different kinds of chemicals

The CYP are in fact a series of isoenzymes that, in theory, specifically catalyze the oxidation of different types of organic molecules. In mammals there are three main families of CYP implicated in biotransformation of xenobiotics: CYP1, CYP2, and CYP3. Each family may present subfamilies termed CYP2A, CYP2B, etc., and each subfamily may be further divided into distinct proteins coded by different genes. For instance, cytochromes CYP1A1 and CYP1A2 are coded by the corresponding *CYP1A1* and *CYP1A2* genes (note the italics for genes). The families and subfamilies are defined on the basis of similarity in DNA sequence.

Other phylogenetic groups greatly differ in CYP variability. For example CYP 1 to 3 families are present in sea urchins but absent in bivalves and other protostomates.[5]

The induction of the synthesis of a certain CYP is mediated by the specific binding of the chemical inducer to a protein receptor that interacts with the regulatory element of the gene and triggers its transcription. For example, coplanar polyaromatic molecules such as the TCDD dioxin bind to the Aryl hydrocarbon receptor (AhR) in the cytoplasm. The TCDD-AhR complex enters the nucleus and, mediated by a specific nuclear translocator protein, interacts with the *CYP1A1* to promote its transcription into mRNA. The latter travels back to

---

[a] One of the few nonmicrosomal oxidation enzymes are dehydrogenases that transform primary and secondary alcohols, including ethanol into aldehydes and ketones, and the latter into organic acids. These enzymes are cytosolic and require NAD or NADP.

the cytoplasm, and activates CYP1A1 synthesis in the rough endoplasmic reticulum.[6] The binding strength of a ligand to the AhR is thought to be directly proportional to the enhanced gene transcription and associated toxicity. TCDD is the agonist with the strongest affinity to AhR. Other less potent inducers via AhR binding are other dioxins, dibenzofurans, coplanar PCBs and PAHs, and thus toxicity equivalents may be calculated for all those molecules on the basis of their relative affinity to AhR compared to TCDD (see also Section 8.1). In mammals, similar roles have been proposed for the constitutive androstane receptor (CAR) in relation to CYP2 induction, and the pregnane-X-receptor in relation to CYP3 induction. The CAR receptor is absent in fish, which may explain the lack of CYP2 inducibility despite fish do have the *CYP2* gene.[7]

Therefore, the induction of biotransformation pathways caused by exposition to chemical pollutants can be monitored at transcriptional level, by quantifying the amount of the specific mRNA that results from the expression of the gen, or at posttranscriptional level by quantifying the amount (or activity, if the protein is an enzyme) of the protein coded by the mRNA. The earlier may provide a better indication of exposure, if additional pollutants interfere with protein synthesis or enzyme activity at posttranscriptional level.

Induction of biotransformation caused by exposition to organics can be monitored at mRNA or protein levels

12.4 PHASE II: CONJUGATION WITH GLUTATHIONE OR GLUCURONIC ACID

Phase II reactions consist of the addition to the foreign compound of an endogenous metabolite readily available in vivo and added to a suitable functional group present on the original xenobiotic or introduced by Phase I metabolism. The resulting conjugate is more polar, thus facilitating excretion, and less toxic. The most common endogenous metabolites used in Phase II reactions are the **glucuronic acid**, the **sulfate**, and the **glutathione (GSH)**, a tripeptide formed by glutamic acid, cysteine, and glycine. Glucuronic acid and sulfate show more affinity for hydroxyl radicals (see Fig. 12.2), whereas GSH is frequently conjugated to epoxide and ketone groups (see Fig. 7.2).

Phase II aims at increasing polarity and buffering toxicity ...

Many Phase II reactions involve the transfer of glucuronic acid in its activated form, bound to the energy storing nucleotide uridine diphosphate (UDP) (Fig. 12.2). The reaction is catalyzed by the UDP-glucuronosyltransferase (UDPGT), an enzyme present in the SER of hepatocytes and other tissues. Unlike the GSH conjugation, where the activated compound is the xenobiotic (type II conjugations), here the break-down of the endogenous metabolite provides the energy for the reaction (type I conjugations).

...by conjugation mediated by glucuronyltransferase ...

GSH is a tripeptide with cysteine very abundant in hepatocytes. The thiol group of the cysteine confers GSH a double role in cellular defense against toxicants. First, the GS^- chemically reacts with electrophiles protecting cells by removing reactive metabolites. Second, GSH is the substrate for energy demanding Phase

...or glutathione-S-transferases

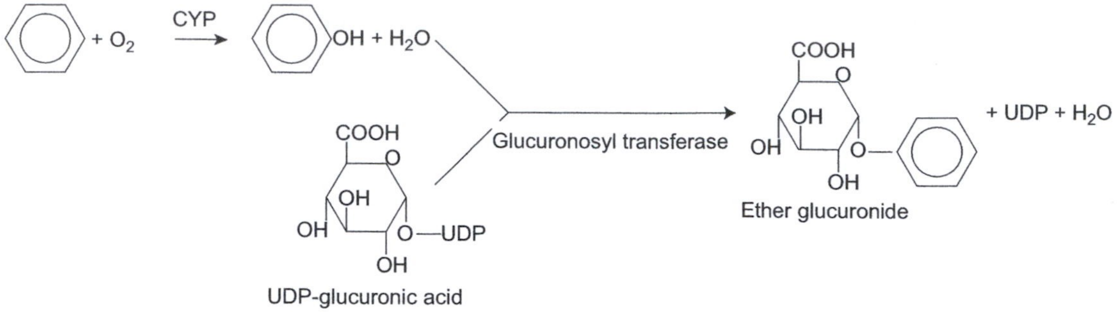

FIGURE 12.2

Biotransformation of benzene, including an oxidation by a cytochrome P450-dependent monooxygenase (CYP) (Phase I), and a conjugation with glucuronid acid (Phase II). The conjugation is catalyzed by a transferase, and the energy needed is provided by the cleavage of the bond with UDP.

II conjugation reactions catalyzed by the **glutathione-S-transferases (GST)**, which mediates the formation of conjugates with the potent electrophilic metabolites originated in Phase I biotransformation of PAHs, PCBs,[8] and other xenobiotics. GST is particularly relevant in the detoxification of the carcinogenic metabolite 7,8-dihydrodiol-9,10-epoxide formed in the Phase I biotransformation of BaP (see Fig. 7.2), since diol epoxides are poor substrates for epoxide hydrolases and cannot be detoxified by this enzyme.[9]

GST are very abundant (up to 0.7 percent of total soluble proteins) and ubiquitous enzymes primarily found in the cytosol of many tissues. Again, like for CYP, there is a number of different GST isoenzymes theoretically substrate-specific. This time the different functional isoenzymes result from the different arrangements of the dimer subunits than constitute the GST.[10]

Phase II enzymes are also inducible in the presence of environmental xenobiotics, and thus susceptible to be used as biomarkers of chemical pollution. However, the inducibility of both GST and UDPGT in fish has been reported to be moderate (up to twofold) in comparison with the inducibility of EROD (up to 100 fold).[11] The use of GST activity as chemical pollution biomarker in bivalves and other invertebrates has shown more promising results (see Section 16.2).

Phase III: catabolism and/or excretion of the conjugates   After Phase II, the resulting conjugates may either be excreted, usually into the bile, or catabolized through further degradative steps. These processes are sometimes termed phase III metabolism. GSH conjugates are degraded to mercapturic acids present in the urine.[12] This is the case of Naphthalene biotransformation products.[13] Excretion of conjugates out from hepatocytes involves a transmembrane pump and requires energy provided by ATP. The common endogenous metabolite conjugated during Phase II allows that a single type of pump can excrete many different biotransformed xenobiotics.[14]

12.5 OXIDATIVE STRESS METABOLISM

Although we usually think of oxygen as the essential element for life, the release of uncoupled high-energy electrons from its molecule may also cause serious injury to cells and tissues. That is why those electrons are handled with inside special cell compartments such as mitochondria and chloroplasts. Many chemicals (e.g., paraquat, cytotoxic quinones, transition metals) produce ROS, which are strong electron donors. This electrons are donated to molecular oxygen to yield superoxide ($\bullet O_2^-$), which may then be metabolized to hydrogen peroxide by the enzyme **superoxide dismutase (SOD)**.

Reactive oxygen species (ROS) may cause cellular injury if not counteracted by antioxidant enzymes

$$2\bullet O_2^- + 2H^+ \rightarrow H_2O_2 + O_2 \qquad \text{(Eq. 12.2)}$$

The hydrogen peroxide is then removed by **catalase (CAT)**

$$2H_2O_2 \rightarrow 2H_2O + O_2 \qquad \text{(Eq. 12.3)}$$

Reduced GSH can also play a role reducing hydrogen peroxide to water, a reaction catalyzed by **glutathione peroxidase (GPx)**.

$$2GSH + H_2O_2 \rightarrow GSSG + 2H_2O \qquad \text{(Eq. 12.4)}$$

When the amount of superoxide formed overwhelms the capacity of available SOD, then hydroxyl radicals ($\bullet OH$) are formed.

$$\bullet O_2^- + H_2O_2 \rightarrow \bullet OH + OH^- + O_2 \qquad \text{(Eq. 12.5)}$$

Metals such as iron can also be responsible for the formation of hydroxyl radicals through the Fenton reaction:

$$H_2O_2 + Fe^{2+} + H^+ \rightarrow \bullet OH + Fe^{3+} + H_2O \qquad \text{(Eq. 12.6)}$$

Hydroxyl radicals can also be involved in the production of further active oxygen species such as singlet oxygen ($^1O_2$), an electronically excited state of molecular oxygen.

Superoxide anions ($\bullet O_2^-$), hydroxyl radicals ($\bullet OH$), and singlet oxygen ($^1O_2$) are termed **ROS**, and they are cytotoxic because they cause membrane damage by peroxidation of its lipids. The lysosome membrane is affected by this mechanism of toxicity and its permeability becomes increased. ROS may also cause damage to proteins (including inactivation of enzymes) and DNA. The metabolic condition caused by all these ROS is known as **oxidative stress**, and the enzymes SOD, CAT, and GPx are known as **antioxidant enzymes**. Antioxidant enzymes, along with low molecular weight ROS scavengers such as GSH or vitamin E, constitute the antioxidant defense system of the organisms. The synthesis of antioxidant enzymes can be induced in the presence of oxidative stress conditions, and hence their activity serves as biomarker of environmental oxidative stress. Increased permeability due to impaired membrane stability in

antioxidant enzymes are induced upon exposure to oxidative stress

the lysosomes is also used as a biomarker, particularly in organisms prone to accumulation of chemicals, such as bivalves.[15]

GSH is the main cytosolic ROS scavenger

Scavengers neutralize ROS by direct reaction with them, thus being temporarily oxidized before being reconverted by specific reductases to the active form. Polar scavengers act as antioxidants in the cytoplasm, while liposoluble scavengers arrest the propagation of lipid peroxidation reactions on the membranes. The most abundant cytosolic scavenger is reduced **GSH**, a tripeptide that directly neutralizes several reactive species through its oxidation to the oxidized form of glutathione (GSSG). The main lipid soluble scavengers are **carotenoids** and **vitamin E**.[16]

12.6   METALLOTHIONEINS AND STRESS PROTEINS

The synthesis of stress proteins and metallothioneins is induced by many sources of cellular stress

The induction of proteins synthesis as a response to any kind of environmental stress (temperature, salinity, hypoxia, chemicals, etc.), has been demonstrated in all organisms examined to date, from prokaryotes to humans. These **stress proteins** were also termed heat shock proteins because the first one was discovered in *Drosophila* flies submitted to heat shock.[17] Stress proteins are classified according to their molecular weights that cluster around 7−8, 16−24, 60, 70 and 90 kDa, resulting the names Stress90, Stress70, and so on. Proteins around 7−8 kDa are called ubiquitin, and the 60 KDa family has been renamed chaperon 60 (cpn60). Their roles are related with the homeostasis of protein levels, structure, and function, including facilitation of three-dimensional folding needed for normal protein function, and preservation and continuous repair of protein structure upon adverse conditions.[18] Despite their ubiquity, for some of them there appears to be a great deal of homology in their aminoacid sequence across taxonomic groups. Thus, ubiquitin, Stress70, and cpn60 seem good candidates as general cellular stress biomarkers because they are inducible, their background levels are low, and their sequence is highly conserved among taxonomic groups.[19]

Metallothioneins (MT) are heat-stable soluble proteins of low molecular weight (below 7 KDa) containing about 30 percent of cysteine residues preferentially placed in Cys-Cys, Cys-X-Cys, or Cys-X-X-Cys sequence. They are involved in the transport of essential metals present in dissolution as divalent cations such as Cu^{2+} and Zn^{2+}. Due to this regulatory role basal levels of MTs are present in unpolluted environments. However, MTs are inducible upon exposure to xenobiotic metals such as Cd, Hg, and Ag in both vertebrates and invertebrates. The divalent cations of these metals are sequestered by the MTs, thus preventing metal toxicity. The MT-metal complexes are synthesized in the liver and transported to the kidney, and show a rapid turnover in the lysosomes. The protease degradation of the complex in the lysosomes of the kidney glomerulal cells releases free Cd and explains the nephrotoxicity that this metal causes in vertebrates.

The induction of MT synthesis upon metal exposure has been used as biomarker of metal pollution. However, unfortunately many other chemical and even physiological sources of stress induce MT synthesis, and they are considered as general stress proteins. Since other chemicals, oxidative stress, and physiological stimuli can also induce MT synthesis, this induction is not by itself diagnostic of metal exposure.[20]

KEY IDEAS

- Biotransformation of organic xenobiotics normally proceed in a first step of oxidation (Phase I) that introduces a reactive group in the xenobiotic, and a second step of conjugation (Phase II) with a nontoxic endogenous metabolite that yields an excretable product.

- Phase I of biotransformation is mediated by inducible cytochrome P450 dependent monooxygenases (CYP).

- In principle, different CYP forms, coded by different genes, are specifically responsible for the degradation of certain families of chemicals.

- The synthesis of the Phase I and Phase II enzymes involved in the detoxification metabolism is induced by the presence in the environment of chemical pollutants. This can be used as a biological tool for monitoring chemical pollution, and the inducible molecules are considered biomarkers.

- Exposure to some organics and transition metals cause oxidative stress through the formation of ROS that affect the lysosomes membrane increasing its permeability. This can be used as a quantitative and rather unspecific biomarker of chemical pollution.

- The antioxidant defense of the organism is composed by inducible antioxidant enzymes (SOD, CAT, and GPx) that metabolize cytotoxic ROS and low molecular weight oxidable molecules (GSH, vitamin E) that directly react with ROS.

Endnotes

1. e.g. Darriba S, Sánchez-Marín P. Lead accumulation in extracellular granules detected in the kidney of the bivalve *Dosinia exoleta*. Aquat Living Resour 2012;26(1):11−17.

2. Marris ME, Levy G. Renal clearance and serum protein binding of acetaminophen and its major conjugates in humans. J Pharm Sci 1984;73(8):1038−1041; Landrum PF, Lydy MJ, Lee H. Toxicokinetics in aquatic systems: Model comparisons and use in hazard assessment. Environ Toxicol Chem 1992;11:1709−1725.

3. EC. Opinion on 4-methylbenzylidene camphor (4-MBC); 2008.

4. Timbrell J. Principles of biochemical toxicology. 3rd ed. London: Taylor & Francis; 2000.

5. Hahn ME. In: Newman MC, editor. Fundamentals of ecotoxicology. 4th ed. 2015. pp 186−192.

6. Whitlock JP. Induction of cytochrome P4501A1. Annu Rev Pharmacol Toxicol 1999;39: 103−125.

7. Hahn (2015) op. cit.

8. Bakke JE, Bergman ÅL, Larsen GL. Metabolism of 2,4′,5-Trichlorobiphenyl by the mercapturic acid pathway. Science 1982;217(13):645−647.

9. Foureman GL. Enzymes involved in metabolism of PAH by fishes and other aquatic animals: hydrolysis and conjugation enzymes (os Phase II enzymes). In: Varanasi U, editor, Metabolism of polycyclic aromatic hydrocarbons in the aquatic environment. Boca Raton: CRC Press; 1989. pp. 185−202.

10. Sheehan D, Meade G, Foley VM, et al. Structure, function and evolution of glutathione trans-ferases: implications for classificatuon of non-mammalian members of an ancient enzyme superfamily. Biochem J 2001;360:1−16.

11. Foureman GL. Enzymes involved in metabolism of PAH by fishes and other aquatic animals: hydrolysis and conjugation enzymes (os Phase II enzymes). In: Varanasi U, editor. Meta-bolism of polycyclic aromatic hydrocarbons in the aquatic environment. Boca Raton: CRC Press; 1989. pp. 185−202.

12. Foureman (1989) op. cit.

13. Timbrell (2000) op. cit. (p. 99).

14. Sheehan D. Applications of in vitro techniques in studies of biomarkers and ecotoxicology. In: Mothersill C, Austin B. In vitro methods in aquatic toxicology. Chichester: Springer; 2005. pp. 55−76.

15. Regoli F. Total oxyradical scavenging capacity (TOSC) in polluted and translocated mussels: a predictive biomarker of oxidative stress. Aquat Toxicol 2000;50:351−361.

16. Regoli F, Giuliani ME. Oxidative pathways of chemical toxicity and oxidative stress bio-markers in marine organisms. Mar Enviorn Res 2014;93:106−117.

17. Wright DA, Welbourn P. Environmental toxicology. Cambridge: Cambridge University Press; 2002. (pp. 123-125).

18. Schüürmann G, Markert B. editors. Ecotoxicology. Ecological fundamentals, chemical expo-sure, and biological effects. New York: Wiley; 1998. (pp. 543−546).

19. Newman MC. Fundamentals of ecotoxicology. 4th ed. Boca Raton: CRC Press; 2015.

20. Roesijadi G. Metallothioneins. In: Newman MC. Fundamentals of ecotoxicology. 4th ed. Boca Raton: CRC Press. 2015; pp. 195−200.

Suggested Further Reading

- Regoli F, Giuliani ME. Oxidative pathways of chemical toxicity and oxidative stress biomarkers in marine organisms. Mar Environ Res 2014;93:106−17.

- Sheehan D, Meade G, Foley VM, Dowd CA. Structure, function and evolution of glutathione transferases: implications for classification of non-mammalian members of an ancient enzyme superfamily. Biochem J 2001;360:1−16.

- Timbrell J. Principles of biochemical toxicology. 4rd ed. London: CRC Press; 2008.

- Whyte JJ, Jung RE, Schmitt CJ, Tillit DE. Ethoxyresorufin-O-deethylase (EROD) activity in fish as a biomarker of chemical exposure. Crit Rev Toxicol 2000;30(4):347−570.

CHAPTER 13

Theory and Practice of Toxicology: Toxicity Testing

13.1 BASIC PRINCIPLES OF TOXICOLOGY

Although one was tempted to divide biologically active chemicals into poisons and medicines, in fact the relevant question is the amount, or **dose**, of the chemical reaching the sensitive targets in our body. The notion that only the dose makes the poison comes from the heterodox Renaissance physician Paracelsus who claimed that no substance is inherently a poison, but at a high enough dose, even apparently innocuous substances may cause toxic, even lethal, effects. We can kill a rat by orally administering to it 30 g of sugar, 2 g of vitamin A, or 1 g of aspirin per Kg of body weight. It is fair to consider those substances as nontoxic if we compare their lethal doses with those of nicotine (1 mg/Kg) or the toxin of botulism (<1 µg/Kg). Dose is thus the key concept when studying the biological effects of a chemical substance, and Paracelsus knew this first hand, since he used toxic substances such as mercury at low doses as medicines.

Any chemical at a high enough dose is a poison

Toxicity can be defined as a negative effect that impairs a biological function beyond the limits of compensation or other forms of rapid recovery. For a substance to elicit a toxic effect on an organism it must be incorporated into the organism and reach the **site of action**. The toxicity may be due to the interaction of the substance with the exposed surface (local toxicity) or more frequently with an internal organ (systemic toxicity) (See Fig. 11.2). Depending on the organ or system targeted we can refer to neurotoxicity, hepatotoxicity, nephrotoxicity, etc. Sublethal toxicity is specially feared when it targets reproduction, DNA (genotoxicity, see Section 16.4), or the normal functioning of hormones (endocrine disruption, see Section 14.3).

Toxic effect is a quantitative response to the amount of chemical reaching a site of action

The relationship between the dose at which the organism is exposed and the dose reaching the target organ is controlled by the processes of uptake, distribution, metabolism, and excretion. This set of processes is termed toxicokinetics (Fig 13.1). The next step, involving the relationship between the internal dose received by the target organ and the toxic effect is called toxicodynamics, and encompasses interactions with molecular receptors, potential antagonisms, or synergisms, etc. In the words of the prominent toxicologist W.J. Hayes, toxicokinetics deals with the question: What does

215

Marine Pollution. https://doi.org/10.1016/B978-0-12-813736-9.00013-1

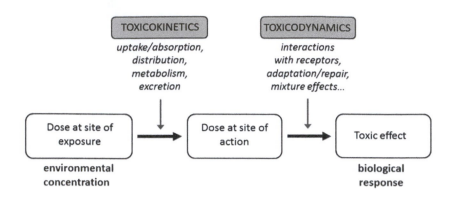

FIGURE 13.1

According to the fundamental principles of toxicology there is a causal relationship between the exposure to a chemical and the toxic effect. The quantitative aspects of this relationship are determined by the toxicokinetic and toxicodynamic processes mediating the amount of chemical reaching the site of action, and the elicited biological response, respectively. Aquatic ecotoxicology normally pays attention to environmental (exposure) concentrations and biological responses, and the intermediate processes are frequently overlooked. *Modified from Renwick AG. Chapter 4. Toxikokinetics. In: Ballantyne B, Marrs T, Syversen T, editors. General and applied toxicology, vol. 1. London: Macmillan Reference Ltd.; 1999. pp. 67–95.*

the organism do to a chemical? Toxicodynamic studies, however, deal with the question: What does a chemical do to the organism?[1]

In aquatic toxicology studies concentration replaces dose.

In aquatic animals, chemicals are very rarely administered by oral ingestion. Most often they are dosed by bathing the organism in water, where a known amount of the chemical is dissolved. Thus, in aquatic toxicology, exposure **concentration** is a practical and ecologically relevant surrogate for dose.

The effect depends on concentration and duration of exposure

The same substance may exert different effects on different sites of action located in different organs. Also, both toxicokinetic and toxicodynamic processes are time dependent, but the timescales of the different processes may be very different. Classic toxicology makes the difference between acute (short-term) and chronic (long-term) effects, and the same chemical may target different organs at short and long term, such as ethanol, which acutely affects the nervous system but chronically damages the liver. Therefore, the overall effect at organismic level may be more difficult to predict than reflected in the scheme of Fig. 13.1.

Despite this complexity, we can summarize some general principles of toxicology: (1) There is a **causal relationship** between exposure to a toxicant and harmful effects due to toxicant reaching the action site; (2) this relationship is **quantitative**: the magnitude of effect depends on the amount reaching the **site of action**; (3) the overall effect depends not only on the amount (dose or concentration) reaching the target but also on the **duration** of exposure. The latter is reflected in Haber's rule: for a given poisonous gas $C \times T = k$, where C is the concentration, T is the time necessary to produce a given toxic effect, and k is a constant. This law has been extended to many other toxicological models and holds true as far as toxicokinetics keep steady state (continuous exposure, uptake equals elimination) and no repair mechanisms (e.g., induction of synthesis of detoxification enzymes) appear. In practice departures from those simple conditions do take place, but the law is still useful in risk assessment as a worst-case scenario.

The most basic and universal mechanism of toxicity is shown by chemicals that target biological membranes and cause a disturbance in their structure and functioning. These chemicals are called **narcotics** since when administered to humans cause sleep at low doses and lack of sensation at higher doses, and they were thoroughly investigated due to their applications in anesthesia. This mechanism of toxicity does not involve chemical reactions, and it is thus termed **nonreactive toxicity** or **narcosis**. Since biological membranes, including those of neurons, consist of a lipid bilayer, hydrophobic chemicals may cause membrane dysfunction by changing its fluidity and permeability to ions.[3,a]

Nonreactive toxicity or narcosis targets biological membranes

The toxicity of narcotics can be modeled as a function of the K_{ow}, which describes the hydrophobicity and thus the affinity for membrane lipids of the molecules. This is an example of a QSAR model, which allows prediction of biological properties (in this case the toxicity) as a function of some physicochemical characteristic of the substance (in this case the K_{ow}). When this is done two different equations result for **nonpolar** and **polar narcotics**, the latter about 5−10 fold more toxic. Polar narcotics show a higher affinity for membrane phospholipids than nonpolar ones. When this is taken into account by using a nonpolar hydrocarbon, n-hexane, rather than octanol as neutral lipid surrogate, and two lipidic phases—the polar lipids of the membrane and the neutral storage lipids—are considered, then the estimated membrane critical residuals are the same for nonpolar and polar narcotics.[4] It seems that the differences in QSAR models for both kinds of narcotics are an artifact due to the limitations of octanol as membrane lipid surrogate, and these differences disappear when a more realistic surrogate of biological membranes, i.e., artificial liposomes made of actual membrane phospholipids, is used.[5]

Chemicals may show toxicity mechanisms that involve effects on biomolecules such as proteins or DNA by direct interaction with those biomolecules or indirectly through interaction with a molecular receptor which may trigger a cascade of biochemical effects. These instances of **reactive toxicity** may target universal metabolic pathways such as uncoupling of oxidative phosphorylation, DNA binding and dysfunction of gene expression, inhibition of basic enzymes, or oxidative stress.

Reactive toxicity may target universal metabolic pathways such as ATP synthesis ...

Hydrophobic compounds carrying ionogenic groups, such as substituted phenols, may have, in addition to narcosis, additional toxicity by interfering

---

[a] The exact mechanism thought is not known, and maybe simply by direct interaction with the membrane lipids, or indirectly by more specific interactions with amphiphilic (with both hydrophobic and hydrophilic properties) regions of membrane proteins, as proposed by Franks NP, Lieb WR, Mechanisms of General Anesthesia. Environmental Health Perspectives 1990;87:199−205.

with the ion gradients that are the basis for fundamental processes such as ATP synthesis. These chemicals, weak organic acids, are called **uncouplers** because they can transport protons across membranes short-cutting the transmembrane ATP synthase and thus uncoupling the respiratory chain from the ATP synthesis.[6]

... or very specific metabolic pathways, such as the very selective pesticides

Reactive toxicants may also target sites of action in metabolic routes very specific for certain taxonomical groups. This **selective toxicity** is the basis for the development of pesticides, intended to be toxic only for a limited set of target species and innocuous for the remaining nontarget organisms.

With independence of the metabolic pathway targeted, toxicokinetic issues, specific for the various patterns of organization, impose dissimilar effects of the same toxicants with the same toxicity mechanisms on different systematic groups and life stages within each group.

13.2   TESTING AQUATIC TOXICITY; CONCENTRATION: RESPONSE CURVES, MEDIAN EFFECTIVE CONCENTRATION AND TOXICITY THRESHOLD

Toxicity tests may follow concentration: response or time-to-death designs

There are two fundamental practical questions a toxicologist is interested to answer. Which of a series of chemicals is more toxic? Which is the safe dose for that chemical? These questions are best answered experimentally by toxicity testing using in vivo or in vitro biological models, and estimation from the tests results of two toxicity parameters, the median effect concentration, and the toxicity threshold (TT). We will deal with the methods to estimate these parameters from a toxicological data set but first we must introduce the main toxicity testing designs.

Ecotoxicology departed from human toxicology when aiming at the protection of the wild organisms, rather than human health, but—perhaps unfortunately—inherited the standard methods of human toxicology, including toxicity testing protocols with emphasis on lethal effects and short exposure times. As stated in the basic principles of toxicology (Section 13.1), toxicity depends not only on dose (or concentration, for aquatic exposures) but also it is directly proportional to the duration of exposure. Hence, toxicity test design may follow two different patterns, the most common **concentration:response** experiments, where exposure duration is arbitrarily fixed to a constant value, or the **time-to-death** curves (also called mortality curves), which are experiments where exposure concentration is fixed and individual deaths, or other discrete events, are recorded through time.

Concentration: response curves are fitted to regression models to derive the median effect concentration

In **concentration:response** tests the toxicity of a substance is experimentally described by the quantitative relation between the exposure concentration and a biological response recorded at a fixed exposure time, typically 48—96 h. Groups of individuals of a test species are exposed for that period to fixed concentrations of the substance, some biological response, typically mortality, is recorded, and the so-obtained data are fitted to different mathematical models termed concentration:response curves. The equations of those curves allow calculation of a simple parameter, the **median effective concentration (EC$_{50}$)**, corresponding to the model-derived theoretical concentration causing a 50% reduction in the biological endpoint recorded. When that endpoint is survival the parameter obtained is the **median lethal concentration (LC$_{50}$)**. The toxic potency of a chemical on the test species increases as the EC$_{50}$ decreases, which answers the first question we addressed, and allows chemicals to be ranked according to their toxicity for the test species. For example, Table 13.1 shows acute EC$_{50}$ values of different chemicals to early life stages of bivalves, ranked from less toxic (higher EC$_{50}$) to the most toxic (lowest EC$_{50}$).

In aquatic toxicology, acute toxicity is frequently measured in 48 h tests for invertebrates and 96 h for fish, while for chronic toxicity, less frequently reported, exposure time must last a significant portion of the life-span of the biological model, typically weeks or months. The term subchronic is occasionally used when exposure time is longer than acute but less than one-third of the time to sexual maturity of the test organism. The purpose of chronic exposures is to offset the effect of exposure time and provide time-independent toxicity parameters. Acute to chronic ratios show values around 100-fold increased sensitivity in chronic exposures, although the value for metals seems to be consistently higher than for organic toxicants.[7]

Aquatic acute toxicity tests normally take 2−4 days and chronic tests a few months

To cope with the effect of time on the results of short-term concentration: response tests it was suggested to repeat the tests at increasing exposure times, plot logarithm of EC$_{50}$ versus logarithm of duration, and calculate a time-independent EC$_{50}$ by extrapolation to infinite duration. This approach

Table 13.1 Median Effect Concentration (48 h-EC$_{50}$) of Different Substances Dissolved in Seawater for Bivalve Embryos, Using Normal Embryo Development as Endpoint

Substance	EC$_{50}$ (mg/L)
Acetone	1000
Crude oil	100−1000
Fe	30
fuel-oil	1−10
Chloroform	1−10
Parathion	1
DDT	1
SDS (surfactant)	0.7
Hydrogen sulfide	0.3
Copper sulfate	0.03
Cl	0.01−0.1
Hg	0.01
Sodium cyanide	0.01
TBT	0.001

though is time-consuming and requires much animal testing. An alternative approach is to conduct time-to-death tests, where concentration is fixed and mortality (or any other discrete sublethal response) is recorded for each individual through time. The resulting data set allows calculation of the median lethal time (LT_{50}), or median time needed for 50% of the tested individuals to die out at a given exposure concentration. This approach is sometimes advocated as more ecologically meaningful than the classical short-term concentration:response approach.[8]

The effect of exposure time on a given endpoint is very relevant for toxicity testing design, but it must not be mistaken with the potential mechanistic differences in the toxic effects of a given chemical after short-term and long-term exposures (See Section 13.1). Acute and sustained exposures may target separate metabolic routes, physiological processes, or even different life stages. In fact, the common derivation of separate values for acute and chronic environmental quality criteria (EQC) stands as a useful strategy of environmental protection as long as both sets of EQC values were funded on solid and independent ecotoxicological information (See Chapter 19.4).

In any case, the LC_{50}, i.e., the concentration causing 50% mortality in the exposed population, is the toxicological parameter most frequently reported in aquatic toxicology, and thus the one chosen to compare the toxicity of different chemicals or to classify substances according to their toxicity. Following the assessment criteria used for the prioritization of hazardous substances,[9] chemicals with short-term LC_{50} below 1 mg/L can be considered as highly toxic, whereas those with short-term LC_{50} above 100 mg/L can be regarded as showing low toxicity (see also Table 13.1). A median (50%) level of effect is favored over other effect levels because in regression models the variance is lowest at the middle of the toxicity curve,[10] and thus this parameter can be estimated with less uncertainty than lower or higher levels of effect.[11]

Toxicity thresholds (TT) can be estimated from no observed effect concentration (NOEC)/lowest observed effect concentration (LOEC) or from $EC_{05/10}$

Having a quantitative parameter to compare toxicants, the EC_{50}, we can address now the second fundamental question. Which is the safe dose for a chemical? Obviously a 50% mortality, or a 50% reduction in any biological trait relevant for the fitness of the individual (growth, development, reproduction) is far too high to be environmentally acceptable, and the parameter we actually wish to know is the TT, i.e., the concentration above which deleterious effects begin to take place. Totally safe concentrations would be those causing 0% effect. In toxicity testing this is termed **NOEC**, and the lowest concentration tested showing a significant effect is termed **LOEC**. Between both experimental parameters lays the theoretical TT. In practice, calculation of NOEC and LOEC is strongly dependent on experimental design (tested concentrations, number of replicates) and the power of the statistical test used. A more robust estimate of the TT may be obtained from a dose:response curve by using a regression model and calculating a lower effect level such as a 10% or 5% reduction in the endpoint, resulting in the EC_{10} or EC_{05}, respectively. The level of effect selected to estimate the TT depends on both ecological and statistical constraints. On the one hand, it should correspond to an ecologically acceptable reduction of the biological response, which can vary depending on the biological significance of the response measured. On the other hand, the parameter must meet basic statistical requirements of regression models. First, regression curves should not be extrapolated beyond the range of effect levels obtained experimentally, and second, confidence intervals of the parameter will enlarge, enhancing the chance to obtain a parameter not significantly different to zero, as the level of effect decreases.

Strengths and limitations of the NOEC/LOEC (hypothesis testing approach) and EC_x (regression approach) have been thoroughly discussed from a statistical standpoint.[12,13] However, the quality of the data set, in terms of a proper experimental design, the technical quality of the experiments, and the use of healthy and homogeneous biological material, is the key aspect to obtain reliable toxicological inferences. If the quality of the experiments and reproducibility of results are not assured the parameters inferred would be poor disregarding the statistical treatment chosen. Low background effects, moderate biological variability, presence of no-effect and partial-effect treatments, and narrow confidence intervals of the EC_x parameters are features that support the quality of toxicity testing data.

Also, the concept of a universally safe concentration such as NOEC clashes with the biological variability in susceptibility within a natural population, which can be easily explained taking into account the number of biological processes mediating between exposure and effect (See Fig. 13.1). The regression approach in contrast conforms with this natural variability, and it is in harmony with the empirical finding that dose triggering a discrete biological effect in a group of individuals is not fixed but follows a log-normal distribution (see next Section). Therefore, despite the appeal of safe or no-effect values, TTs estimated from low levels of effect derived from regression models result more useful for environmental management.

As we have just discussed, toxicity of chemicals to aquatic organisms is commonly quantified in terms of water concentration causing a certain level of effect, typically the LC_{50}. However, we have seen (Section 13.1) that effect is determined by the dose at site of action, not at site of exposure, and the fraction of exposure dose that reaches the site of action is dependent on the toxicokinetic processes that in turn are markedly influenced by biochemical, physiological, and anatomical variability. Thus, a more accurate but also more complicated description of a chemical's toxicity is achieved by measuring the internal concentration in an organism at or near the target site, at the time of mortality (or some other sublethal endpoint). This is the case of the **critical body burden** (CBB), which has received many different names (lethal body burden,[14] critical body residue, internal lethal concentration,[15] etc.).

Unlike LC_{50}, lethal body burden may be independent of exposure time

When only baseline or nonreactive toxicity is studied, then the CBB for a given chemical becomes independent of exposure time,[16] overcoming an important limitation of the LC_{50}. Furthermore, when the toxicity of structurally similar chemicals is compared, a common value is found as long as the chemicals were hydrophobic, nondegradable in the time required for bioaccumulation to occur, and with limited chemical reactivity.[17] For example, for three polar narcotics tested on fish, A.P. van Vezel et al.[18] found CBB values ranging from 1.1 ± 0.3 to 1.8 ± 0.6 mmol/kg, disregarding pH that affects the degree of ionization of these compounds and fixed time LC_{50} values. Ionized forms are taken up more slowly, but after a longer period a similar CBB is reached in the fish. Unfortunately, the constancy of CBB values does not hold for reactive chemicals with additional mechanisms of toxicity, such as uncoupling of ATP synthesis or enzyme inhibition.

QSAR models predict toxicity on the basis of chemical properties of related molecules

When experimental data are not available for a particular chemical but experimental toxicity data sets for similar molecules on the same test species are available, the toxicity of that particular chemical may be modeled as a function of some structural or physicochemical characteristic of the substance, typically its K_{ow} partition coefficient, using **QSAR models**. A wide range of computer programs facilitate the implementation of more complex QSAR models taking into account different properties of the molecule such as molecular weight, molecular topology, steric qualities, water solubility, partition coefficients, nucleophilicity, electrophilicity, acid dissociation, etc. The need of toxicity data to meet statutory requirements for new chemicals, the pressure for reduction of animal testing, and reasons of cost have led to an abuse of these models. However, modeled toxicity parameters are only reliable when based on sufficient experimental data, and experimental values will always be more accurate provided testing follows adequate standard procedures.

13.3   LOG-NORMAL DISTRIBUTION OF INDIVIDUAL SUSCEPTIBILITY; PROBIT ANALYSIS AND OTHER CONCENTRATION: RESPONSE MODELS

Toxicological practice found two regularities:

After more than a century of toxicological practice in the laboratory with very different biological models, common patterns of response have been observed that support two general conclusions relevant for the experimental design of toxicity tests and the mathematical calculation of toxicity parameters.

… logarithmic increases in dose cause linear increases in the biological response …

1. Logarithmic increases in dose cause linear increases in the biological response. This can be illustrated with the classical experiments of A.W. Greenwood et al.,[19] who recorded the growth of the comb, a secondary sexual character in male chicken, as a function of the dose of testosterone administered by injection in castrated cockerels. The response was linear only once the dose of testosterone administered was expressed in log scale (Fig. 13.2). In fact, this is related with Fechner's law, developed in the 19th century in the field of human physiology, which in its simplest form states that sensation is proportional to the logarithm of the stimulus intensity. This empirical

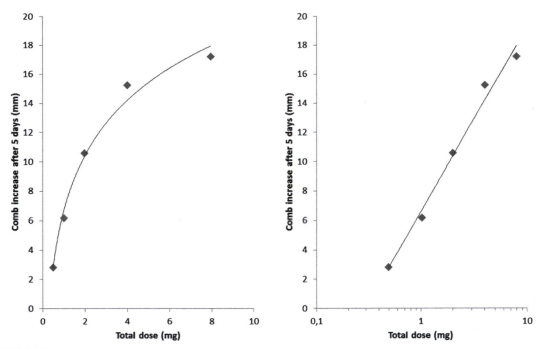

FIGURE 13.2

Pattern of response of a biological endpoint—the growth of the comb in castrated cockerels—against the dose of testosterone administered, represented either in linear (left) or log scale (right). It can be noted that linear increments in the biological response are obtained only as a result of logarithmic increments of the dose. *Redrawn from Greenwood et al. (1935) op. cit.*

law seems to be universal for any biological response elicited by an exogenous chemical. A rather overlooked implication of this law is that when we wish to average biological variables the geometric and not the arithmetic mean must be used.[20]

2. The variation in the sensitivity of individuals within a population to a given chemical follows a normal distribution, provided that the dose is expressed in a log scale: i.e., sensitivity to a chemical follows a log-normal distribution within a population. Again, this was observed already in the early 20th century, for example by C.I. Bliss and coworkers[22] who found that the individual lethal dose of a natural drug, the extract of *Digitalis*, on cats followed a log-normal distribution (Fig. 13.3). Thus, even though all of the cats came from the same source and doses were corrected by body weight, sensitivity to this drug showed in the cat population a normally distributed biological variability that can be described similarly to any other quantitative trait by a mean and a standard deviation.

... and individual susceptibility to a chemical in a population follows a log-normal distribution

Now, the total number of cats killed by a given dose of *Digitalis* will be all the cats whose individual lethal dose is lower or equal to that dose, i.e., the cumulative-frequencies normal distribution curve, which shows the typical

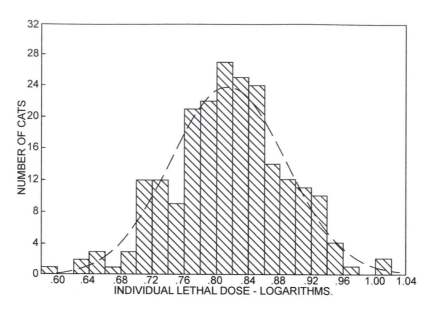

FIGURE 13.3

Frequency distribution of the variable number of cats plotted against the logarithm of the individual lethal dose. The dashed line curve is that expected from the theoretical normal distribution. *From Bliss CI. The U. S. P. Collaborative Cat Assays for Digitalis. J Am Pharm Assoc 1944; 33(8):225–245.*

sigmoid shape so commonly found in toxicological experiments (Fig. 13.4 left; see also Fig. 1.2). The changes in slope of this curve reflect the differences in relative frequencies of the typical bell-shaped normal curve. Values close to the mean have higher frequencies, which reflects on a steeper slope in the middle of the sigmoid curve that decreases as we depart toward both extremes due to the corresponding decrease of frequencies.

The percentage of cats killed by a given dose of *Digitalis* would be the area below the curve to the left of the abscissa corresponding to the dose administered. We are interested in the equation of that curve since it relates dose and response, and allows the quantification of the potency of a toxicant in terms of EC_{50} or some other effect level. Calculation of that area involves the integration of the rather complex equation of the normal distribution, nonlinear fitting, and the need of additional tests to check the goodness of fit. Although this can be computed with the help of statistical software, in the past it was achieved by an ingenious transformation that linearizes the sigmoid shape of the normal curve of cumulative frequencies: the probit analysis.[24] In brief, probit analysis takes advantage of the normal distribution of the individual susceptibilities in log scale and replaces the experimental doses (x) by their normal equivalent deviates (the value minus the mean divided by the standard deviation) plus five (to avoid negative values): $Y = 5 + (x - \mu)/\sigma$. The resulting units are called **normal probability units** or "**probits.**"

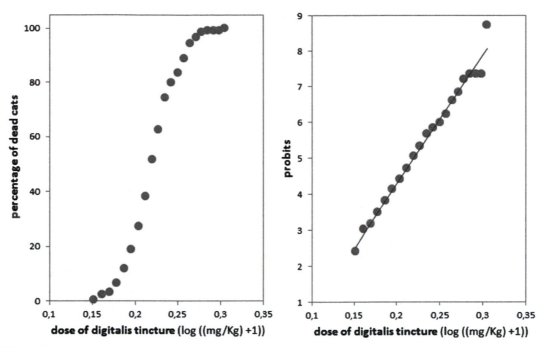

FIGURE 13.4

Percentage of cats killed by a given dose of *Digitalis* expressed in linear scale (left) or in probability units (probit) scale (right) versus log dose. Note that probit transformation linearizes the response and allows calculation of LC_{50} using linear regression. *Data from Bliss (1944) op. cit.*

Now, for each proportion or percentage of response P the corresponding value in probits is calculated as the abscissa that corresponds to the same value P of probability in the transformed normal distribution function with mean 5 and variance 1.

From the integration of that function for doses corresponding to different percentages of effect, tables were obtained relating percentage of response to probits. In the precomputer era those tables allowed transforming toxicity data, in terms of percentage of mortality, to probits.[b] Once data are expressed as probits (Y), they fit to linear functions of log of dose (D): $Y = a + b \log D$ (Fig. 13.4), and toxicity parameters and their confidence levels could be obtained simply using linear regression. The EC_{50} will thus be the value of D corresponding to a $Y = 5.00$, or the EC_{10} that corresponding to $Y = 3.72$.

Because of the complexity of the cumulative normal distribution function, during the precomputer era other transformations equally useful to linearize the response (P) but not requiring the use of complex statistical tables were advocated, such as the angular transformation (arcsine of square-root of P),

---

[b] In addition to analytical solutions, probability scale paper, were Y axis was printed in probit scale, allowed to plot cumulative normal distribution curves as straight lines.

the logit transformation ($\ln (P/(1 - P))$), the transformed logit (($\text{logit}/2) + 5$), or the Weibull transformation ($\ln[-\ln(1 - P)]$). The latter has the advantage to be modifiable to fit asymmetrical (skewed) datasets. Nonlinear regression models such as the two and three parameter logistic equations yield also good fitting to dose:response curves, and show the practical advantage that one of the fitting parameters is already the EC_{50}. For most toxicological data sets, the EC_{50} values estimated by the different models are in fact quite similar,[26] but only the probit analysis is based on the empirical laws described at the beginning of this section.

13.4   BIOTIC AND ABIOTIC FACTORS AFFECTING TOXICITY

Toxicity tests must standardize all biotic and environmental factors affecting toxicity

Numerous biotic and abiotic factors affect the toxicity of a given substance, and thus the value of the EC_{50} and TT. This must be borne in mind when designing toxicity tests and comparing their results. Standard methods for toxicity testing place emphasis on these factors that interfere with the results. Among the biotic factors, species, size, sex, or life stage, and among the abiotic factors, temperature, salinity, pH, organic matter, and interaction with other chemicals, must be taken into account. Since toxicity tests are conducted in a laboratory, special care must be taken in using individuals acclimated to the testing conditions. Lack of thermal or ionic acclimation usually enhances sensitivity. Inadequate maintenance conditions of the test organisms, in terms of feeding, density, or stress, are also expected to increase sensitivity, and it is common that marine toxicity tests conducted by inexpert laboratories yield lower EC_{50} values than those from quality-assured facilities.

The first choice in toxicity testing is the election of the **test species**. Chemicals showing nonspecific mechanisms of toxicity (narcotics, uncouplers) are expected to show similar effects on systematically distant model organisms. On the other hand, selective chemicals acting on very specific metabolic routes will be much more toxic for target species. Choice of test species is frequently made based on logistic reasons (organisms easy to maintain and reproduce in laboratory), but caution must be taken not to bias the choice toward very robust organisms. To reduce the chance of selective mechanisms of toxicity to pass unnoticed due to the choice of test species, a battery of tests using systematically distant models (e.g., algae, invertebrates, and fish) is commonly advised.

13.4.1   Size
In terrestrial toxicology, size effects on sensitivity are commonly corrected by expressing dose normalized by body mass (dosage). This is less common in aquatic toxicology.

13.4.2   Sex
When using adult stages, sex is a possible confounding factor, sometimes related to the gametogenic cycle that demands differential energy investment

in males and females. In mammalian toxicology, standard methods normally prescribe the use of individuals with the same sex.

13.4.3 Life Stage

Marine invertebrate embryos and larvae are several orders of magnitude more sensitive to chemical pollution than adults, as expected from the body mass difference and naked body surface of embryos, directly exposed to dissolved chemicals. Table 13.2 illustrates this with the case of mercury toxicity to adults and larvae of three common species: a crab, a shrimp, and an oyster. Those early life stages are the weakest link in the life history of marine annelids, arthropods, mollusks, and echinoderms, and their high susceptibility bear important ecological implications for pollution events whose effects can pass unnoticed if we only pay attention to adult populations.

Early life stages of marine invertebrates are orders of magnitude more sensitive than adults

13.4.4 Temperature

In ectotherms (both invertebrates and fish) metabolic rates are directly related to temperature. This affects not only active uptake but also biotransformation and excretion rates, and thus generalizations cannot be made across classes of contaminants with different limiting toxicokinetic and toxicodynamic processes. Many studies found an increase in the acute toxicity of both metals and organic chemicals with temperature. J.B. Sprague, who also reported this effect, argues that it may be due to a reduction in the time-to-death at higher temperatures rather than a true increase in the intrinsic toxicity of the chemical. He reviews instances of no effect or even an increase of toxicity at low temperature when tests are carried on long enough to suppress the effect of slower uptake kinetics, and concludes that no universal assumptions should be made about temperature effects on toxicity.[27]

Acute toxicity increases with temperature, but this may be merely due to faster uptake

Regarding long-term exposures, increased toxicity at higher temperatures may be due to other factors apart from kinetics. Temperature enhances the metabolic costs of maintenance and less energy reserves may remain available for active detoxification processes. On the other hand, physiological acclimation must

Table 13.2 Acute Toxicity of Mercury to Adult and Larvae of Three Marine Species. Larvae Show an Approximately 100—1000-Fold Higher Sensitivity Than Adults

Sp	Hg 48-h LC$_{50}$ (mg/L)		
	Adult	Larva	A/L Ratio
Carcinus maenas	1.2	0.014	86
Crangon crangon	5.7	0.01	570
Ostrea edulis	4.2	0.003	1400

Data from Connor PM. Acute toxicity of heavy metals to some marine larvae. Mar Pollut Bull 1972;3(12): 190—192.

also be taken into account when studying the long-term effects of temperature on toxicity. As a result of the physiological plasticity of ectotherms, median effective concentrations increase (i.e., toxicity decreases) when test organisms are previously acclimated to different exposure temperatures.

13.4.5   Salinity and pH

Trace metal toxicity decreases with salinity ...

In oceanic waters, salinity and pH values show little variation in the water column, but this is not the case for shallow estuarine waters where tidal cycles, riverine inputs, and rain conditions may greatly affect those variables that in turn change the chemical speciation of polar organic and inorganic pollutants. For similar reasons, salinity and pH may vary also in pore waters from coastal sediments.

It is well known that toxicity of trace metals present in solution as divalent cations decreases as salinity increases, within the range of salinities tolerated by the test species.[28] This is consistent with the increasing dilution effect caused by nontoxic Ca^{2+} and Mg^{2+} competing for the sites of action with the trace metal ions, as predicted by the FIAM model (see Section 10.4).

For polar organics the unionized form is more readily taken up, and thus more toxic, since it diffuses through biological membranes, and water pH is the major factor influencing ionization. Similarly, pH greatly affects toxicity of ammonium, since toxicity is due to the unionized form, NH_3, more abundant at high pH values where the equilibrium in Eq. (13.1) is displaced to the left.

$$NH_3 + H^+ \leftrightarrow NH_4^+ \hspace{3cm} \text{(Eq. 13.1)}$$

13.4.6   Organic Matter

... and dissolved organic matter

As discussed in Section 10.4, dissolved organic matter may bind metal ions and render them not bioavailable to filter feeders. Humic substances in particular show a remarkable capacity to sequester metals and low molecular weight organics, and since they are too large to be taken up through the gills may play a role in suppressing the biological effects of potentially toxic contaminants.[29]

13.4.7   Interactions With Other Chemicals: Mixture Toxicity

Toxicity of a chemical can be enhanced (synergism), reduced (antagonism), or unaffected (additivity) by the presence of another

Additivity allows prediction of joint toxicity by summing the toxic units

The toxicity of a chemical can be enhanced (**synergism**), reduced (**antagonism**), or unaffected by the presence of another toxicant. If two chemicals in a mixture act through the same mechanism and do not affect the toxicity of each other, simple similar joint action, also termed concentration addition, is present. The combined effects of two chemicals with simple similar joint action are additive, i.e., they may be predicted by the summation of the concentrations (C) of the individual chemicals after adjustment for the differences in toxic potencies, i.e., their **toxic units (TU)**.

$$TU = C/EC_{50} \hspace{3cm} \text{(Eq. 13.2)}$$

For any pair of chemicals A and B, **additivity** can be experimentally investigated by testing the chemicals both in combination and individually, and calculating the interaction factor (IF), that is:

$$IF = \left(EC_{50(A)m} / EC_{50(A)i}\right) + \left(EC_{50(B)m} / EC_{50(B)i}\right) \qquad \text{(Eq. 13.3)}$$

where EC_{50m} and EC_{50i} are the concentrations of each chemical causing a 50% effect in the mixture and individually, respectively. Combined toxicity is additive when IF = 1.

Plotting the fraction of the individual $EC_{50(A)}$ versus the fraction of the individual $EC_{50(B)}$ that tested in combination produce the same toxicity generates the so-called isobole (a line denoting combinations with the same overall toxicity) graphics (Fig. 13.5). Combinations resulting from additive, or concentration addition, effects generate a straight isobole joining 100% $EC_{50(A)}$ with 100% $EC_{50(B)}$, whereas synergistic interactions produce curved isoboles bent toward the origin (concave), and antagonistic interactions are denoted by convex isoboles.

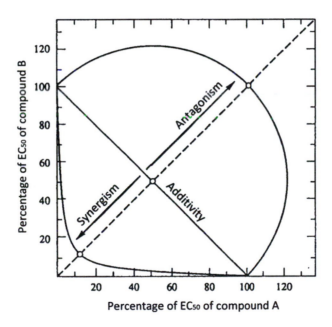

FIGURE 13.5

Isoboles graphic representing, in terms of percentages of the individual EC_{50} values mixed, the combinations of two chemicals A and B showing the same combined toxicity. If there is no interaction and the effects are additive then any combination along the straight line joining 100% $EC_{50(A)}$ and 100% $EC_{50(B)}$ cause a overall 50% effect. Positive interactions (synergism) produce concave isoboles since the same overall effect needs mixing lower fractions of each chemical's EC_{50}, while negative interactions (antagonism) cause the opposite effect. The dashed line represents combinations of the same fraction of each EC_{50}. From Rozman KK, Doull J, Hayes Jr WJ. Dose and time determining, and other factors influencing, toxicity. Chapter 1, pp 3–101. In: Krieger R, editor. Hayes' handbook of pesticide toxicology. 3rd ed. Elsevier; 2001.

Investigating the interaction between two toxicants may be complicated by the fact that this may vary with the proportions of the chemicals used in the mixture, particularly when the individual toxicity curves are not parallel. Therefore, it is customary to test mixtures of equal proportions of the individual EC_{50} values (dotted line in Fig. 13.5).

I.L. Marking and V.K. Dawson[30] conveniently transformed the IF to obtain values of 0 for no interaction (additivity), negative values for negative interactions (antagonism), and positive values for positive interactions (synergism), by calculating an index (I_{MD}) that takes the form $I_{MD} = (1/IF) - 1$ when $IF \leq 1$, and $I_{MD} = -IF + 1$ when $IF \geq 1$.

For n chemicals with additive effects, the overall toxicity of the mixture may be predicted simply from the sum of TU for each chemical present in the water:

$$TU = TU_{(A)} + TU_{(B)} + ... + TU(n) \hspace{2cm} \text{(Eq. 13.4)}$$

In fact, for complex mixtures slight departures from additivity in both senses— positive and negative interactions—tend to compensate one another, and the concentration addition model provides useful estimates of the total effect. This is particularly relevant for the management of wastewater effluents since the presence of multiple chemicals may cause overall toxicity in the receiving waters even when each individual chemical was present at a concentration below the respective TT.

KEY IDEAS

- There is a causal relationship between exposure to a toxicant and harmful effects due to toxicant reaching the action site. This relationship is quantitative: the magnitude of effect depends on the amount reaching that site. The overall effect depends on the amount reaching the target and the duration of exposure.

- Mechanisms of toxicity include unspecific alteration of cell membranes (narcosis), uncoupling of oxidative phosphorylation or interaction with biomolecules (proteins, DNA, etc.) that cause alteration in their structure and function (e.g., enzyme inhibition).

- Toxicity tests may follow concentration:response or time-to-death designs. The first produce estimates of the median effective concentration (EC_{50}), termed LC_{50} when the endpoint is survival, and the TT.

- The TT may be estimated by testing the significance of differences with control (NOEC/LOEC approach) or by calculating a concentration causing a low level of effect using regression models.

- Unlike LC_{50}, lethal body burden may be independent of exposure time.

- QSAR models, useful in risk assessment, predict toxicity on the basis of chemical properties of related molecules such as solubility or partition coefficients.

- Toxicological practice found two empirical laws: effect linearly increases as dose geometrically increases, and the individual susceptibility follows a log-normal distribution in a population.

- Early life stages of marine invertebrates are orders of magnitude more sensitive to chemicals than adults.

- Salinity and organic matter decrease the toxicity of metals.

- Toxicity of a chemical can be enhanced (synergism), reduced (antagonism), or unaffected (no interaction) by the presence of another. The combined toxicity of chemicals with the same mechanism and no interaction is predicted by summing the TUs, calculated by dividing the concentration of each chemical by its individual EC_{50}.

Endnotes

1. Rozman KK, Doull J, Hayes Jr WJ. Dose and time determining, and other factors influencing, toxicity. Chapter 1, pp 3−101. In: Krieger R, editor. Hayes' handbook of pesticide toxicology. 3rd ed. Amsterdam: Elsevier; 2010.

2. Renwick AG. Chapter 4. Toxikokinetics. In: Ballantyne B, Marrs T, Syversen T, editors. General and applied toxicology, vol. 1. London: Macmillan Reference Ltd.; 1999. pp. 67−95.

3. Reviewed by Escher BI, Schwarzenbach RP. Mechanistic studies on baseline toxicity and uncoupling of organic compounds as a basis for modeling effective membrane concentrations in aquatic organisms. Aquat Sci 2002;64:20−35.

4. van Vezel AP, Punte SS, Opperhuizen A. Lethal body burdens of polar narcotics: chlorophenols. Environ Toxicol Chem 1995;14:1579−1585.

5. Vaes WHJ, Urrestarazu Ramos E, Verhaar HJM, Hermens JLM. Acute toxicity of nonpolar versus polar narcosis: Is there a difference? Environ Toxicol Chem 1998;7(7):1380−1384.

6. Terada H. Uncouplers of oxidative phosphorylation. Environ Health Perspec 1990;87:213−218.

7. Lange R, Hutchinson TH, Scholz N, Solbé J. Analysis of the ECETOC aquatic toxicology (EAT) database. II. Comparison of acute to chronic ratios for various aquatic organisms and chemical substances. Chemosphere 1998;36(1):115−127.

8. Newman MC, Dixon PH. Ecologically meaningful estimates of lethal effects on individuals. In: Ecotoxicology: a hierarchical treatment; 1996. pp. 225−2453.

9. Hedgecott S. 16: Priorization and standards for hazardous chemicals. In: Calow P, editor. Handbook of ecotoxicology, vol. 2, Oxford: Blackwell. pp. 368−393.

10. Finney DJ. Statistical method in biological assay. London: Griffin; 1978.

11. Smith EP, Cairns J. Extrapolation methods for setting ecological standards for water quality: Statistical and ecological concerns. Ecotoxicology 1993;2:203−219.

12. OECD. Current approaches in the statistical analysis of ecotoxicological data: a guidance to application. OECD Environmental Health and Safety Publications Series on Testing and Assessment, No 54, Paris: Environment Directorate; 2006. 147 pp.

13. Newman MC, Clements WH. Ecotoxicology, a comprehensive treatment. Boca Raton: CRC Press; 2008.

14. de Bruijn J, Yedema E, Seinen W, et al. Lethal body burdens of four organophosphorous pesticides in the guppy (*Poecilia reticulata*). Aquatic Toxicol 1991;20:111−122.

15. Chaisuksant Y, Yu Q, Connell DW. The Internal Critical Level Concept of Nonspecific Toxicity. Rev Environ Contam Toxicol 1999;162:1−41.

16. de Bruijn et al. (1991) op. cit.

17. See Endnote 15.

18. See Endnote 4.

19. Greenwood, AW, Blyth, JSS, Callow, RK. Quantitative studies on the response of the capon's comb to androsterone. Biochem J Jun 1935;29(6):1400−1413.

20. Galton F. The geometric mean, in vital and social statistics. Proc R Soc London Ser B 1879;29: 365−367.

21. Greenwood et al. (1935) op. cit.

22. Bliss CI, Hanson JC. Quantitative estimation of the potency of Digitalis by the cat method in relation to seculr variation. J Am Pharm Assoc 1939;28(8):521−530.

23. Bliss CI. The U. S. P. Collaborative Cat Assays for Digitalis. J Am Pharm Assoc 1944;33(8): 225−245.

24. Finney DJ. Probit analysis. 3rd ed. Cambridge Univ Press; 1971. 333 pp.

25. Bliss (1944) op. cit.

26. Newman MC. Quantitative methods in aquatic ecotoxicology. Lewis Publ. 1994. (p. 120).

27. Sprague JB Measurement of pollutant toxicity to fish. II. Utilizing and applying bioassay results. Water Res 1970;4:3−32.

28. Mclusky DS, Bryant V, Campbell R. The effect of temperature and salinity on the toxicity of heavy metals to marine and estuarine invertebrates. Oceanogr Mar Biol Ann Rev 1986;24: 481−520.

29. Ortego LS, Benson WH. Effects of dissolved humic material on the toxicity of selected pyrethroid insecticides. Environ Toxicol Chem 1992;11:261−265.

30. Marking, LL, Dawson, VK. Method of assessment of toxicity or efficacy of mixtures of chemicals. US Fish Wildlife Service Invest. Control 1975;67:1−8.

Suggested Further Reading

• Finney DJ. Probit analysis. 3rd ed. Cambridge Univ Press; 1971. 333 pp.

• Newman MC. Fundamentals of ecotoxicology. The science of pollution. 4th ed. Boca Raton: CRC Press; 2015. 654 pp.

• OECD. Current approaches in the statistical analysis of ecotoxicological data: a guidance to application. OECD environmental health and Safety Publications Series on Testing and assessment, No 54. Paris: Environment Directorate; 2006. 147 pp.

• Rozman KK, Doull J, Hayes Jr WJ. Dose and time determining, and other 5 factors influencing, toxicity. In: Krieger R, editor. Hayes' handbook of pesticide toxicology. 3rd ed. Elsevier; 2010 [Chapter 1], pp. 3−101.

CHAPTER 14

Sublethal Toxicity at the Level of Organism

14.1 ENERGY COSTS OF DETOXIFICATION AND REPAIR

Sublethal toxicity of chemicals can be investigated at various levels of organization. We will deal here only with those that are recorded on organisms, descending to the molecular mechanisms only to explain the basis of the variables measured in the individuals. Also, the key aspect of a sublethal effect on an individual, from the standpoint of long-term ecosystem protection, is its relevance for the **biological fitness** of the individual that determines its capacity to reproduce and thus to contribute to maintain the size of the population. Indeed, population size is determined by the balance between **mortality** and **reproduction** (birth) rates. The other components of fitness are secondary, and their relevance lies on their indirect effects on mortality and reproduction rates. Impaired embryo development or delayed growth of immature stages decrease the chance of the individual to reach reproductive age. Lack of energy reserves may reduce fecundity or delay spawning to a season unsuitable for larval survival. Reduced motility may increase the chance to be predated or decrease the ability for matching. When pollution affects these fitness-related traits, population sizes of the sensitive species are expected to decline leading to long-term changes in community structure, and thus in ecosystem functioning.

> **Ecologically relevant sublethal effects are those related to fitness because they affect population size**

Toxicity testing has proved that the sublethal effects of pollutants on traits that affect the individual's fitness, such as reproduction, growth, development, motility or physiological energetics, frequently take place at concentrations much lower than the LC_{50}. For example the 21-d LC_{50} of Cd for adult *Crassostrea virginica* oysters is approximately 100 times higher than the Cd concentration inhibiting growth after 20-d exposure in the same species.[1] Many marine invertebrates show complex life cycles that include planktonic larval stages and settlement after metamorphosis onto benthic habitats. Timing and success of crucial life cycle events such as gametogenesis, spawning, or larval settlement is affected by environmental factors, including chemical pollution. Therefore, the investigation of the Toxicity Threshold (TT) of pollutants for those

> **Sublethal endpoints may be orders of magnitude more sensitive than LC_{50}**

233

Marine Pollution. https://doi.org/10.1016/B978-0-12-813736-9.00014-3

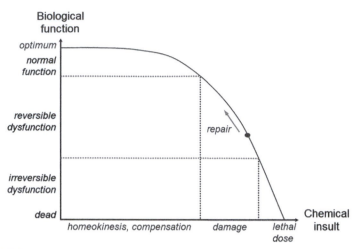

FIGURE 14.1

Schematic representation of the effect of increasing chemical insult on some vital biological function of the organism. Within low ranges homeokinesis and compensation mechanisms allow normal functioning at near optimum level. Above a certain threshold dysfunction that may be reversible thanks to repair mechanisms arises. Compensation and repair demand increasing energy investment.

fitness-related sublethal endpoints is more relevant for the protection of the marine ecosystems than lethal toxicity studies.

Sublethal toxicity demands compensation and repair mechanisms that cause energy expenditure

Sublethal toxicity causing a biological dysfunction arises when the dose at the target site exceeds the capability of the physiological mechanisms to maintain the function within normal values. If we represent the response of the function to increasing chemical concentrations (Fig. 14.1), we will see that in the low range of concentrations potential damage is efficiently buffered by the mechanisms of **homeokinesis** (that maintain the function within normal values) and **compensation** (that reverse the disturbed function to normal values) of the metabolism. Homeokinesis is defined as the capacity of the living organisms to keep constant some activities, i.e., physiological functions, in a variable environment, in analogy to the capacity to keep constant the internal state, i.e., the composition, termed homeostasis.[2] However, when a threshold is exceeded, those mechanisms cannot cope with the increased chemical insult and a sharp decrease in the levels of the biological function (dysfunction) arises. This impairment can be either repaired when the level of damage is reversible, or persist and lead to the dead of the individual. Compensation and repair mechanisms involve energy expenditure.

The energy invested in those mechanisms is subtracted from competing physiological processes such as motility, gametogenesis, or somatic growth. For example, in rainbow trout, dietary copper exposure up to 200 mg/Kg did not result in any observable behavioral effect.[3] Homeostatic adjustments are supposed to buffer that moderate chemical insult. However, exposure to 500 mg/Kg disturbed swimming activity in terms of reduced time spent swimming, although maximum swimming speeds remained unaffected.[4] Cu was accumulated in the intestine and the liver, the latter chelated by metallothioneins, and growth after 3 months was unaffected. In contrast, when the Cu dose increased to 730 mg/Kg the detoxification ability was apparently exceeded, the efficiency of food conversion decreased and growth rate was reduced.[5]

Therefore, the ability of an organism to stay active, grow up, and reproduce may be compromised by the environmental pollutants. There seems to be a hierarchy built up by evolution in the energetic allocation trade-offs. Essential somatic maintenance related to vital functions ranks above reproduction and growth, which in turn results more critical than the maintenance of normal levels of activity. All those sublethal effects are much subtler than mortality, but as long as they depress the individual's fitness they have the potential to affect the **size of populations**, and thus the structure of the biological communities that compose the ecosystems.

14.2 EFFECTS ON PHYSIOLOGICAL ENERGETICS, GROWTH, AND DEVELOPMENT

Living systems seem in contrast with the second law of thermodynamics, which states that any spontaneous process involving energy proceeds at the expense of irreversibly increasing entropy, a state variable inversely related with the order of the system components. Living organisms manage to maintain and develop into increasingly complex structures. In fact, organisms do not escape from the universal thermodynamic constraints, they simply maintain their structure and functioning by consuming external energy and increasing the entropy of their environment. Thus, energy consumption is the common trait that allows any organism to stay alive. For heterotrophic organisms, this energy is obtained from food, and its partition is described by the so-called **energy budget**:

Sublethal levels of chemicals decrease the energy budget, growth performance, and developmental success

$$G = I - (F + E + R) \tag{Eq. 14.1}$$

where G is the surplus of energy available for gametogenesis and somatic growth, I is ingested food energy, F is energy loss through the feces, E is energy loss through excretion of N products (urine, ammonia), and R is metabolic expenditure (usually assessed through respiration).

The energy expenditure demanded by the detoxification processes enhances the negative term R in the energy budget. Other substances may impair the feeding ability or reduce the performance of the digestion and absorption of the food, thus affecting also Eq. (14.1).

Growth performance is one of the most sensitive responses to exposure to contaminants. For example, oyster larval growth is already retarded at 4 µg/L of Hg, whereas normal embryogenesis and larval swimming are affected at 8 µg/L, larval survival at 32 µg/L, and metamorphosis at 64 µg/L.[6] Fast-growing early life stages such as larvae and juveniles are particularly suitable to study these effects, whose ecological relevance is obvious considering that free-living immature stages of marine organisms are submitted to heavy mortalities, and reduced growth rate markedly decreases the chance of survival until reproductive stage. Growth and other sublethal responses may show stimulation by low levels of toxicants, a phenomenon termed hormesis that greatly complicates calculation of toxicity thresholds.[7]

Scope for growth (SFG) in filter feeders provides a sensitive and relevant response to pollution

In adults, measuring growth rates may demand long exposure times. A practical alternative is to measure the SFG, a physiological parameter that consists of the balance in terms of rate functions between the energy gains (feeding, absorption) and losses (excretion, respiration), i.e., the energy available for growth and reproduction. This parameter can be calculated by recording the rates involved in the energy balance (Eq. 14.1), taking into account that in marine invertebrates energy loss associated to excretion of N is frequently considered as negligible, and converting all ratios into the common currency of energy (Joules/h). Thus,

$$SFG = I - F - R = (I \cdot AE) - R \qquad \text{(Eq. 14.2)}$$

where AE is the food absorption efficiency.

In filter feeders these rates can be rapidly recorded in laboratory. Assuming filtering activity causes an exponential decrease in the concentration of suspended particles in a closed system of volume V, then the filtering rate (FR), in L/h, will be,

$$FR = [\log_e(C_0/C_t)] \, V/t \qquad \text{(Eq. 14.3)}$$

where C_0 and C_t are the particle concentration (number per L) initially and after t hours, respectively. For moderate decreases in C, the Ingestion Rate (IR, J/h) may be then estimated as,

$$IR = FR \cdot C \cdot J \qquad \text{(Eq. 14.4)}$$

where C is the particle concentration in the water, in mass units (g/L), and J is the caloric equivalent (J) of the food particles (23.5×10^{-3} J/g) for POM.[8]

AE, in turn, is estimated from the organic matter percentage of food (f) and feces (e) according to the expression.[9]

$$AE = f - e/(100 - e)f \qquad \text{(Eq. 14.5)}$$

Finally, R is assessed by measuring the oxygen consumption in a closed container and applying the oxycaloric equivalent, 0.465 J/μmole O_2.

SFG decreases as chemical pollution increases

An additional advantage of SFG is that it can be recorded in laboratory under standard conditions and still reflects the original energy balance of the organism in the wild, responding to many environmental factors, including pollution. SFG in mussels from both mesocosms and natural environment was inversely related to the PAH body residues according to a semilogarithmic inverse relationship (r = −0.87) without an apparent threshold of effect (Fig. 14.2). In fact, in many cases the SFG response depends mainly on the feeding rate response,

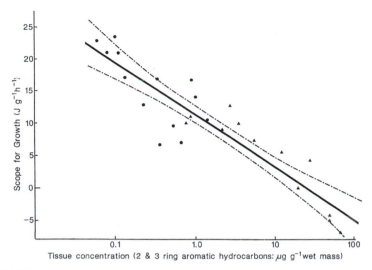

FIGURE 14.2

Relationship between scope for growth (SFG) and the log concentration of two and three ring aromatic hydrocarbons in the tissues of *Mytilus edulis* from mesocosms (triangles) and field studies (circles). SFG is standardized for 1 g mussel and a common ration level, and all measurements were carried out during active growth period. *Modified from Widdows J, Donkin P. The application of combined tissue residue chemistry and physiological measurements of mussels (Mytilus edulis) for the assessment of environmental pollution. Hydrobiologia 1989;188(189):455–461.*

and measuring the latter provides enough sensitivity to detect environmental changes. This is due to the high sensitivity of the gill ciliary activity to both nonspecific narcosis and neurotoxic action of metals and other chemicals.[10]

Thus, the SFG in bivalves has been proposed by J. Widdows and coworkers[11] as a nonspecific, highly sensitive, and ecologically relevant biological tool for the assessment of marine pollution, and it has been advocated as a practical tool in coastal pollution monitoring programs worldwide. The combination of the SFG and chemical analyses in mussels has allowed the identification and quantitative assessment of regions and sites significantly altered by the presence of pollutants.[12] However, caution must be taken when SFG values of mussels from geographically heterogeneous regions or recorded in different seasons are compared, since mussel condition and natural factors such as food availability affect SFG and mask the effect of contaminants on this parameter.[13] In those broad monitoring networks chemical pollutants affecting SFG may be different in each area, and tight relationships between SFG and pollutant bioaccumulation, such as those depicted in Fig. 14.2, must not be expected (see Fig. 17.2).

As illustrated in Table 13.2, early life stages of marine invertebrates are particularly sensitive to chemical pollution, since they are naked life forms with large surface to volume ratio exchanging dissolved substances through the whole body surface. In addition, embryogenesis, and particularly illustrated, is a very sensitive period of the life cycle since differentiation of tissues and organs of the individual is

Teratogenic chemicals interfere with normal embryo development

occurring. In fish, late blastula and early gastrula stages are more sensitive than earlier and later stages, although most research focused on more advanced phases because morphological abnormalities in spine, head or eye are more conspicuous and easy to identify. In sea urchin embryos, frequent models for embryological studies, the onset of gastrulation is the most sensitive period to external factors.[14] Gastrulation and formation of the shell gland are the most sensitive stages to trace metals in the embryogenesis of bivalves.[15] In human and by extension vertebrate toxicology a chemical agent prone to disturb the normal development of the embryo and causing malformations is called a **teratogen** and the effect is termed teratogenicity. In oviparous vertebrates, chemicals may be transferred to embryos during egg formation. In mammals, birth defects may be caused by exposure via the mother, by placental transfer.

14.3 EFFECTS ON REPRODUCTION; ENDOCRINE DISRUPTION

Endocrine disrupting compounds (EDCs) are synthetic chemicals that interfere with the metabolism of hormones or act as their mimics

Sexual differentiation, gametogenesis, and reproduction of animals are controlled by the hormones synthetized by the endocrine system. Foreign compounds, called **endocrine disrupting compounds (EDCs)** can either interfere with the metabolism of hormones or act as their mimics and bind to their receptors, thus impairing these fundamental processes for life. According to the World Health Organization, "an endocrine disrupter is an exogenous substance or mixture that alters function of the endocrine system and consequently causes adverse health effects in an intact organism, or its progeny, or (sub)populations."[16] Many natural compounds are known to have abortive properties, and farmers have long known for instance that grazing sheep on new growth clover reduce pregnancies. We know now this is due to the action of phytoestrogens that can disrupt the normal metabolism of sexual hormones. However, only recent concern has been raised about the potential endocrine disrupting effects of synthetic chemicals such as detergent, cosmetics components, or plastic additives increasingly present in our environment. Since the effects of hormones on gene expression are modulated through interaction with protein receptors, the chemicals that show affinity with those receptors and trigger similar effects are termed agonists, and those blocking the receptor antagonists. Unfortunately, an abuse of these terms without the needed fundamental research on the mode of action is frequent in experimental ecotoxicology literature.

Tributyl-tin (TBT) inhibits the synthesis of estradiol causing imposed male traits in gastropod females

Although sex is genetically determined in nonhermaphroditic species, sexual differentiation is influenced by the conversion rates of the steroid hormones: estrogens (female) and androgens (male). Mori and coworkers[17] demonstrated about half a century ago that exposure of *Crassostrea gigas* oysters to an estrogen during the undifferentiated phase of the gametogenic cycle increased the female sex rate.

Both testosterone, the main male hormone, and estradiol, the main female hormone, originate from the same precursor, cholesterol, throughout a metabolic pathway that includes the transformation of testosterone into 17β-estradiol, a reaction catalyzed by a CYP19 aromatase enzyme. The inhibition of this enzyme by chemicals such as TBT and other organotin compounds may cause **imposex**, the masculinization of females in marine gastropods, a syndrome that can eventually cause the sterilization of the affected individuals (see also Sections 9.8 and 16.5). The mechanism of inhibition may not be a direct interaction with the enzyme but a down-regulation of the aromatase gene expression[18] upon binding of TBT to the retinoid X receptor, whose natural agonist, the retinoic acid, caused imposex when injected to the gastropod *Thais clavigera*.[19]

The massive use of TBT in boat hull paints caused a decline of gastropod populations[20] that recovered after the ban of this product. Fig. 14.3 illustrates this recovery for the case of the Portuguese coast, based on the degree of imposex in wild *Nassarius reticulatus*, a common gastropod present in soft substrates. The

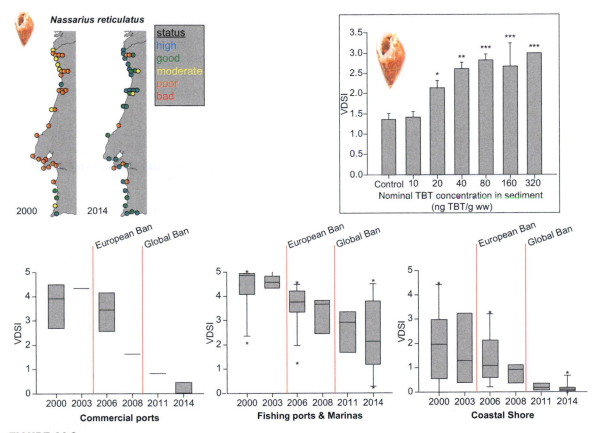

FIGURE 14.3

Recovery in the levels of imposex recorded in wild populations of the soft-substrate gastropod *Nassarius reticulatus* along the Portuguese coast from 2000 to 2014. Data are also separately presented for commercial ports, fishing ports and marinas, and coastal shore. Dates of the European and global ban of tributyl-tin (TBT) for boat hull paints are also shown. The insert shows a dose:response induction of imposex on healthy *N. reticulatus* females conducted in laboratory. *Modified from Laranjeiro FMG. The use of Imposex/Intersex as a tool to assess the ecological quality status of water bodies. (PhD Thesis). Universidade de Aveiro; 2015.*

validation of this method was conducted in laboratory by exposing healthy *N. reticulatus* to sediments spiked with increasing amounts of TBT, and recording imposex—assessed through the VDSI index (see Section 16.5)—after 28 days.

Xenoestrogens cause vitellogenin (VTG) synthesis and other female traits in male fish

The female hormone estradiol (See Figure 8.7) acts through binding to specific protein receptors in the cell. The activated receptor triggers the expression of estrogen-responsive genes related with sex differentiation and gametogenesis. Many synthetic chemicals called **xenoestrogens** have proved to be estrogenic, meaning they show affinity to estrogen receptors (ER) and thus trigger the biochemical pathways activated by the female hormone, though at levels thousands to million times lower than the natural hormone estradiol.

The estrogenic activity may be tested through several in vitro tests, including the use of recombinant yeast strains and vertebrate or fish cell cultures. The first test uses yeast whose genome has integrated the DNA sequence of the human ER, and which contains also plasmids with a reporter gene responsive to the activated receptor. When a synthetic chemical has affinity by the ER, it binds and activates the receptor, which induces the expression of the reporter gene: an enzyme that degrades a chromogenic substrate whose product is measured by absorbance.[22] The affinity is quantified through a dose:response (concentration of the foreign chemical vs. absorbance) curve.

In human toxicology, human breast cancer cells are used to assess estrogenicity. The endpoint measured may be cell proliferation, or more specifically light production, by using a cell line stably transfected with an ER-controlled luciferase reporter gene.

Other in vitro tests use cultured hepatocytes, including hepatocytes from some fish species. The cells are exposed to the suspected xenoestrogen and the production of vitellogenin (VTG) is measured by immunoassay.

Finally, other in vivo and in vitro assays evaluate the induction or suppression of the expression of a specific gene, such as that encoding vitellogenin, through quantification of the corresponding mRNA or protein levels.[23] Omics techniques allow this evaluation to be expanded to the whole mRNA (transcriptomics) or protein (proteomics) pool in the cell. Unlike receptor binding or cell proliferation assays, expression assays can be used to detect both agonists and antagonists of the hormone receptor.[24]

Among the synthetic substances showing estrogenic activity in in vitro tests are pharmaceuticals (e.g., ethinyl estradiol, used in contraceptive pills), components of personal care products (e.g., triclosan), plastics monomers and additives (bisphenol-A, phthalates, benzophenones), the cleaning agent nonylphenol, and some pesticides (metoxychlor, DDT) (See also Section 8.3). However, the in vivo activity of the EDCs cannot be accurately predicted by the in vitro assays, and the predictive value of these tests for adverse effects at individual scale, especially of those concerning human health, is extremely controversial.[25]

Exposure of fish to xenoestrogens has been associated with decreased fertility, decreased hatching success, demasculinization and feminization, and alteration of immune functions. In oviparous species, the production of VTG, an egg yolk protein precursor, is critical for successful reproduction in females. Normally, only mature females produce enough endogenous estrogen to induce vitellogenesis, but exposure to xenoestrogens in the external environment can trigger this response in male and juvenile fish as well.

The simultaneous presence of male and female gonadal tissue in individuals of a species with separate sexes is called **intersex**.[26] The most frequently reported manifestation of intersex is the presence of oocytes within the male gonads, a phenomenon observed in fish as early as 1931 by Goldschmidt. Intersex can be induced by exposure to EDC in laboratory, but in field studies it is difficult to distinguish between natural intersex and intersex due to stressors, and evidence that EDCs were actually impacting the reproductive health of natural fish populations is less convincing.[27]

14.4 BEHAVIORAL EFFECTS WITH IMPLICATIONS ON BIOLOGICAL FITNESS

Behavioral effects have been repeatedly reported as the most sensitive endpoints responding to pollutants. Unfortunately, as sensitivity increases ecological relevance of the measured biological response may decrease. Variables related to motility in its different forms (swimming activity or swimming speed, avoidance, burrowing activity in infaunal animals, changes in taxis), or continuous recording of physiological rates (ventilation rates, heart beating, valve closure in bivalves) have been proposed as sensitive indicators of contaminant effects, and the latter proved useful for real time monitoring of water quality.

Behavioral endpoints are the most sensitive responses to pollutants

Fish ventilatory patterns are conveniently monitored by remote electrodes that detect gill opercular movement, and the frequency or amplitude of certain events such as gill purges (reversed circulation of water through the gills to clean excess mucus) can be continuously recorded and processed by a computer. Provided the normal range of variation in the parameter being monitored was statistically determined, the system can be programmed to deliver a real time alarm when the signal exceeds a given threshold. This is the basis of the **biomonitors**, also called Biological Early Warning Systems (BEWS), which take advantage of some sensitive biological endpoint to enable the operation of an electronic device for the automatic detection of impaired water quality. Other biological responses proposed for use as biomonitors of fresh water environments are fish movement patterns within a tank, activity levels, or rheotaxis (the ability to swim against a current), and valve movements in *Dreissena polymorpha* mussels. Regarding marine environment, valve closure in bivalves, or mussel heart beating rates have provided biological endpoints for similar devices at experimental level.

Patterns of fish gill ventilation or bivalve shell closure allow continuous monitoring of water quality

A different issue is to derive ecologically relevant information from those responses, which requires proof of the relevance of those behavioral changes on the biological fitness of the individual, an exercise very rarely done. Animal behavior integrates genetic, biochemical, and physiological attributes, and behavioral changes do not necessarily lead to the associated increase in energy expenditure discussed in Section 14.1, nor are they easily related to quantitative fitness-related traits. Thus, some behavioral responses may be very useful as early warning signals and at the same time anecdotic from an ecological standpoint.

The clam reburial bioassay provides a simple and ecologically relevant tool to assess sediment toxicity

An interesting exception is provided by the clam reburial experiments. Several studies used borrowing behavior of clams as a biological tool to detect chemical pollutants in natural or spiked sediments.[28] Clam reburial was reported to be up to 20 times more sensitive as endpoint than short term mortality. By counting the number of clams remaining on the sediment surface at short time intervals and adjusting the resulting ratios of buried animals to log-logistic or log-probit functions of time, a ET_{50} value (time needed for complete burial of a half of the population) that provides an objective estimate of burrowing speed can be obtained. The duration of the test is often 24 h, although ET_{50} values in control sediments may be as low as a few minutes. The test detected Cu and Zn in the sediment down to tens of ppm, Cd down to ppm units, and Hg and nonylphenol down to tenths of ppm, and seemed to respond to the pore water concentration rather than the total metal content. The response recorded is clearly relevant for survival, since infaunal organisms must bury into the sediment to avoid predation, and W.H. Pearson et al.[29] demonstrated this in the field by recording a higher consumption of clams by crabs in artificially oiled sediments associated to a shallower vertical distribution of the clams in the contaminated sand. Interference of biological factors (clam size) and grain size are challenges for the routine use of this sediment bioassay.

Lack of standardization and field validation are in fact typical shortcomings of behavioral tests that have limited so far their application to risk assessment or to the derivation of water quality criteria.

KEY IDEAS

- Toxicants cause biochemical, physiological, and behavioral sublethal effects on the organisms at concentrations below those causing mortality. When these effects decrease the biological fitness of the individual, i.e., the ability to grow, reproduce, and survive, then they have ecological relevance because they will potentially decrease the population size.

- From a physiological standpoint, sublethal toxicity demands compensation and repair mechanisms that cause increased energy expenditure. Energy allocated to these mechanisms cannot be invested in somatic growth or production of gametes.

- The feeding rate and energy budget (the difference between the energy gains and losses) of mussels are affected by chemical pollution, and this can be used to monitor water quality, although natural factors such as food availability, temperature, or gametogenic stage interfere with these rates.

- Some chemicals termed EDC alter the normal functioning of the endocrine system. The best-known case in the marine environment is the masculinization of female gastropods induced by TBT, which interferes with the aromatase enzyme that transforms testosterone into estradiol.

- Xenoestrogen EDC chemicals such as some plastic additives induce the synthesis of VTG in male fish. Xenoestrogenicity can also be tested in vitro using genetic engineering techniques.

- Behavioral responses related to swimming activity, bivalve shell closure, or fish ventilation rates may be continuously recorded to provide real-time signs of deteriorated water quality. These devices are termed biomonitors.

Endnotes

1. Taylor D. A summary of the data on the toxicity of various materials to aquatic life. vol. 2. Cadmium. 2nd ed. Brixham: Brixham Laboraotory; 1981.

2. Prosser CL. Adaptational biology. Molecules to organisms. New York: John Wiley & sons; 1986.

3. Handy RD, The assessment of episodic metal pollution. II. The effects of cadmium and copper enriched diets on tissue contaminant analysis in rainbow trout (*Oncorhynchus mykiss*). Arch Environ Contam Toxicol 1992;22:82—87.

4. Handy RD, Sims DW, Giles A, et al. Metabolic trade-off between locomotion and detoxification for maintenance of blood chemistry and growth parameters by rainbow trout (*Oncorhynchus mykiss*) during chronic dietary exposure to copper. Aquat Toxicol 1999;47:23—41.

5. Lanno RP, Slinger SJ, Hilton JW. Maximum tolerable and toxicity levels of dietary copper in rainbow trout (Salmo gairdneri Richardson). Aquaculture 1985;49:257—268.

6. Beiras R, His E. Effects of dissolved mercury on embryogenesis, survival, growth and metamorphosis of *Crassostrea gigas* oyster larvae. Mar Ecol Prog Ser 1994;113:95—103.

7. Stebbing ARD, Hormesis — the stimulation of growth by low levels of inhibitors. Scie Total Environ 1982;22:213—234.

8. Smaal AC, Widdows J. The scope for growth of bivalves as an integrated response parameter in biological monitoring. In: Kramer KJM, editor. Biomonitoring of coastal waters and estuaries. Boca Raton: CRC Press; 1994. pp. 247—267.

9. Conover RJ Assimilation of organic matter by zooplankton. Limnol Oceanogr 1966;11: 338—354.

10. Smaal & Widdows (1994) op. cit.

11. Bayne BL, Moore MN, Widdows J, et al. Measurement of the responses of individuals to environmental stress and pollution: studies with bivalve molluscs. Phil Trans R Soc Lond B 1979; 286:563—581; Widdows J, Staff HJ. Biological effects of contaminants: measurement of scope for growth in mussels. ICES techniques in marine environmental sciences no 40. Int Council Exploration Sea; 2006.

12. Widdows J, Donkin P (1989) op. cit.; Widdows J, Donkin P, Staff FJ, et al. Measurement of stress effects (scope for growth) and contaminant levels in mussels (*Mytilus edulis*) collected from the Irish Sea. Mar Environ Res 2002;53:327–356.

13. González-Fernández C, Albentosa M, Campillo JA, et al. Influence of mussel biological variability on pollution biomarkers. Environ Res 2015;137:14–31.

14. Vlasova GA, Khristoforova NK. The effect of cadmium on early ontogénesis of the sea urchin Strongylocentrotus intermedius. Sov J Mar Biol 1982;8(4):210–215.

15. For sea-urchins see: Zhadan PM, Vashchenko MA, Medvedeva LA, Gareyeva RV. The effect of environmental pollution, hydrocarbons and heavy metal son reproduction of sea urchins and bivalves. In: Ilyichev VI, Anikiev VV, editors. Oceanic and anthropogenic controls of life in the Pacific Ocean. The Netherlands: Kluwer Academic Publishers; 1992. pp. 267–268. For fish see: von Westernhagen H. 4 Sublethal effects of pollutants on fish eggs and larvae. In: Hoar WS, Randall DJ, editors. Fish physiology, Vol. XIA, Academic Press; 1988. pp. 253–346.

16. International Programme on Chemical Safety, World Health Organization.

17. Effect of steroid in oyster III Sex reversal from male to female in *Crassostrea gigas* by estradiol-17B. Bull Jap Soc Sci Fisheries 1969;35:1072–1076.

18. Saitoh M, Yanase T, Morinaga H, et al. Tributyltin or triphenyltin inhibits aromatase activity in the human granulosa-like tumor cell line KGN. Biochem Biophys Res Commun 2001;289: 198–204.

19. Nishikawa J, Mamiya S, Kanayama T, et al. Involvement of the retinoid X receptor in the development of imposex caused by organotins in gastropods. Environ Sci Technol 2004;38: 6271–6276.

20. Bryan GW, Gibbs, PE, Hummerstone LG, et al. The decline of the gas-tropod *Nucella lapillus* around the south-West England: evidence for the effectoftributyltin from antifouling paints. J Mar Biol Assoc. UK 1986;66:611–640.

21. Laranjeiro FMG. The use of Imposex/Intersex as a tool to assess the ecological quality status of water bodies. (PhD Thesis). Universidade de Aveiro; 2015.

22. Routledge EJ, Sumpter JP. Estrogenic activity of surfactants and some of their degradation products assessed using a recombinant yeast screen. Environ Toxicol Chem 1996;15(3): 241–248.

23. Knoebl I, Blum JL, Hemmer MJ, et al. Temporal Gene Induction Patterns in Sheepshead Minnows Exposed to 17β-Estradiol. J Exp Zool 2006; 305A:707–719.

24. Gross TS, Arnold BS, Sepúlveda MS, et al. Endocrine disrupting chemicals and endocrine active agents. Chapter 39. In: Hoffman DJ, et al, editors. Handbook of ecotoxicology. 2nd ed. Lewis Publishers; 2003.

25. Segner H, Navas JM, Schäefers C, et al. Potencies of estrogenic compounds in in vitro screening assays and in life cycle tests with zebrafish in vivo. Ecotoxicol Environ Safety 2003;54: 315–322. DeMott RP, Borgert CJ. Reproductive toxicology. Chapter 11. In: Williams PL, James RC, Roberts SM, editors. Principles of toxicology. Environmental and industrial applications. 2nd ed. New York: Wiley; 2000.

26. Bahamonde PA, Munkittrick KR, Martyniuk CJ. Intersex in teleost fish: are we distinguishing endocrine disruption from natural phenomena? General Comp Endocrionol 2013;192: 25–35.

27. Mills LJ, Chichester C. Review of evidence: are endocrine-disrupting chemicals in the aquatic environment impacting fish populations? Sci Total Environ 2005;343:1–34.

28. Reviewed by Beiras R. Clams as biological tools in marine ecotoxicology. In: Da Costa González, F, editor. Clam Fisheries and aquaculture. New York; Nova Science Publishers; 2013.

29. Pearson WH, Woodruff DL, Sugarman PC, Olla BL. Effects of oiled sediment on predation on the Littleneck Clam, Protothaca staminea, by the Dungeness Crab, Cancer magister. Estuar Coast Shelf Sci 1981;13:445–454.

Suggested Further Reading

- Beiras R. Clams as biological tools in marine ecotoxicology. In: Da Costa González F, editor. Clam Fisheries and aquaculture. New York: Nova Science Publishers; 2013.
- Laranjeiro FMG. The use of Imposex/Intersex as a tool to assess the ecological quality status of water bodies. PhD Thesis. University of Aveiro; 2015.
- Sumpter JP. Xenoendocrine disrupters—environmental impacts. Toxicol Lett 1998;102—103: 337—42.
- Widdows J, Donkin P. The application of combined tissue residue chemistry and physiological measurements of mussels (*Mytilus edulis*) for the assessment of environmental pollution. Hydro-biologia 1989;188(189):455—61.

CHAPTER 15

Effects of Pollution on Populations, Communities, and Ecosystems

15.1 CHANGES IN THE PRESENCE AND ABUNDANCE OF A POPULATION: THE CONCEPT OF INDICATOR SPECIES

If the sensitivity to a certain type of pollution across the different species in a community follows a normal distribution, then we could identify a number of species particularly sensitive and thus prone to quickly disappear after a pollution event, and others from the other end of the distribution, far more tolerant than average, that will stand and even thrive because of lack of resources competition to eventually dominate the community under polluted conditions. For example, certain epiphytic lichens are very sensitive to atmospheric SO_2, and they are absent in the vicinity of air pollution sources.[1] On the other hand, chironomid larvae dominate benthic communities in the sediments of eutrophic lakes because they are more resistant to hypoxic conditions than other insect larvae. Both sets of species are useful as indicators of pollution by absence or occurrence (and increase in abundance) respectively. In marine environments, Reish[2] was the first to report that certain taxa such as the polychaete *Capitella* proliferated in organically polluted marine sediments, while other polychaetes or bivalves such as *Tellina* disappeared. A species whose absence, presence, or abundance reflects a specific environmental condition is termed an **indicator** species. The limited mobility and relative long life cycle of benthic fauna and flora make them more suitable for these studies than planktonic or nektonic organisms. Table 15.1 summarizes some data on benthic marine taxa reported as indicative of organic pollution effects. Despite biogeographic variability some general patterns can be observed. In soft substrate habitats, proliferation of certain polychaetes and oligochaethes (the latter rarely identified to genus) and disappearance of small bivalves and echinoderms are indicative of organic pollution. In rocky shores macroalgae are more useful as indicators, and replacement of some red and brown algae (e.g., *Gelidium*, *Pelvetia*) by green (*Ulva*, *Enteromorpha*) and ceramiaceous or calcified red algae

> Indicator species are those whose presence or absence informs on pollution status

Marine Pollution. https://doi.org/10.1016/B978-0-12-813736-9.00015-5

Table 15.1 Marine Benthic genera of Soft and Hard Substrates Reported as Characteristic of Clean and Polluted Environments, and Thus Potentially Useful as Indicators of Organic Pollution by Absence or Presence, Respectively.

	Indicator by Absence	Indicator by Presence
Soft substrate	Scoloplos[3], Tellina[1,11], Nucula[8,11], Abra[5], Ampelisca[8,9], Equinocardium[6], Amphiura[12]	Capitella [1], Scolelepis[2], Chaetozone [3], Polydora[2,6], Corophium[2], Hydrobia [2]
Hard substrate	Pelvetia, Gelidium[7], Patella[10], Chthamalus[10]	Ulva [13], Enteromorpha[13], Ceramium[7,13], Corallina[4], Polydora[10]

Data from: [1] Reish (1955), [2] Pearson & Rosenberg (1978), [3] Davies et al. 1984, [4] Bellan (1985) and Soltan et al. (2001), [5] Gray et al. (1990), [6] Gray (1992), [7] Gorostiaga & Díez (1996), [8] Grall & Glémarec (1997), [9] Gómez Gesteira & Dauvin (2000), [10] Pagola-Carte & Saez-Salinas (2001), [11] Je et al. (2003), [12] Rosenberg et al. (2004), [13] Correa et al. (1999) and Juanes et al. (2008).[3]

(e.g., *Ceramium, Corallina*) are indicative of organic pollution and hypereutrophication, although strong seasonal variability in macroalgal populations must be taken into account.

Capitella is the most frequently reported indicator of pollution in marine sediments

Most studies summarized in Table 5.1 targeted pollution caused by organic wastes and urban sewage. Those indicators are thus species particularly sensitive or resistant to the specific changes caused by excessive organic enrichment and hypereutrophication, namely deficit of oxygen, increased water column turbidity, decreased sediment pH and redox potential, and production of reduced toxic chemicals such as H_2S and NH_3. We know from laboratory toxicity tests that remarkable differences in sensitivity to toxic chemicals—even to those acting through universal mechanisms of toxicity—among taxa may exist, and thus caution must be taken to extend the indicative value from sensitivity to organic inputs to other kinds of pollution. However, some polychaete taxa identified as indicators in organic enrichment studies are also apparently selectively resistant to toxic chemicals. J.M. Davies and coworkers[4] reported a succession of dominant polychaetes across a strong oil pollution gradient, with maximum abundance of *Capitella* in the vicinity of the oil well, replaced by *Chaetozone* and *Goniada* at intermediate distances and the occurrence of *Scoloplos* in the most distant sites (Fig. 15.1). A.B. Josefson et al. found similar selective resistance to metal pollution concentration in the polychaetes from sediments of a Greenland fjord system affected by Pb–Zn mining.[5] Therefore, the widely distributed *Capitella* seems to be a pollution indicator of almost universal value.

The concept of indicator species has been frequently criticized since it supports a static vision where shifts in tolerance within a region due to acclimation, or between biogeographic regions due to genetic differences are not considered.[6] Despite this well-funded criticism, indicator species are frequently advocated for the assessment of atmospheric and aquatic pollution, including marine ecosystems.

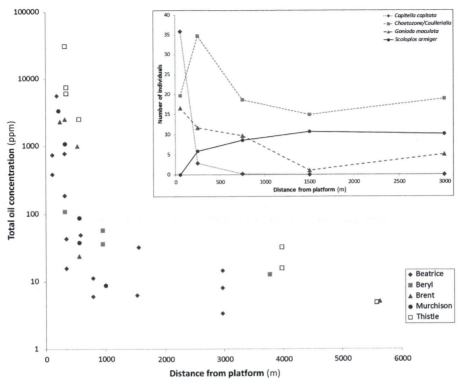

FIGURE 15.1

Changes in the oil concentration in the sediment, and (insert) benthic polychaete abundance as a function of the distance to the oil platform. Notice the different tolerance to oil of the polychaetes studied, with *Capitella* as the most resistant and *Scoloplos* as the most sensitive, which give these organisms' oil pollution indicators value within that area. *Modified from Davies et al. (1984) op. cit.*

The generalization of the indicator species concept, attributing a given fixed value of tolerance to each species in the community allows the calculation of benthic macrofauna **biotic indices**, advocated to assess the environmental status of aquatic ecosystems. This approach stems from observations in lakes and freshwater streams, relating the abundance of certain insect taxa with the dissolved oxygen and organic matter values of the water. Those observations led to the classification of all common species of macrofauna into discrete groups of tolerance. Since the beginning of the 20th century, the **saprobic index** based on tolerance to organic enrichment has been successfully applied to evaluate the quality of freshwater streams. The assignation of species to tolerance groups, from the very sensitive or **oligosaprobic** (typically plecoptera and ephemeroptera) to the very resistant or **polysaprobic** (typically chironomids or oligochaetes), was based on observations at different distances from a source of wastewaters. A score is arbitrarily assigned to each tolerance group, and the index is computed for each sampling site taking into account the species tolerance score weighted by the species abundances. W.H. Clemments and coworkers[7]

In fresh waters, composition of benthic macrofauna, summarized in biotic indices, respond to organic enrichment

developed a similar index for freshwater benthic fauna using toxicity tests with copper to objectively quantify the sensitivity of each taxon and successfully applied this index to assess pollution gradients in streams caused by trace metals.

In marine waters, biotic indices are useful to track temporal changes but not to compare among geographical areas

However, attempts to extend the use of biotic indices to marine ecosystems face multiple limitations. In marine environments the community composition is determined by natural factors such as depth and salinity that are nearly invariable in most freshwater habitats, and others such as hydrodynamism or bottom texture that show a much broader range of variability. In addition, the linear pollution gradients used to assign species to tolerance groups in streams are not found in coastal areas submitted to variable sea currents and tidal cycles. Thus, biotic indices are not expected to identify differences in pollution status across different geographical areas. Still, applications such as identification of temporal trends or impact after an anthropogenic event are not hampered by these limitations, and similar approaches to the freshwater saprobic index were followed to build up biotic indices based on composition of marine benthic infauna, either using the whole community (e.g., ITI,[8] BI,[9] AMBI,[10] BRI,[11] BENTIX[12]) or just a subset of species (e.g., BOPA,[13] see also[14]). These indices use from two to six discrete groups of tolerance, occasionally termed "ecological groups," except for BRI, which computes a "pollution tolerance score" for each species by using ordination multivariate analysis. Unfortunately, a central element in the rationale of those indices, such as the quantification of the tolerance, is frequently ill-defined using general terms such as endurance to "stress," "disturbance," or even "change." Criteria used for the assignation of species to tolerance groups include presence in organically enriched bottoms, deposit versus suspension feeding, opportunistic life history, or occurrence at a given distance from sewage outfalls. In the case of other benthic indices (ISI,[15] BQI[16]) quantification of tolerance is based on the species diversity of the communities where each species occurs. In practice, these indices have produced conflicting results,[17] and failed to discriminate pollution status among different estuaries[18] or to identify sites polluted by toxic chemicals,[19] although coherent results were also reported when applied to temporal changes or strong pollution gradients in a small uniform area.[20]

The generalized use of abundance data for all or most infaunal species made by many benthic indices has somewhat obscured the initial strength of the indicator species concept, i.e., the use of a single or a small group of species not necessarily abundant in the community whose tolerance to a specific source of pollution departs widely from average. The computation within those indices of the abundances corresponding to ubiquitous species with little or none discriminant value introduces noise in the value of the index and reduces its discriminant power.

15.2 CHANGES IN COMMUNITY STRUCTURE; DIVERSITY INDICES AND SPECIES: ABUNDANCE CURVES

Pollution may alter the environmental factors that determine the distribution and abundance of the species. Alterations may involve a displacement of environmental conditions beyond the range of tolerance of certain species in the community, which will disappear. Organic pollution, though, is also a food resource exploitable by detritivorous species whose abundance, and that of their predators, may increase. Abundance of nutrients will benefit species with high reproductive output and short life cycles that may replace the efficient competitor but slow growing organisms that dominate under low nutrient scenarios. These changes in the composition of the benthic fauna can be tracked using the classical **indices of diversity**. Among them the most frequently used is the Shannon index (H_S),[21] that in bits takes the expression:

Several indices quantify the diversity of a community

$$H_S = -\sum_{i=1}^{S} p_i \log_2 p_i \qquad \text{(Eq. 15.1)}$$

where p_i is the proportion of total number of individuals, or biomass, corresponding to species i in a total of S species.

Species diversity actually has two components, **richness**, i.e., total number of species present (S), and **evenness**, i.e., similar share of individuals (or biomass) among species. Given a sample with a total of N individuals belonging to S species, richness may be quantified using the Margalef index[22] (d_M), that allows a quantification of richness less sample size dependent than simply taking S:

$$d_M = (S - 1)/\ln N \qquad \text{(Eq. 15.2)}$$

When diversity is quantified using H_S, then the evenness index (J),[23] ranging from 0 to 1, will be:

$$J = H_S/H_{Smax} \qquad \text{(Eq. 15.3)}$$

where maximum diversity, $H_{Smax} = \log_2 S$.

Although these indices have been largely applied to taxonomic studies that identify species or genus, it is worth noting that they are applicable to any other taxon or functional unit in the community.

After a comprehensive review on the effects of organic inputs, from urban waste water to paper-mill effluents or hydrocarbons, on the structure of marine benthic communities, T.H. Pearson and R. Rosenberg[24] proposed their well-known paradigm according to which anthropogenic organic enrichment at low-to-moderate levels increase biomass and number of species, but eventually triggers a peak of abundance of a few opportunistic species (typically polychaetes), and the disappearance of all species sensitive to the changes in the sediment due to decomposition of organic material (reduced dissolved oxygen,

In soft bottom marine environments organic enrichment causes proliferation of opportunists and reduction of diversity

pH, redox potential, production of H_2S and NH_3) at high levels of organic pollution. Dissolved oxygen at the seabed is an important condition that shapes the structure of the benthic community.[25] For example, most infaunal bivalves, echinoderms, and crustaceans die at oxygen saturation levels below 10%, whereas the polychaetes *Capitella* and *Polydora* tolerate conditions below 5%. However, sulfides may promote the proliferation of certain bivalves with endobiont chemosynthetic bacteria that use reduced S as energy source, such as *Thyasira flexuosa*. In anaerobiosis only a few nematodes and *Beggiatoa* bacteria survive.[26] At high levels of organic pollution the combined effects of disappearance of sensitive species plus a decrease in evenness caused by the dominance of a few opportunistic species will cause a marked decrease in diversity. Therefore, this paradigm predicts a bell-shaped response of diversity to organic pollution (Fig. 15.2A, dotted line), with reductions in diversity occurring rather late in the sequence of organic pollution impact.[27] Ecological effects of hypoxia were further discussed in Section 2.4.

Natural factors as fluctuating salinity may also explain low diversity

The Pearson and Rosenberg paradigm was developed to study successional changes in benthic fauna prompted by organic pollution and recovery after pollution abatement, not to analyze patterns of diversity from different geographical areas. Background values of marine benthic diversity are highly dependent on natural factors such as depth or bottom texture. Furthermore, in many estuarine environments submitted to fluctuating conditions due to tidal cycles species diversity is naturally low, and the careless application of assessment criteria derived from pollution gradients under stable conditions leads to incorrect assessments of human impact. In the Minho estuary (Northwest Iberian Peninsula), for example, a comprehensive integrative monitoring showed absence of pollution from the chemical and ecotoxicological standpoints, but moderate or poor ecological status when standard benthic macrofauna indices that included diversity were applied[28] (see also Section 19.2).

Physical disturbance or toxic chemicals are not expected to cause the same pattern of response than organic enrichment

Physical disturbance (e.g., dredging activities) or toxic chemicals are not expected to cause the peak of opportunistic species described by Pearson and Rosenberg, derived from the use of organic matter as food resource.[29] The effects would rather be restricted to a progressive decrease in species richness as the most sensitive species begin to disappear. Thus, diversity should be maximal at minimum levels of pollution, and decrease as a threshold related with the potency of the toxic pollutant is exceeded (Fig. 15.2A, dotted line). This theoretical pattern was confirmed by the benthic diversity recorded in an oil field (Fig. 15.2B), and in metal polluted fjords (Fig. 15.2C; see also[30]) where, in addition, no clear change in the distribution of individuals among species, i.e., in evenness, was described.[31]

Effects of pollution on the structure of algal and plant communities were less frequently studied. Polluted sites typically show a lower structural diversity with dominance of crustose or filamentous algae and absence of a dense canopy. Turbidity is a key aspect as far as effects on photosynthetic species are concerned.[32] Hypereutrophication has been linked to shifts from seagrass- to

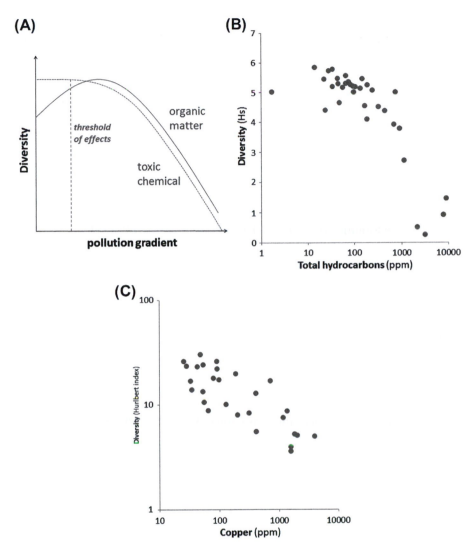

FIGURE 15.2
Patterns of response of the species diversity in benthic communities along a pollution gradient caused by organic matter (A, continuous line) or a toxic chemical (A, dotted line). Since organic matter may be used as food, diversity increases at low to moderate level of impact. In contrast, for a toxic chemical diversity is maximal at the minimum levels of pollution and decreases when the concentration of the chemical exceeds a toxicity threshold. Actual relationships between benthic diversity and hydrocarbon (B) and copper (C) pollution are also shown. *Own elaboration. (B): data from Gray (1989) op. cit.; (C): data from Rygg (1985) op. cit.*

algae-dominated communities.[33] The removal of structure-building species such as seagrass will certainly affect also the composition of the fauna.

Depressed diversity index values do not inform on how pollution may specifically affect to each component of diversity: richness and evenness. In fact, in terrestrial communities it has been described that the application of a pesticide reduces the species richness but increases evenness, which partially buffers the

Species:abundance curves better describe changes in community structure

overall effects on diversity because the pesticide kills many of the dominant species.[34] Species richness seems to be the most sensitive index to contaminants in marine communities also.[35] Moreover, remarkable changes in community structure may not be reflected in the values of those indices. Decreased Shannon diversity was restricted to a few hundred meters around oil wells, whereas increased hydrocarbon contents in the sediment extended for more than 1 km and changes in abundance of polychaete indicator species were observable along at least 2 km transects[36] (See also Fig. 15.1).

A graphical way to depict the composition of the community, more useful for an analytical assessment of the anthropogenic impact on the community structure, stems from the pioneer work by F.W. Preston,[37] who found out that when abundance was considered in a logarithmic scale the number of species belonging to a given class of abundance followed a normal distribution, although the left side of the Gaussian curve corresponding to the rarest species was veiled due to limitations of the sampling methods.

Comparisons in the shape of the species:abundance curves among sites due to shifts in the number of rare or abundant species are more informative than simple values of community indices. Fig. 15.3A shows the curves resulting from data obtained in an oil field at the North Sea.[38] Sites 30 and 37, located within 100 m of the oil platform, show disturbed patterns compared to control Site 1, more than 3 km away from the platform. Disturbed assemblages showed the following changes: (1) the number of rare species (abundance classes 1 and 2) decreased, and (2) a few species spectacularly increased their abundance, and abundance classes 7 to 10, absent in the control, appear. However, these effects were restricted to sites very close to the oil activities, and no significant differences in Shannon diversity were observed between sites less than 500 m and more than 500 m apart from the platform. The study reflected other interesting conclusions. Four indicator species (belonging to intermediate abundance classes) showed patterns of abundance inversely related to distance to platform, and thus were more informative than traits of the whole community. On the other hand, when sites were clustered according to benthic community structure by using multivariate analysis, percent of mud and barium content (present in the drilling muds) of the sediment explained the grouping better than the total hydrocarbon concentration.

Fig. 15.3B shows similar curves comparing a copper polluted fjord (Orkdalsfjord) with a control fjord. The benthic community inhabiting the sediments polluted with Cu (205 mg/kg DW) showed a much sharper reduction in species richness that affected not only the low abundance classes but all except the highest abundance class. As a result the slope of the species:abundance curve decreases.[39]

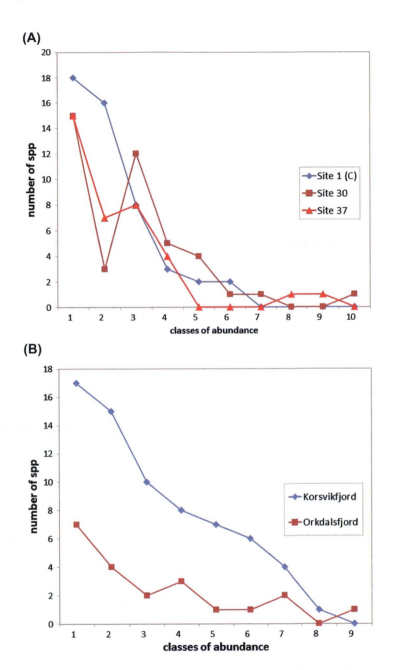

FIGURE 15.3

Number of species belonging to different classes of abundance, in log 2 scale, for benthic fauna in a North Sea oil field (A) and two fjords with different levels of Cu pollution (B). Control sites are shown in blue. Notice the reduction of species belonging to classes of low and (in the case of Cu pollution only) medium abundance, and, in the case of the oil field, the appearance of a few species with extremely high abundance in the polluted sites. *Own elaboration. (A) data from Gray et al. (1990) op. cit.; (B) data from Rygg (1986) op. cit.; Rygg & Skei (1984) op. cit.*

15.3   EFFECTS OF POLLUTION AT ECOSYSTEM LEVEL

We may wish to know how changes on community structure affect ecosystem functioning

In previous chapters we have studied the effects of pollution at molecular, physiological, and organismic levels, with focus on the effects to traits of individuals that are relevant for their survival, reproduction, and growth, and thus to the size of their populations. We have seen that pollutants either directly interfere with vital functions of the organisms or change the environmental conditions that shape the biological communities, the living component of the ecosystems. The special case of imbalances in community structure caused by oil spills has been summarized in Section 7.5. We wish to turn our attention now to the effects of pollution on ecosystem functioning.

The interest of studying the effects of pollution on ecosystems is manifold. From a theoretical standpoint we may wish to understand how the changes in the community composition described in previous sections may prompt also changes in the functional attributes of the ecosystems related to the cycling of matter and the energy flow, and thus try to predict the ecological relevance of those changes. If a sensitive species disappearing because of pollution is functionally redundant then the ecosystem process where the missing species is implicated will not be altered. Moreover, the relationship between ecological diversity and **stability** in its different forms (resistance to change under anthropogenic pressure, capability for returning to the same prestress conditions, rapidness of recovery) is a classical theme in ecological theory[40] that bears important practical implications in the conservation of valuable ecosystems. However, for the reason just stated, stability is more related to functional diversity than to structural diversity.

since vital ecosystem services may be at risk

We are also interested in assessing the risk that pollution poses for the many **ecosystem services** provided by marine ecosystem to humankind, including provisioning services such as fisheries, regulating services such as fixation of atmospheric CO_2 through primary production, and cultural services such as recreational opportunities. All those services are based on functional attributes such as efficient nutrient recycling that avoids turbid hypereutrophic waters, adequate balance between production and consumption of plant biomass, or efficient channeling of energy through long food chains that end up in large size species of high commercial value. Purification of air and water, mitigation of droughts and floods, detoxification and decomposition of wastes, maintenance of biodiversity, or provision of aesthetic values are just some of the services ecosystems provide to human kind. Since these services are provided for free they are frequently taken for granted and absent from economic budgets, although some efforts to put figures yield estimations above US$ 33 trillion annually, about 63% contributed by marine systems, and most of this coming from coastal systems.[41]

However, effects on ecosystem functioning are difficult to predict

The undeniable relevance of normal ecosystem functioning to humankind prompted attempts to identify functional attributes useful to monitor "ecosystem health," in an analogous way to the assessment that a doctor can conduct on our health from our blood pressure, glucose in the urine, and so on. Community respiration or primary production have been proposed as potential indicators of environmental status related to functioning in an aquatic ecosystem.[42]

However, attempts to use these attributes for monitoring aquatic pollution have generally failed. Eugene Odum already observed that when lakes were affected by large acid or nutrient inputs, primary productivity and other functional aspects were remarkably homeostatic, despite wide alteration of the composition of fish and plankton species.[43] The idea of an assessment of ecosystem health, like the assessment that a doctor can conduct after a medical check, no matter how appealing it may result, is hampered by the intricate multiple connections among ecosystem components that provide the homeostasis of the major state variables in an ecosystem, and these connections bear a much higher level of complexity than those operating on an individual. Whereas organisms are highly centralized systems whose organs are controlled by direct chemical (hormones) and electrical (nerves) signals, the control mechanisms in ecosystems are diffuse, and consist of weak but numerous feedback signals running along large and complex routes and acting just on certain components of the system.[44] Examples of those diffuse control mechanisms are the microbial subsystems that regulate the storage and release of nutrients, or the behavioral traits that govern species interactions and thus population abundance. As a consequence, to predict the effect of a pollution event on an ecosystem as a whole is a very difficult task.

Hypereutrophication is supposed to shift ecosystem functioning from dominance of large species with low turnover rates to blooms of fast-growing species that overwhelm the consumption capacity of grazers and divert increasing amounts of energy into detrital pathways. This is the case of the replacement of seagrass meadows, with higher diversity, longer food chains, and a higher percentage of the energy stored in biomass from high trophic levels, by communities dominated by green macroalgae with a higher relevance of detritivores.[45] However, the shift is not linear, and the system may pass through **bifurcations** that give raise to new stability points, with no return to the original state even when the intensity of disturbance—e.g., input of nutrients—is reduced. In addition, the intermittent supply of pollutants in combination with the random variability of environmental conditions may generate aperiodic fluctuations that follow **chaos dynamics**, which are very difficult to predict.

In summary, while community structure has been used as indicator of water pollution for almost a century, little progress has been made in developing simple and practical tools relating the effects of pollution on functional traits at ecosystem level. This is due to the complex network of regulation mechanisms that govern the functioning of natural ecosystems.

and functional attributes generally failed as pollution indicators

KEY IDEAS

- If the sensitivity to a certain type of pollution across the different species in a community follows a normal distribution, then we could identify a number of species particularly sensitive that disappear after a pollution event, and others far more tolerant than average that will stand and even thrive because of lack of resources competition to eventually dominate the community under polluted conditions.

- Pearson and Rosenberg proposed their well-known paradigm according to which organic enrichment at low-to-moderate levels increases biomass and the number of species, but eventually triggers a peak of abundance of a few opportunistic species (typically polychaetes) and the disappearance of all species sensitive to the changes in the sediment due to decomposition of organic material at high levels of organic pollution.

- From the Pearson and Rosenberg paradigm follows that species diversity is maximal at medium levels of organic enrichment. In contrast, for toxic chemicals diversity is maximal at minimum concentrations of the toxicant. However, many natural factors determine the composition of the benthic community, and diversity may be low for natural reasons, for example in fluctuating environments such as estuaries.

- Biotic indices based on the classification of species into groups of tolerance and/or associated to diversity have been used to track anthropogenic change. Since they rely on the sensitivity to organic enrichment they must be validated for physical disturbance (e.g., dredging activities) or toxic chemicals.

- Due to the influence of natural factors (salinity, depth, hydrodynamism, bottom texture) with strong geographical variability on the composition of marine benthic communities, it is not advisable to use community structure beyond assessment of temporal trends or strong pollution gradients within a single area.

- The changes in shape and slope of species:abundance curves may provide more information on changes in community structure than simple community indices.

- Ecosystem services provided by marine ecosystems, including provisioning of food through fisheries, fixation of atmospheric CO_2, and recreational opportunities depend on the undisturbed functioning of natural ecosystems. The effects of pollution on ecosystem functioning, and thus on those ecosystem services, are difficult to predict due to the complex dynamics at ecosystem level.

Endnotes

1. See Table 4.7 in: Wright DA, Welbourn P. Environmental toxicology. Cambridge: Cambridge University Press; 2002. (pp. 128−129).

2. Reish DJ. The relation of polychaetous annelids to harbor pollution. Public Health Rep 1955; 70(12):1168−1174.

3. Data from: Reish (1955) op. cit.;
    Pearson & Rosenberg (1978) op. cit.;
    Davies et al. (1984) op. cit.;
    Bellan G, Effects of pollution and man-made modifications on marine benthic communities in the Mediterranean: A review. In: Moraitou-Apostolopoulou M, Kiortsis V, editors. Mediterranean marine ecosystems; 1985. pp. 163−194. New York: Plenum Publishing Co.;
    Soltan D, Verlaque M, Boudouresque CF, et al. Changes in macroalgal communities in the vicinity of a Mediterranean sewage outfall after the setting up of a treatment plant. Mar Pollut Bull 2001;42:59−70.; Gray et al. (1990) op. cit.;
    Gray (1992) op. cit.; Gorostiaga JM, Díez I. Changes in the sublittoral benthic marine macroalgae in the polluted area of Abra de Bilbao and proximal coast (Northern Spain). Mar Ecol Prog Ser 1996;130:157−167.;
    Grall & Glémarec (1997) op. cit.;
    Gómez Gesteira JL, Dauvin J-C. Amphipods are good bioindicators of the impact of oil spills on soft-bottom macrobenthic communities. Mar Pollut Bull 2000;40(11):1017−1027.;
    Pagola-Carte S, Saiz-Salinas JI. Cambios en el macrozoobentos de sustrato rocoso del abra de Bilbao: 14 años de seguimiento de la recuperación biológica. Bol Inst Esp Oceanogr 2001; 17(1 y 2):163−177.;
    Je J-G, Bellan T, Levings C, et al. Changes in benthic communities along a presumed pollution gradient in Vancouver Harbour. Mar Environ Res 2003;57:121−135.;
    Rosenberg et al. (2004) op. cit.;
    Correa JA, Castilla JC, Ramírez M, et al. Copper, copper mine tailings and their effect on marine algae in Northern Chile. J Applied Phycol 1999;11:57−67.;
    Juanes JA, Guinda X, Puente A, et al. Macroalgae, a suitable indicator of the ecological status of coastal rocky communities in the NE Atlantic. Ecol Indic 2008;8:351−359.

4. Davies JM, Addyu JM, Blackman RA, Blanchard JR, Ferbrache JE, Moore DC, Somerville HJ, Whitehead A, Wilkinson T. Environmental effects of the use of oil-based drilling muds in the North Sea. Mar Pollut Bull 1984;15(10):363−370.

5. Josefson AB, Hansen JLS, Asmund G, et al. Threshold response of benthic macrofauna integrity to metal contamination in West Greenland. Mar Pollut Bull 2008;56:1265−1274.

6. Zettler ML, Proffitt CE, Darr A, et al. On the myths of indicator species: issues and further consideration in the use of static concepts for ecological applications. PLoS One 2013; 8(10):e78219.

7. Clemments WH, Cherry DS, Van Hassel JH. Assessment of the impact of heavy metals on benthic communities at the Clinch River (Virginia): Evaluation of an index of community sensitivity. Can J Fish Aquat Sci 1992;49(8):1686−1694.

8. Word JQ. The infaunal trophic index. Pages 19−39. In Bascom W, editor. Southern California Coastal Water Research Project annual report. El Segundo, California, USA; 1978.

9. Grall J, Glémarec M. Using biotic indices to estimate macrobenthic community perturbations in the Bay of Brest. Estuar Coast Shelf Sci 1997;44:43−53.

10. Borja A, Franco J, Pérez V. A marine biotic index to establish the ecological quality of soft-bottom benthos within European estuarine and coastal environments. Mar Pollut Bull 2000;40(12):1100−1114.

11. Smith RW, Bergen M, Weisberg SB, et al. Benthic response index for assessing infaunal communities on the Mainland Shelf of Southern California. Ecol Appl 2001;11:1073−1087.

12. Simboura N, Zenetos A. Benthic indicators to use in ecological quality classification of Mediterranean soft bottom marine ecosystems, including a new biotic index. Mediterranean Mar Sci 2002;3/2:77−111.

13. Dauvin JC, Ruellet T. Polychaete/amphipod ratio revisited. Mar Pollut Bul 2007;55:215−224.

14. Roberts RD, Gregory MR, Foster BA. Developing an efficient macrofauna monitoring index from an impact study - A dredge spoil example. Mar Pollut Bull 1998;36(3):231−235.

15. Rygg B. Indicator species index for assessing benthic ecological quality in marine waters of Norway. Norwegian Institute For Water Research. Report No 40114; 2002. pp. 1−32.

16. Rosenberg R, Blomqvist M, Nilson HC, et al. A. Marine quality assessment by use of benthic species-abundance distributions: a proposed new protocol within the European Union Water Framework Directive. Mar Pollut Bull 2004;49:728−739.

17. Labrune C, Amouroux JM, Sarda R, Dutrieux E, Thorin S, Rosenberg R, Grémare A. Characterization of the ecological quality of the coastal Gulf of Lions (NW Mediterranean). A comparative approach based on three biotic indices. Mar Pollut Bull 2006;52:34−47. Pinto R et al. Review and evaluation of estuarine biotic indices to assess benthic condition. Ecol Indic 2009;9:1−25.

18. Puente A, Diaz RJ. Is it possible to assess the ecological status of highly stressed natural estuarine environments using macroinvertebrates indices? Mar Pollut Bull 2008;56:1880−1889.

19. Marín-Guirao L, Cesar A, Marín A, et al. Establishing the ecological quality status of soft-bottom mining impacted coastal water bodies in the scope of the Water Framework Directive. Mar Pollut Bull 2005;50:374−387.

20. Muxika I, Borja A, Bonne W. The suitability of the marine biotic index (AMBI) to new impact sources along European coasts. Ecol Indic 2005;5:19−31. Josefson et al. (2008) op. cit.

21. Shannon CE, Weaver W. The mathematical theory of communication. University of Illinois Press; 1949.

22. Margalef, R. Information theory in ecology, Gene Sys 1958;3:36−71.

23. Pielou EC. The measurement of diversity in different types of biological collections. J Theor Biol 1966;13:131−144.

24. Pearson TH, Rosenberg R. Macrobenthic succession in relation to organic enrichment and polution of the marine environment. Oceanogr Mar Biol Annu Rev 1978;16:229−311.

25. E.g. González-Oreja JA, Saiz-Salinas JI. Exploring the relationships between abiotic variables and benthic community structure in a polluted estuarine system. Wat Res 1998;32(12):3799−3807.

26. Gray JS. Eutrophication in the sea. In: Colombo G, Ferrari I, Ceccherelli VU, Rossi R, editors. Marine eutrophication and population dynamics. Olsen & Olsen. 1992; pp. 3−15.

27. Gray JS. Effects of environmental stress on species rich assemblages. Biol J Linnean Soc 1989; 37:19−32.

28. Beiras R. Assessing ecological status of transitional and coastal waters; current difficulties and alternative approaches. Front Mar Sci 2016. https://doi.org/10.3389/fmars.2016.00088.

29. Marín-Guirao et al. (2005) op. cit.; Quintino V, Elliot M, Rodrigues AM. The derivation, performance and role of the univariate and multivariate indicators of benthic change: Case studies at differing spatial scales. J Exper Mar Biol Ecol 2006;330:368−382.

30. Josefson et al. (2008) op. cit.

31. Rygg B. Effect of sediment copper on benthic fauna. Mar Ecol Prog Ser 1985;25:83−89.

32. Heck KL. Community structure and the effects of pollution in sea-grass meadows and adjacent hábitats. Mar Biol 1976;35:345−357.

33. E.g. McClelland JW, Valiela I. Mar Ecol Prog Ser 1998;168:259−271.

34. Barrett GW. The effects of an acute insecticide stress on a semi-enclosed grassland ecosystem. Ecology 1968;49(6):1019−1035.

35. Johnston EL, Roberts DA, Contaminants reduce the richness and evenness of marine communities: A review and meta-analysis. Environ Pollut 2009;157:1745−1752.

36. Davies JM, Addyu JM, Blackman RA, Blanchard JR, Ferbrache JE, Moore DC, Somerville HJ, Whitehead A, Wilkinson T. Environmental effects of the use of oil-based drilling muds in the North Sea. Mar Pollut Bull 1984;15(10):363−370.

37. Preston FW. The commonness, and rarity, of species. Ecology 1948;29(3):254−283.

38. Gray JS, Clarke KR, Warwick RM, et al. Detection of initial effects of pollution on marine benthos: an example from the Ekofisk and Eldfisk oilfields, North Sea. Mar Ecol Prog Ser 1990;66:295−299.

39. Rygg B. Heavy-metal pollution and log-normal distribution of individuals among species in benthic communities. Mar Pollut Bull 1986;17(1):31-36.;
Rygg, B, Skei, J. Proceedings of the International Workshop on Biological Testing of Effluents and Receiving Waters. OECD/US-EPA/Environ Canada; 1984. pp. 153−183.

40. MacArthur RH. Fluctuations of animal populations, and a measure of community stability. Ecology 1955;36:533−536.

41. Costanza R, d'Arge R, de Groot R, Farber S, Grasso M, Hannon B, Limburg K, Naeem S, O'Neill RV, Paruelo J, Raskin RG, Sutton P, van den Belt M. The value of the world's ecosystem services and natural capital. Nature 1997;387:253−260.

42. Odum EG, Cooley JL. Ecosystem profile analysis and performance curves as tolos for assessing environmental impact. In: Biological evaluation of environmental impacts: the proceedings of a symposium. Office of Biological Services, Fish and Wildlife Service, USDI-FWS/OBS-80/26. US Government Printing Office; 1980. pp. 94−102.

43. Odum EG. Trends expected in stressed ecosystems. Bioscience 1985;35:419−422.

44. Patten BC, Odum EP. The cybernetic nature of ecosystems. American Naturalist 1981;118: 886−895.

45. E.g. Marques JC, Nielsen SN, Pardal MA, et al. Impact of eutrophication and river management within a framework of ecosystem theories. Ecol Model 2003;166(1−2):147−168.

Suggested Further Reading

- Johnston EL, Roberts DA. Contaminants reduce the richness and evenness of marine communities: a review and meta-analysis. Environ Pollut 2009;157:1745−52.

- Pearson TH, Rosenberg R. Macrobenthic succession in relation to organic enrichment and pollution of the marine environment. Oceanogr Mar Biol Ann Rev 1978;16:229−311.

- Zettler ML, Proffitt CE, Darr A, et al. On the myths of indicator species: issues and further considerations in the use of static concepts for ecological applications. PLoS One 2013;8(10): e78219.

Part III: Monitoring and Abatement of Marine Pollution

CHAPTER 16

Biological Tools for Monitoring: Biomarkers and Bioassays

16.1 USING NATIVE ORGANISMS: BIOMARKERS OF EXPOSURE AND EFFECTS

As discussed in Chapter 17, current approaches to pollution monitoring combine chemical and biological tools. The first group of biological tools available to monitor environmental pollution is based on studies conducted on individuals transplanted to, or more commonly native from, the site under investigation, and exposed to local environmental conditions for long periods of time. These are generically called biomarkers. In this context, a **biomarker** is a biological response measured in an organism naturally exposed to the site under study that serves as indicator of the presence and/or the effect of environmental pollutants. As early as 1964, C.J. Dawe and coworkers found enhanced occurrence of hepatic neoplasms in benthic fish from a lake in Maryland (US), and speculated that these tumors "may prove to be useful as indicators of environmental carcinogens."[1] The response may be measured at biochemical, cellular, physiological, anatomical, or behavioral level. Cause-effect relationships are better identified at low levels of biological organization, while ecological relevance in turn increases as level of organization increases. Responses measured at the molecular level often provide sensitivity and specificity, which allows an insight into mechanisms of action, but the biological significance for the overall fitness of the individual is often unclear.[2] Therefore a comprehensive and integrated monitoring program should combine a suite of biomarkers measured at different levels of organization, as well as other biological tools such as bioassays and population or community indicators (See Fig. 17.1).

In environmental monitoring, biomarkers are measured in individuals native from the study site or transplanted to that site. The requirements for an organism to be used as biomarker sentinel are common to those discussed for biomonitors (Section 17.3) except that some biochemical biomarkers demand the ability to biotransform pollutants. In fact, good bioaccumulators such as mussels show low or null activities for some enzymes used as biomarkers, such as ethoxyresosufin-O-deethylase (EROD). In addition, the biological response measured must be (1) **quantitative**, showing a robust dose:response pattern and a known time-dependence between exposure and effect; (2)

Biomarker: a biological response indicating the presence of a pollutant

The ideal biomarker should be specific, sensitive, and quantitative

265

Marine Pollution. https://doi.org/10.1016/B978-0-12-813736-9.00016-7

sensitive to low levels of pollution, to provide an early warning signal; and (3) **specific** to a single type of pollutant. The three requirements are rarely met at once, but the inhibition of the aminolevulinic acid dehydratase (ALAD) activity (an enzyme with a role in hemoglobin synthesis) by Pb is an example of a sensitive, quantitative, and specific biological measurement useful in humans and wildlife as biomarker of exposure to that trace metal. Still, it also illustrates the limitations of biomarkers. Decreased ALAD activity, which showed significant correlation with lead content, was not related though to any of the fitness variables (number of eggs laid, number of young fledged, and prefledging body weight) measured in birds exposed to lead from automotive emissions.[3]

The reproductive syndrome called imposex in marine gastropods, specifically caused by the biocide tributyl-tin (TBT), formerly used in antifouling paints, provides another example of an almost ideal biomarker. In fact, anatomical measurements of imposex in affected female snails better track TBT pollution than the costly chemical analysis of the TBT itself. Unfortunately, many biomarkers initially proposed as very specific to certain groups of chemicals (metallothioneins for trace metals, acetylcholinesterase (AChE) for organophosphate and carbamate insecticides, etc.) have been later found to respond to a much broader range of sources of stress.

Biomarkers may indicate simply exposure or also harmful effects

Biomarkers are classified as those of **exposure** and those of **effect**. The latter include responses that imply harmful effects on the biological fitness of the organism, i.e., in its chance to survive and reproduce. Common biomarkers of exposure are analysis of PAH metabolites in bile, induction of phase I and phase II biotransformation enzymes, induction of antioxidant enzymes, or metallothionein synthesis. Biomarkers of effect used in aquatic environments are lysosomal alterations, AChE inhibition, endocrine disruption (VTG synthesis in males, imposex), reduced scope for growth, DNA aberrations, and histopathology and/or neoplasia. While the rationale for this classification may be easily understood, in practice the presence of a fitness-relevant effect in experimentation with biomarkers is very rarely investigated, and most of the so-called exposure biomarkers can also be used as effect biomarkers provided the levels of chemical insult are high enough. For example, PAHs can be traced in the environment by the increased cytochrome P450 monooxygenase (CYP) activity in exposed organisms, and this is traditionally considered an exposure biomarker in fish. However, metabolic activation of certain high molecular weight PAH by CYP causes increased levels of carcinogenic PAH metabolites associated to the development of liver neoplasia in fish, certainly an effect biomarker.

In fact a well-known but poorly investigated limitation of chemically inducible molecular biomarkers is the bell-shaped response to exposure levels[4] (see Fig. 16.2). While the biomarker levels increase within the low range of environmental concentrations, it eventually saturates and can even sharply decrease at higher levels. Thus, the values of the response are the same for very clean and very polluted areas. This is, for instance, the case for detoxification enzymes that are induced by organic xenobiotics in moderately polluted sites but are inhibited by high concentrations of trace metals, which frequently occur in highly polluted sites. Again, this remarks the complementary nature of biological and chemical information, since each one is needed to shed light on the results of the other.

16.2 ENZYMATIC BIOMARKERS: ETHOXYRESOSUFIN-O-DEETHYLASE, GLUTATHIONE TRANSFERASES, ACETYLCHOLINESTERASE

Detoxification enzymes involved in Phase I and II of the biotransformation of xenobiotics, such as EROD and GST, respectively, are induced by the presence in the environment of chemical pollutants taken up by the organisms. The quantification of the induction (Fig. 16.1) may be assayed at **transcriptional level** by quantifying the corresponding mRNA using quantitative reverse transcription polymerase chain reaction (qRT-PCR), at translational level by detecting and quantifying the **amount of protein** by western blot, immunochemistry, or enzyme protein determined by enzyme-linked immunosorbent assays (ELISA), or at enzymatic level by assaying the **activity** of the enzyme. In the first

> Induction of enzymatic detoxification may be studied at transcriptional, protein, or enzyme activity levels

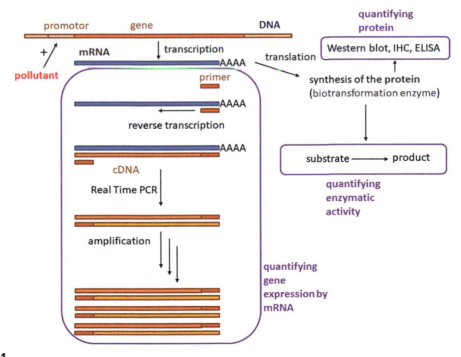

FIGURE 16.1

Molecular tools to quantify the potential induction of detoxification pathways caused by exposure to a pollutant. Quantification may be performed at transcriptional (mRNA), translational (protein) or enzymatic activity levels. *IHC*, immunohistochemistry.

case, when we wish to study the effect of a pollutant at transcriptional level on the expression of a gene, total RNA is extracted. Then complementary DNA (cDNA) corresponding to the mRNA whose transcription we wish to study is synthetized by using reverse transcriptase and primers specific to the target gene sequence. Finally, the cDNA is amplified and quantified using real-time PCR that allows comparing the levels of expression between control and exposed individuals and quantifying the increase or decrease of expression due to the presence of pollutants. A gene whose expression is not affected by the pollutant is used as reference.

Omics technologies go one step further and allow screening of all macromolecules of a certain type within a given kind of cells. Very briefly, **genomics** screen for all genes that are being expressed in a cell type by means of the analysis of the mRNA pool. While genes are identical in all cells, gene expression depends on live stage, cell type, and environmental conditions. Actively expressing genes are transcribed to mRNA. In genomics, the whole RNA pool is used as a template to synthesize labeled cDNA through a two steps process that produces first single strand and then double strand cDNA. This labeled cDNA is then hybridized using microarrays, which are silicon chips that contain, attached to known spots, gene fragments designed with the complementary sequence to the genes we wish to study. After hybridization the signal in each spot, proportional to the expression of the corresponding gene, is quantified by measuring absorbance at the wavelength of the label. Thus, the expression of each gene can be compared to a control and "upregulation" or "downregulation" instances can be identified.

In **proteomics** the target of the analysis is the protein pool. After cell lysis, the proteins are separated by electrophoresis in a polyacrylamide gel and digested with trypsin, and the amino acid sequence of the resulting peptides analyzed by mass spectrometry. Bioinformatics play a key role then to identify the protein fragments.

EROD is a biomarker of fish exposure to coplanar hydrocarbons

More conventional methods quantify enzymatic activities by colorimetric methods using light or fluorescence microwell readers. For example, the **ethoxyresosufin-O-deethylase** activity can be easily assayed in vitro in the microsomal fraction of fish hepatocytes, using 7-ethoxyresorufin as substrate.[5] The EROD is an enzyme belonging to the family of cytochrome P450 1A1 monooxygenases, involved in the Phase I of the biotransformation of organic xenobiotics in vertebrates (see Section 12.3). EROD is induced by exposure to chemicals with a coplanar polycyclic structure such as certain biocides, PAHs, PCBs, dioxins, and furans.[6]

Advocated by OSPAR, MEDPOL, and HELCOM (See Section 17.4), EROD is frequently used as exposure biomarker in regional monitoring programs using epibenthic fish such as dab (*Limanda*), *Callionymus*, and other species.[7] Unfortunately, background levels are very different among species, with OSPAR Background Assessment Criteria (BAC) values of 10 pmol/min/mg of protein for

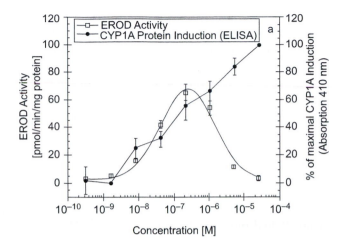

FIGURE 16.2

Comparison of the induction of ethoxyresorufin-O-deethylase (EROD) activity and immunodetected cytochrome P450 1A (CYP1A) enzyme protein determined by enzyme-linked immunosorbent assays (ELISA) upon exposure to benzo[k]fluoranthene. *Source: Fent & Bätscher (2000) op. cit.*

plaice, 24 for flounder, and 147—178 for dab. This complicates monitoring at broad geographical scales. Furthermore, both environmental (temperature, oxygen) and biological factors (age, development stage) have been found to interfere in those EROD background levels.[8] Fish sampling is more complicated and costly than collecting sessile invertebrates, but unfortunately, bivalves do not show appreciable levels of EROD activity. Other hydrocarbon induced monooxygenases, such as the aryl-hydrocarbon hydroxylases have been proposed as an alternative biomarker of PAH exposure in bivalves. Thus, the measurement of benzo-a-pyrene hydroxylase (BaPH) activity, an enzyme also suitable to monitor hydrocarbon exposure in fish,[9] has been proposed as a useful Phase I biomarker in mussels,[10] but simple and safe standard methods and assessment criteria are still to be developed.

In laboratory, EROD induction is concentration and exposure time-dependent. Using fish hepatoma cell lines, K. Fent and coworkers[11] found a concentration-related induction of EROD activity with medium and high molecular weight PAHs, with maximum induction at lower concentrations for the most potent inducers in the order dibenz[a,h]anthracene, dibenz[a,i]pyrene, benzo[k] fluoranthene, benzo[a]pyrene, benzo[a]anthracene, and pyrene. Induction decreases at high concentrations resulting in bell-shaped concentration activity curves (Fig. 16.2). The same pattern of response was found in experiments with male *Xiphophorus* fish exposed to the organochlorine biocide triclosan.[12] For a fixed exposure time, the response is bell-shaped, but the peak of maximum induction is displaced to lower concentrations as exposure time increases.

The bell-shaped response to pollutant concentration exhibited by the EROD activity is in contrast with the amount of inducible CYP1A, which can be quantified by immunoassay, and elicits a monotonic response with maximum values at maximum pollutant concentrations.[14] This suggests that at high concentrations, parental compounds or their residues may cause an inhibition in EROD catalytic activity despite increased CYP1A levels, which is a relevant limitation of in vivo EROD induction as a biomarker in highly polluted scenarios.

glutathione transferase (GST) activity is induced by exposure to chemical pollutants

The **glutathione transferases (GSTs)** are cytosolic enzymes involved in the Phase II of the biotransformation of xenobiotics by catalyzing the conjugation of electrophilic substrates to glutathione (see Section 12.4). The GST activity is unspecifically induced in marine organisms (including invertebrates) by exposure to a broad series of contaminants such as reactive oxygen species,[15] metals,[16] hydrocarbons,[17] and pesticides.[18] Unlike other proposed enzymatic biomarkers, GST provides a robust response relatively independent of seasonal factors.[19] GST is also the subject of comprehensive medical research because of its potential role protecting against chronic diseases that arise from oxidative tissue damage.[20]

Many chemicals inhibit AChE activity

Signals propagate along the nervous fibers by means of an electrical mechanism: a depolarization of the neuronal membrane. However, neurons are not in physical contact; a gap, the synapse, separates the axon of the presynaptic neuron from the dendrites of the postsynaptic neuron, and prevents the transmission of this electrical signal. Thus, the nerve impulse travels across the synapses thanks to the secretion by the presynaptic neuron of an endogenous chemical called a neurotransmitter that specifically interacts with receptors in the postsynaptic neuron and propagates the electrical signal to the latter. The most common and ubiquitous neurotransmitter is acetylcholine, present in the central and parasympatic nervous systems and in the neuromuscular junctions, and the synapses that use this neurotransmitter are called cholinergic. **AChE** is an enzyme involved in the correct transmission of the nervous signal across cholinergic synapses. Once the depolarization is triggered in the postsynaptic neuron, the AChE hydrolyzes the acetate group of the neurotransmitter to yield inactive choline, interrupting the nervous signal (Fig. 16.3). Synthetic chemicals such as organophosphate and carbamate insecticides selectively bind to AChE and inactivate the enzyme. As a result, the nervous signal is not correctly interrupted, causing in poisoned individuals sustained muscle contraction, paralysis, and occasionally death. The organophosphate methylparathion and other AChE inhibitors are responsible for many accidental poisoning particularly among agricultural workers in developing countries.[21] The mechanism of toxicity involves the irreversible binding of the insecticide metabolite paraoxon to the AChE active site.[22] Unlike organophosphates, the binding of carbamates with AChE is reversible and a quick reactivation of AChE after overexposure takes place, rendering carbamate poisoning less severe.[23]

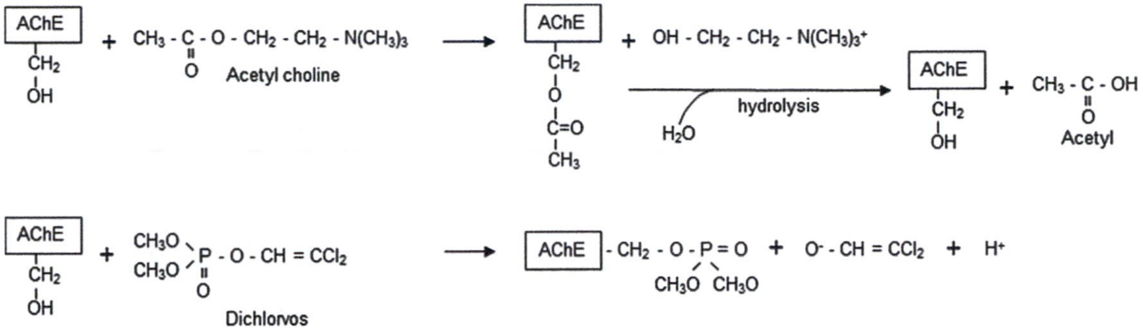

FIGURE 16.3
The normal function of acetylcholinesterase (AChE) involves the reversible binding of acetate to a serine group of the active site and reactivation of the enzyme by hydrolysis. The AChE inhibition by organophosphates is due to the irreversible binding of the chemical, here dichlorvos, to the active site.

AChE activity in a wide range of species is unselectively inhibited by many other types of chemical pollutants, such as metals, surfactants, and PAHs, and it is currently considered as a useful method to detect any contaminant with neurotoxic effects. AChE activity has been demonstrated in a variety of tissues of marine organisms including fish muscle and brain, adductor muscle and gills of shellfish, and abdominal muscle of crustaceans. Thus, the inhibition of AChE activity in molluscs, crustaceans, or fish has been advocated as a general indicator of pollution to be included in environmental monitoring,[24] and standard methods were developed.[25]

As an example of use in monitoring studies, F. Galgani and coworkers showed a decrease in AChE activity measured in dab muscle as fish were caught at sites closer to the Elbe outlet, in the North Sea (Fig. 16.4).[26] Background AChE activities are relatively independent of sex, age, and reproductive stage but temperature is an interfering factor.[27] Assessment criteria for Atlantic and Mediterranean mussels and some benthic fish species are available; OSPAR BAC values are 235–335 nmol/min/mg of protein for flounder, 150 for dab, and 15 to 30 for mussels, and Environmental Assessment Criteria (EAC) values range between 67 percent and 73 percent of BAC values.[28]

16.3 LYSOSOMAL STABILITY

Lysosomes are cellular organelles that contain hydrolytic enzymes playing a central role in intracellular digestion, resorption, and excretion processes. The impermeability of the lysosomal membrane is essential for its normal functioning, but a variety of environmental conditions have been described to reduce lysosomal stability. These include chemical stress but also physical, nutritive, or even emotional stress in rats. Therefore, **lysosomal membrane stability** (LMS) can be considered as a general, nonspecific biomarker of effect that responds to a broad range of pollutants, and its use has been proposed in a

Lysosomal membrane stability is a nonspecific biomarker of pollutant effects

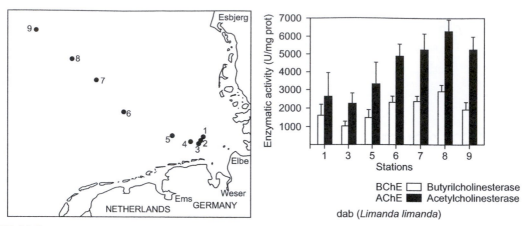

FIGURE 16.4

Cholinesterase activities in the benthic fish *Limanda limanda* along a pollution gradient in the German Bight (North Sea). Unlike butyrilcholinesterase (white bars), AChE (black bars) showed depressed values at the nearshore sites, close to the Elbe and Wesser river mouths, compared to exterior sites. *Modified from Galgani et al. (1992) op. cit.*

tiered approach as part of a Tier 1 screening to identify sites where more costly and sophisticated assessment tools should subsequently be applied.[30]

LMS can be measured by histochemistry in frozen digestive cells, recording the labilization period required to give maximum staining intensity for hydrolase enzymes (e.g., N-acetyl-glucosaminidase),[31] or in blood cells of living mussels (hemolymph granulocytes) and other invertebrate species by the neutral red retention time (NRRT).[32] After 15 min incubation, the neutral red dye is sequestered into the lysosomes of the granulocytes. If the lysosome membranes are damaged the dye leaks out into the cytosol where it can be visualized under the microscope. Observations are made at fixed time intervals, and for each sample NRRT will correspond to the last time period recorded when there was no evidence of dye loss in more than 50 percent of the cells. This ranges from 15 min in much damaged cells up to >1 h in healthy individuals. Standard protocols for measurement of LMS in both frozen tissue sections and live cells from molluscs and fish are available.[33] OSPAR BAC and EAC (value that if exceeded indicates deleterious effects) for mussel NRRT are 120 and 50 min, respectively.

Lysosomal stability is inversely related to chemical body burden

The LMS in wild mussels (*Mytilus galloprovincialis*) sampled from 14 sites with different degrees of chemical pollution along the Iberian Mediterranean coast was inversely correlated to the Chemical Pollution Index (CPI, see Section 1.5), an integrative index of chemical pollution calculated for each mussel population on the basis of the bioaccumulation data. Thus, NRRT was a negative linear function of CPI according to the expression: NRRT = −5.186 CPI + 23.62 ($r^2 = 0.61$) (Fig. 16.5).

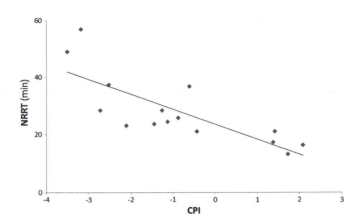

FIGURE 16.5
Neutral red retention time (NRRT) in lysosomes from hemocytes from wild mussel populations sampled along the Mediterranean coast of Spain. NRRT was an inverse function of the Chemical Pollution Index (CPI), an integrative index of the total chemical pollutant body burden. *Data from Martínez-Gómez et al.*

16.4   GENOTOXICITY BIOMARKERS: DNA ADDUCTS, MICRONUCLEUS, COMET ASSAY

Agents causing deleterious changes in the DNA, the macromolecule carrying the genetic information in each cell, are termed **genotoxic**. Changes in the DNA sequence as a consequence of interferences with correct DNA replication are called mutations, and chemicals inducing these changes are **mutagenic**. The first identified agent of genetic alterations was radiation, but some chemicals can also be genotoxic by interacting with the DNA itself or the cellular apparatus involved in its replication. These changes are of particular concern because they can be inherited by the progeny, even when they do not manifest in the parents. Furthermore, although the genetic dotation is identical in all the cells of the body, gene expression is regulated and limited to certain developmental stages or cellular types. Interference with correct gene expression may lead to cancer, and chemicals that may induce cancer are called **carcinogenic**. A chemical with the potential to damage the chromosome structure is called a **clastogen**. Early life stages with high rates of cellular division and differentiation are especially susceptible to DNA damage.

Therefore, the term **genotoxicity** encompasses deleterious changes at different molecular scales. A point mutation or microlesion is a change in the base-pair sequence of the DNA (insertions, deletions, substitutions). A chromosome alteration or macrolesion is a structural alteration of the chromosomal integrity (breaks, deletions, rearrangements) or even a change in the number of chromosomes. Because the earlier are not microscopically visible and much less frequent than the latter, most genotoxicity tests in eukaryotes focus on macrolesions. It is interesting to remark also that cells have a complex DNA repairing machinery, and not all structural changes in the DNA, primary alterations

Genotoxicity: induction of deleterious changes in DNA

Genotoxicity include microlesions (mutations, strand breakages), and macrolesions (chromosome aberrations, micronuclei)

(strand breakages, **DNA adducts**, alteration of bases etc.), persist in the form of secondary alterations: **chromosome aberrations**, **sister chromatid exchange**, or formation of **micronuclei**. The latter are normally used in eukaryotes as markers of exposure to genotoxicants.[35]

Testing mutagenicity: the Ames assay

The detection of mutagenicity, causing changes in the DNA sequence, is normally investigated through in vitro screening assays with prokaryotes, such as the Ames test. This test uses a mutated *Salmonella typhimurium* strain unable to synthesize the essential amino acid histidine, hence only capable to grow in culture media containing this amino acid. A chemical that is a positive mutagen will show the ability to revert the histidine auxotrophia and support a dose-dependent growth of the strain in histidine-free media.

Despite the simplicity and rapidity of prokaryote tests, their utility to predict genotoxicity in eukaryota is limited by the phylogenetic distance. Bacteria lack true nuclei as well as the enzymatic pathways by which most promutagens are activated to form mutagenic compounds. Bacterial DNA has a different protein coat than does eukaryota.

Testing genotoxicity in eukaryotes: DNA adducts, micronuclei, and comet assay

DNA adducts are the result of covalent bonds between carcinogenic chemicals and DNA. They are normally removed by cellular repair mechanisms, but during chronic exposures they may reach steady-state concentrations and thus provide a quantitative measurement of exposure to carcinogenic pollutants. DNA is extracted from liver or analog invertebrate tissues, enzymatically hydrolyzed, and stained with ^{32}P, a radioactive beta emitter that must be handled in special facilities. The radio-labeled adducts can be located by autoradiography and quantified by measuring radioactivity, although the measurement is semiquantitative. Levels of DNA adducts in nonmigratory fish are strongly correlated with the concentration of PAH in sediments, while there is no evidence that age, sex, or environmental factors such as salinity and temperature significantly affect the formation of DNA adducts.

Micronuclei are intracytoplasmic genetic material originated from chromatin lagged in anaphase as a consequence of chromosome breakage or malfunction of the spindle apparatus. A number of protocols for the **micronucleus test** using fish[36] and molluscs,[37] both in vivo (with peripheral blood cells) and in vitro (with cultured pericardial or gill cells), have been described. They are labor intensive and somewhat subjective since they rely on visual inspection, and micronuclei are rare phenomena with low background frequencies. Therefore, thousands of cells must be inspected to detect a few micronuclei. Interindividual variability is usually high. When the micronucleus test in mussel was applied in environmental monitoring, age, water temperature, and season were confounding factors that markedly affected background values.[38] For those reasons the technique, though promising in investigative monitoring, was never incorporated into large-scale monitoring programs. The recent development of automated protocols, such as flow cytometry may overcome some of these shortcomings.

A sensitive and rapid method for DNA strand break detection in individual cells is the single cell gel electrophoresis assay, commonly termed **comet assay**.[39] This method involves embedding individual cells in agarose gel on a microscope slide to be lysed and their DNA unwound under alkaline conditions and subjected to gel electrophoresis. The broken DNA fragments or relaxed chromatin migrate away from the nucleus, and this results in a comet shape that can be visualized with a fluorescent dye. The relative amount of DNA in the comet tail (% tail DNA) is linearly correlated with the number of DNA breaks.

The comet assay was advocated as a suitable technique to monitor genotoxicity in aquatic environments both in vivo[40] and in vitro.[41] It offers considerable advantages over other cytogenetic methods for DNA damage detection, like chromosome aberrations, sister chromatid exchange, and micronucleus test, because the cells studied need not be mitotically active and a small number of cells observed is required. Genotoxicity of metals and pesticides in oysters assessed by the comet assay was found to correlate with teratogenic effects in early life stages assessed by the standard embryo-larval bioassay.[42] For mussel hemocytes, available data suggest a BAC of ca. 10 percent tail DNA, while for dab erythrocytes 4 to 5 percent. Variations in protocols can lead to major differences in results and an inability to directly compare results. Guidelines relating to the use of the comet assay have been published for mammal models, and more recently for bivalves and flatfish.[43]

16.5 IMPOSEX IN FEMALE GASTROPODS AND VITELLOGENIN IN MALE FISH

Imposex can be defined as the superimposition of male characters, such as penis and vas deferens, in females of many marine gastropods. This syndrome, first described in *Nucella lapillus*,[44] is caused by TBT, broadly used in antifouling paints for boat hulls until its ban due to this and other effects on nontarget species or life stages (see Section 9.8). The mechanism underlying this phenomenon is likely to be the downregulation caused by TBT of the aromatase gene expression, and the corresponding reduction in the conversion of androgens to estrogens (see Section 14.3). The resulting anatomical anomalies can be quantified by means of indices such as the **vas deferens sequence index (VDSI)**[45] that can take in *Nucella lapilus* females a discrete enter value ranging from 0 to 6, or the **relative penis length index (RPLI)**[46] that ranges from 0 to 100 according to the length of the imposed penis in females compared to the mean male length. In *N. lapillus*, VDSI values above 2 imply reduced growth and recruitment, and values above 4 imply sterilization of affected females, while values below 0.3 are undistinguishable from zero TBT levels. In TBT-polluted areas the relationship between these anatomical indices and the TBT exposure levels is so tight that they can be used as a simple and cost-effective method for the estimation of TBT concentration in the water.[47] VDSI

Imposex: a biomarker of androgenic chemicals…

discriminates very well among low impact zones, but saturates at TBT concentrations in water around 10 ng Sn/L. In highly impacted zones RPLI, with a more gradual response, has a higher power of discrimination, though its value may show seasonal fluctuations.

...used in the OSPAR coordinated monitoring

In fact, imposex is the only biological effects technique whose measurement is compulsory in the current monitoring schemes of OSPAR (CEMP: Coordinated Environmental Monitoring Programme) and HELCOM. OSPAR assessment criteria for several marine snail species are available to use imposex as a biomarker for TBT pollution[48] (Table 16.1). Equivalences to the five categories of ecological status of the European Water Framework Directive (WFD) have been proposed.[49] After its global ban at the beginning of the century, TBT pollution showed decreasing trends worldwide and affected snail populations have been reported to recover. For example, a long-term study in Aveiro (Portugal) showed a decrease in VDSI mean values for female *N. lapillus* from harbor areas from 4 down to 0.5 in the last 10 years[50] (see also Fig. 14.3).

Vitellogenin: a biomarker of estrogenic chemicals

Vitellogenin (VTG) is a high density fosfolipoprotein, precursor of the main vitellum proteins that provide to the female gametes the energy reserves necessary to undergo embryo development. In egg-lying vertebrates, VTG synthesis takes place in the liver and VTG is then transported through the blood to the ovaries, where it transforms into egg yolk proteins such as lipovitellins and phosvitin. Trouts directly exposed to sewage-treatment effluents showed up to 100,000-fold increases in plasma VTG concentration, and similar effects were obtained by immersion of male trouts in solutions of $1-10$ ng/L of the synthetic estrogen 17α-ethynyl-estradiol, used in contraception pills.[51]

Table 16.1 OSPAR Assessment Criteria for Imposex in Marine Gastropods, Assessed by the Vas Deferens Sequence Index (VDSI) or the Intersex Sequence Index (ISI)

Imposex Assessment Class	VDSI				ISI
	Nucella lapillus	*Neptunea antique*	*Nassarius reticulatus*	*Buccinum undatum*	*Littorina littorea*
A (<BAC)	<0.3	<0.3	<0.3	<0.3	<0.3
B (<EAC)	0.3–2.0	0.3–2.0			
C	2.0–4.0	2.0–4.0	0.3–2.0	0.3–2.0	
D	4.0–5.0	>4.0	2.0–3.5	2.0–3.5	0.3–0.5
E	5.0–6.0		>3.5	>3.5	0.5–1.2
F	population absence				1.2–4.0

BAC, *Background Assessment Criteria, established at VDSI < 0.3 for* N. lapillus *and* N. antique *only; the other species are not sufficiently sensitive to establish this criterion.* EAC, *Environmental Assessment Criteria, established at VDSI > 2.0 for* N. lapillus *and* N. antique *and at VDSI > 0.3 for* N. reticulatus *and* B. undatum; L. littorea *ISI is not sufficiently sensitive.*
Source: *OSPAR 2013, Background document and technical annexes for biological effects monitoring, Update 2013.*

Following these spectacular findings, the induction of VTG synthesis in male fish, normally under control of the female hormone estradiol, was assumed to be a reliable biomarker for assessing the presence of environmental pollutants that act as estradiol mimics, called xenoestrogens (see Section 14.3), and the technique was suggested for incorporation to marine pollution monitoring programs (see also Section 16.4). However, difficulties for methodological standardization, the interference of multiple intrinsic and environmental factors with the background levels of VTG, and lack of assessment criteria for sentinel species have so far precluded the use of this biomarker at large scale.

Methods proposed for quantification of VTG include immunoassays,[52] which require the previous isolation of the VTG antibody for the monitor species, and an indirect assay based on the levels of alkali-labile phosphates (ALP).[53] VTG antibodies are available for a limited number of species. Since VTG are very abundant and they contain significant amounts of phosphates, ALP in plasma proteins have been proposed as an indirect method to measure VTG, but the possible interference of other proteins of similar composition must not be neglected. In bivalve molluscs, VTG and vitellins are both produced directly in the gonads, which render its differentiation more difficult, and proteomic techniques have proved that ALP is not a valid method for quantification of VTG.[54]

16.6 USING LABORATORY ORGANISMS: ECOTOXICOLOGICAL BIOASSAYS

The second group of biological tools available to monitor environmental pollution is based on studies conducted under strictly standardized conditions in the laboratory, or more rarely in the field, using homogeneous biological materials, frequently populations reared in laboratory or aquaculture facilities. These tools are called **ecotoxicological bioassays**. Fig. 16.6 summarizes the main differences between biomarker and bioassay studies applied to environmental monitoring. Biomarkers target biological responses measured in each site under investigation, while bioassays are intended to assess the toxic effects of inert environmental matrices on standard laboratory specimens. In addition, biomarkers frequently use molecular and cellular responses, and they result from long-term exposures of the studied organisms, while bioassays normally record lethal and sublethal responses at organismic level after short-term exposures.

Bioassays test the toxicity of environmental samples according to strictly standardized procedures

Therefore, bioassay methods must be adapted to the type of environmental matrix sampled: surface water, interstitial water, or sediments. Thus, bioassays can be classified into **liquid phase** and **solid phase**, depending on the environmental matrix to be tested, and manipulations needed to obtain the testing media (Fig. 16.7). In all cases sample storage time and manipulation prior to

BIOMARKERS IN MONITORING STUDIES

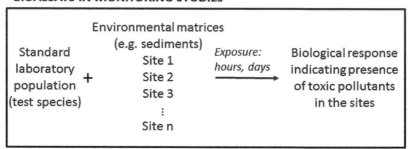

BIOASSAYS IN MONITORING STUDIES

FIGURE 16.6

Conceptual differences between biomarkers and ecotoxicological bioassays in monitoring studies.

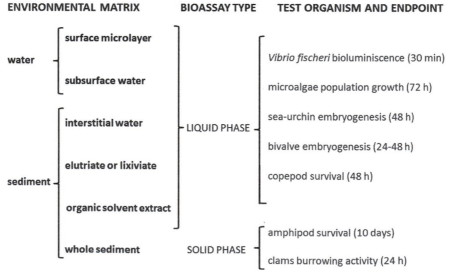

FIGURE 16.7

Types of bioassays according to the environmental matrix to be tested and the main organisms used in marine ecotoxicological tests.

testing must be minimized, since these steps may produce artifacts which constitute the main weakness of laboratory bioassays.

The sea **surface microlayer** can be sampled by using Teflon or glass surfaces or rotating drums. This allows collecting a film of water that adheres to the Teflon surface, representing the first few hundred microns of the water column. This microlayer is the interface for pollutants entering the marine waters from the atmosphere, and it is also enriched with hydrophobic pollutants which dissolve in the oil slick frequent in ports and other coastal waters subject to navigation activities. **Subsurface water** is sampled using conventional oceanographic bottles—e.g., Niskin—lowered by a rope to the desired depth and closed by a messenger. Water samples for toxicity testing must be stored with no air space in glass containers in the dark, kept refrigerated and tested within 24—48 h. If water samples are to be frozen then they must be filtered prior to freezing, but filters may retain contaminants.

> Field sampling, sample storage, and manipulation are critical steps to obtain reliable results

Sediments for use in ecotoxicological studies are frequently homogenized and sieved through 2 mm mesh to remove large particles and macrofauna. When sampling areas with high microgeographical variability such as intertidal mudflats, it is advisable to take 3 subsamples from the vertices of a triangle. For practical reasons these subsamples can be later mixed into a single composite for testing and chemical analysis. Sediment for use in toxicity testing should always be stored at 4°C. Freezing and freeze-drying artificially increase their toxicity.[55] Sediment toxicity has been reported to increase, decrease, or remain unchanged with storage time at 4°C, depending on the type of sediment, categories of contaminants associated with it, and toxicity test method and system applied.[56] Therefore, the establishment of acceptable storage times is quite arbitrary and generally reflects practical aspects of sampling and laboratory procedures. The sooner after sampling toxicity tests are performed, the better they will reflect actual field conditions of the analyzed sediment. ASTM recommends that sediment storage time does not exceed 14 days.[57]

Since sediments are the preferred inert matrix in environmental monitoring studies, but effects on water column organisms are also relevant, methods have been developed to obtain a liquid phase from the sediments. This includes obtaining the sediment pore water, aqueous extraction of the sediment using control seawater (elutriation or lixiviation), and extraction of the sediment with organic solvents. **Interstitial pore waters** can be obtained from the sediment in situ, by divers, using syringes. Laboratory extraction is more common and can be done by centrifugation,[58] vacuum suction,[59] or pneumatic pressure.[60] Sandy sediments are more suitable for pneumatic pressure or vacuum extraction, whereas fine-grained sediments provide large volumes of pore water when centrifuged.

Elutriates (or **lixiviates**) are aqueous extracts of the sediments which reproduce the potential remobilization of contaminants from the bottom sediments to the water column, which may take place in the seafloor due to natural (hydrodynamism, bioturbation) or anthropogenic (dredging, trawling) causes. For obtaining the elutriates, fresh sediments are mixed with filtered seawater of oceanic characteristics. In most protocols sediments are mixed with water at a proportion of 100 g to 500 mL in closed flasks filled to the brim and hermetically sealed, leaving no air space. Elutriation in air-tight conditions has been proved to increase sensitivity compared to stirring in open vessels.[61] Rotatory agitation is provided for 30 min at 60 rpm, followed by overnight decantation.

Extraction of the sediment with **organic solvents** maximize sensitivity and may allow chemical characterization of the toxic agent by testing fractions extracted with solvents of different polarities, but it also decreases the environmental relevance of the findings due to potential toxicity of solvent residues and bioavailability issues.

Salinity, dissolved oxygen, sulfide, or ammonia are confounding factors that may cause false positives in the bioassays

Natural conditions in the environmental matrix to be tested, such as temperature, salinity, pH, dissolved oxygen, sulfide, and ammonia concentrations are potential **confounding factors** for the interpretation of test results, and must be checked to make sure they lay within optimal values for the test species or within the requirements of the standard protocols, otherwise they are a source of false positives. This is particularly relevant for matrices that require manipulation such as pore water or elutriates, but also to surface waters, submitted to low salinity periods, or whole sediments, which may be naturally rich in toxic substances such as sulfides and ammonia derived from the anaerobic decomposition of organic matter or may show artificially enhanced levels due to a too-long storage period. To avoid possible artifacts due to these manipulations of the samples, in situ **bioassays** may be conducted. In this case the standard biological materials are placed in the field at the study sites, using cages or other deployment devices in a similar way to active monitoring transplantation experiments. Since environmental variables affecting the biological responses measured in the bioassay cannot be controlled in the field, they must be recorded to take them into account for the interpretation of the data. Although sample manipulation is not required, in situ bioassays are very labor intensive. It can be useful for investigative monitoring at hot spots but it is generally not suitable for routine monitoring at large geographical scales.

An ideal bioassay should be sensitive, ecologically relevant, and easy to standardize

The main requirements of an ideal ecotoxicological bioassay to be used in routine marine pollution monitoring networks are (1) **sensitivity**, (2) **ecological relevance**, and (3) suitability for **standardization**. Sensitivity allows detection of pollutants at low concentrations in the environment before levels become so high that harmful effects are already done. Use of early life stages such as embryos, larvae, and neonates maximize sensitivity and facilitate microscale testing. However, no single bioassay shows a universal range of sensitivity to unmask the presence in the environmental sample of any toxic chemical, mainly because biocides and other substances may show highly selective toxicity. Therefore, a battery of several bioassays representing the main taxonomic groups of marine organisms—ideally bacteria, algae, bivalves, crustaceans, echinoderms, and fish—should be conducted.

Ecological relevance means that the endpoint used in the test should be relevant for the ecological fitness of the individual, and the test species should be representative of taxa present in the ecosystem to be protected. Suitability for standardization means that the test must be simple, the response robust, and the facilities and equipment needed inexpensive. This requirement is frequently overlooked by researchers. Robustness of results must be assured through a strict **quality control** (QC) of the biological material and experimental procedures. Acceptability criteria concerning control response, and interlaboratory comparisons play a central role in QC, an essential aspect very well developed in analytical chemistry but somewhat neglected in biological measurements.

Internationally accepted standard methods have been developed for ecotoxicological bioassays using a large number of marine organisms and biological endpoints, including bacterial bioluminiscence, microalgal population growth, macroalgae reproduction, rotifer survival and reproduction, polychaete growth, copepod and mysid acute survival and life cycle tests, bivalve and sea urchin embryogenesis and larval development, sea urchin fertilization, amphipod and lugworm survival and reburial in sediment, fish embryo hatching and larval survival, etc. We will briefly review here some of them and refer to the further reading section for additional discussion on advantages and limitations of common marine bioassays.

16.7 LIQUID PHASE BIOASSAYS

16.7.1 *Vibrio fischeri* Bioluminiscence

The heterotrophic marine bacteria *Vibrio fischeri* has a luciferase enzyme coupled to the respiratory chain, and thus it produces light in direct relation to its metabolic rate. Light produced at serial dilutions of the tested sample is recorded in a photometer after 30 min exposure, and an IC_{50} (the dilution of the sample that inhibits light emission to a half) quantifies the toxicity of the sample. The bacteria may be stored freeze-dried, and commercial software is available to calculate the IC_{50}. The standard test is suitable for liquid samples[62] and it has been adapted to cope with solid phases.[63] This is the most rapid and standardized test, but it has been developed to check toxicity to microorganisms in WWTP reactors. The relevance of results applied to environmental samples is limited by the phylogenetic distance between bacteria and most aquatic resources. Thus, the use of this test as screening tool to identify pollution hot spots is not advised.

16.7.2 Microalgae Growth Inhibition

Freshwater and marine[64] (*Phaeodactylum, Skeletonema*) monoalgal cultures are used in static short-term growth inhibition tests conducted in flasks[65] or microplates.[66] Algal density after 72 h incubation is recorded either by manual (counting cells) or automatic (Coulter counter) counts of cell numbers per volume, or using fluorimetry. The ISO standard includes as acceptability criteria a minimum control growth rate of 16-fold increase in density in 72 h and a maximum coefficient of variation among replicates of 7 percent.

16.7.3   Bivalve Embryogenesis

Reproductive stocks of adult bivalves (mainly oysters and mussels) are commonly available from aquaculture facilities. In vitro fertilized eggs of *Crassostrea* spp., *Mytilus* spp., and many other bivalve species have been used to characterize seawater and sediment quality through liquid phase bioassay methods, an approach pioneered by C.E. Woelke in the United States,[67] and later standardized by ASTM,[68] ICES,[69] and IFREMER.[70] Percentage of morphologically abnormal larvae or retarded embryos is recorded as endpoint, and it provides a sensitive and quantitative method for detection of many chemical pollutants. According to OSPAR, toxicity of a liquid sample is classified as "elevated" when abnormalities are >20 percent, and "high concern" when they are > 50 percent.[71]

16.7.4   Sea-Urchin Embryo Test (SET)

Sea urchin embryos of the genera *Paracentrotus, Hemicentrotus, Pseudocentrotus, Strongylocentrotus, Anthocidaris,* and *Arbacia* among others have been used around the globe to assess water quality after the seminal work of D.P. Wilson[72] and M.P. Bougis[73] in Europe, and later N. Kobayashi in Japan.[74] Different echinoid species show similar sensitivity to pollutants, which allows the combined use of species with different breeding seasons. Sea urchins are easily maintained in aquarium, and by suitable thermal and food conditioning they are amenable to achieve mature gonads at any time of the year. In vitro fertilization is very simple and sea urchin embryogenesis is used as endpoint in liquid phase marine toxicity tests standardized by ASTM,[75] Environment Canada,[76] or ICES.[77] Morphology of the larvae is frequently used as an endpoint, but larval length (Fig. 16.8) provides a more sensitive, gradual, and observer-independent response amenable for automatic recording,[78] and it has been adopted as endpoint by OSPAR for the standard Sea-urchin Embryo Test (SET). Assessment criteria to classify sediments according to the toxicity of their elutriates have also been developed.[79] When larval size increase is expressed as a proportion of control response (proportion net response, PNR), a sample is considered as toxic when PNR < 0.7 and highly toxic when PNR < 0.5. Natural or storage-caused ammonia[80] and sulfide[81] concentrations in the sediment are a relevant confounding factor. More robust results are obtained when serial dilutions of the elutriate are tested and toxic units (TU) are calculated as $TU = 1/CE_{50}$, where CE_{50} is here the dilution of the elutriate (in log scale) causing a 50 percent reduction in PNR. A sample is then considered as toxic when TU > 0.3 and highly toxic when TU > 0.9.

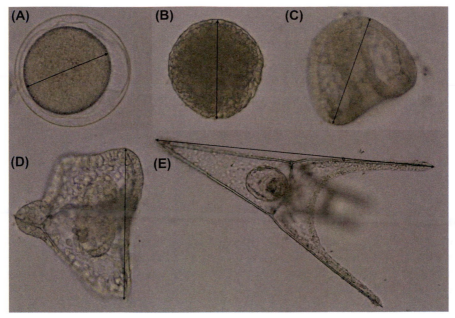

FIGURE 16.8
Embryogenesis and early larval development of the *Paracetrotus lividus* sea urchin. Fertilized egg (A), early embryo (B), gastrula (C), prism larva (D), and pluteus larva (E). The latter is the normal stage achieved after 48 h incubation of fertilized eggs at 20°C. *Black arrows* indicate measurement of size (maximum dimension), the endpoint used in the Sea-urchin Embryo Test (SET). *Photographs: R. Beiras.*

16.7.5 Copepod Survival

Calanoid copepods (e.g., *Acartia* spp.), main components of marine zooplankton, and benthic harpacticoid copepods (e.g., *Tigriopus* spp., *Tisbe battagliai*) are easily cultured in laboratory, requiring minimal space and equipment. Standard methods were developed by ISO and ASTM for their use in acute[82] and chronic[83] bioassays. The acute test records mortality (or lack of motility) after 48 h exposure. Sensitivity may be enhanced by using nauplius larvae.[84]

16.8 SOLID PHASE BIOASSAYS

16.8.1 Amphipod Survival

Amphipods are sediment-burrowing crustaceans present in freshwater, estuarine, and saltwater sediments. Standard methods have been developed for a 10-d survival laboratory test with *Rhepoxynius abronius* or *Ampelisca abdita* in North America[85] and *Corophium* spp. in Europe,[86] among other species.[87] Salinity and grain size are relevant confounding factors in this test, and for this reason the more tolerant species *Leptocheirus plumulosus* replaced *A. abdita* in some US monitoring networks. Although amphipods are generally less tolerant to pollution than polychaetes, the lethal response shows limited

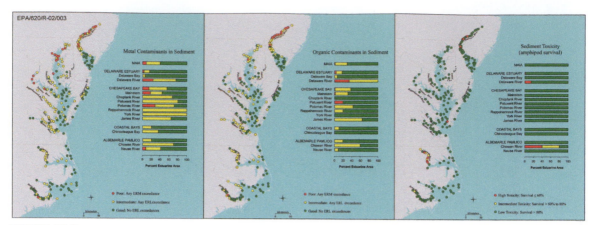

FIGURE 16.9

Classification of estuarine and coastal sediments from the US Mid-Atlantic according to metal contents (left), organic pollutants contents (middle), and toxicity to *A. abdita* amphipods. ERL/ERM criteria were used for the classification of the chemical data. Discrepancies found between sediment chemistry and toxicity were attributed to a too-strict criteria of mortality for a sediment to be classified as toxic, and to bioavailability issues. *From: US-EPA (2002) op. cit.*

sensitivity, and some protocols advise recording reburial capacity at the end of the test or even amphipod growth. Validity criteria for maximum control mortality range from 10 to 15 percent, and samples are considered toxic when mortality is significantly higher than controls after appropriate statistical treatment.

The sediment toxicity bioassay with amphipods is part of the evaluation of coastal pollution in the United States conducted by the US Environmental Protection Agency (US-EPA) (Fig 16.9). Sediment toxicity was considered low when mortality ≤20 percent, medium from >20 percent to ≤40 percent, and high when >40 percent. Subsequent developments reduced the number of categories to two with a benchmark of 20 percent mortality. Sediment toxicity is integrated in an assessment tool termed Sediment Quality Index that includes also sediment chemistry.[88]

16.8.2   Clam Reburial

In the 1980s W.H. Pearson and coworkers proved both in laboratory[90] and in the field[91] that clams took longer to bury themselves in artificially oiled compared to clean sand, and stayed at shallower depths in the sediment. This behavioral response is far more sensitive than acute mortality (See Section 14.4), but interference of biotic (clam size and species) and abiotic (sediment texture) factors precluded the development of standard procedures.

KEY IDEAS

- In environmental monitoring biomarkers are biochemical, cellular, physiological, or anatomical responses measured in native or transplanted organisms that show evidence of exposure and/or effects to pollutants.

- Molecular biomarkers provide early warning signals of the presence of pollution before effects at higher levels of organization took place.

- Molecular responses induced by exposure to environmental chemicals can be studied at transcriptional (individual gene expression, genomics), translational (immunoassays, proteomics), or enzymatic levels.

- Induction of EROD (Phase I detoxification), GST (Phase II detoxification), and other detoxification enzymes are common exposure biomarkers.

- LMS and acetylcholinesterase activity can be considered as general, nonspecific biomarkers of effect that respond to a broad range of pollutants.

- Imposex in gastropods and vitellogenin synthesis in male or juvenile fish are effect biomarkers that specifically detect endocrine disrupting chemicals.

- Ecotoxicological bioassays test the toxicity of environmental samples according to strictly standardized procedures, using sensitive species and life stages representative of the ecosystems to be protected.

- Salinity in water samples, and dissolved oxygen, sulfide, or ammonia in sediments, pore water or elutriates, are confounding factors that may cause false positives in the bioassays. Sediment storage time must be reduced to a minimum, and freezing or freeze-drying is not suitable for toxicity testing purposes.

- Liquid phase bioassays use as endpoints microalgal growth inhibition, bivalve or sea urchin embryo development, and copepod or mysid survival. Assessment criteria and minimum control performance are part of the QC for these tests.

- The 10 days amphipod survival test is the most common whole sediment bioassay.

Endnotes

1. Dawe CJ, Stanton MF, Schwartz FJ. Hepatic Neoplasms in Native Bottom-feeding Fish of Deep Creek Lake, Maryland. Cancer Res 1964;24:1194–1201.

2. McCarthy JF, Shugart LR. Biomarkers of environmental contamination. Boca Raton: Lewis Publishers; 1990. (p. 11).

3. Walker CH, Sibly RM, Hopkin SP, et al. Principles of ecotoxicology 4th ed. Boca Raton:CRC Press; 2012. (p. 182).

4. Fent K, Bätscher R. Cytochrome P4501A induction potencies of polycyclic aromatic hydrocarbons in a fish hepatoma cell line: demonstration of additive interactions. Environ Toxicol Chem 2000;19(8):2047–2058.

5. Galgani F, Payne JF. Biological effects of contaminants: microplate method for measurement of ethoxyre- sorufin-0-deethylase microplate method for measurement of ethoxyresorufin-0-deethylase (EROD) in fish. ICES Techn Mar Environ Sci 1991;13. Copenhagen.

6. Reviewed by van der Oost R, Beyer J, Vermeulen NPE. Fish bioaccumulation and biomarkers in environmental risk assessment: a review. Environmental Toxicology and Pharmacology 2003;13:57–149.

7. Burgeot T, Bocquené G, His E, et al. Monitoring of biological effects of polluants: Field application. In: Garrigues et al, editors. Biomarkers in marine organisms. A practical approach. Elsevier. 2001; pp. 179–213.

8. Andersson T, Forlin L. Regulation of the cytochrome P450 enzyme system in fish. Aquat Toxicol 1992;24:1–20.

9. Payne JF. Field Evaluation of Benzopyrene Hydroxylase Induction as a Monitor for Marine Petroleum Pollution. Science 1976;191:945–946.

10. Michel X, Salaün J-P, Galgani F, et al. Benzo(a)pyrene Hydroxylase Activity in the Marine Mussel *Mytilus galloprovincialis*: A Potential Marker of Contamination by Polycyclic Aromatic Hydrocarbon-Type Compounds. Mar Environ Res 1994;38:257–273.

11. Fent K. Fish cell lines as versatile tools in ecotoxicology: assessment of cytotoxicity, cytochrome P4501A induction potential and estrogenic activity of chemicals and environmental samples. Toxicol Vitro 2001;15:477–488.

12. Liang X, Nie X, Ying G, et al. Assessment of toxic effects of triclosan on the swordtail fish (*Xiphophorus helleri*) by a multi-biomarker approach. Chemosphere 2013;90:1281–1288.

13. Fent K, Bätscher R. Cytochrome P4501A induction potencies of polycyclic aromatic hydrocarbons in a fish hepatoma cell line: demonstration of additive interactions. Environ Toxicol Chem 2000;19(8):2047–2058.

14. Fent and Bätscher (2000) op. cit.

15. Eaton DL, Bammler TK. Concise review of the glutathione-S-transferases and their significance to toxicology. Toxicol Sci 1999;49:156–164.

16. Canesi L. Heavy metals and glutathione metabolism in mussel tissues. Aquat Toxicol 1999;46: 67–76.

17. Lee RF, Keeran WS. Marine Invertebrate Glutathione-S-transferases: Purification, Characterization and Induction. Mar Environ Res 1988; 24:97–100; Cheung CCC, Zheng GJ, Li AMY, et al. Relationships between tissue concentrations of polycyclic aromatic hydrocarbons and antioxidative responses of marine mussels, *Perna viridis*. Aquat Toxicol 2001;52:189–203.

18. Reviewed by van der Oost (2003) op. cit.

19. Vidal-Liñán L, Bellas J, Campillo JA, Beiras R. Integrated use of antioxidant enzymes in mussels, *Mytilus galloprovincialis*, for monitoring pollution in highly productive coastal areas of Galicia (NW Spain). Chemosphere 2010;78:265–272.

20. Eaton and Bammler (1999) op. cit.

21. Wesseling C, Castillo L, Elinder CG. Pesticide poisoning in Costa Rica. Scand J Work Environ Health 1993;19:227–235.

22. Landis WG, Yu M-H, editors. Introduction to environmental toxicology. 2nd ed. Boca Raton: Lewis Publishers; 1998. (p. 108).

23. Britt J. Properties and effects of pesticides. Chapter 17. In Roberts SM et al, editors. Principles of toxicology 3rd ed. New York: Wiley; 2015.

24. e.g. Bocquené G, Galgani F, Truquet P. Characterisation and assay conditions for use of AChE activity from several marine species in pollution monitoring. Mar Environ Res 1990;30:75–89.

25. Bocquené G., Galgani F. Biological effects of contaminants: Cholinesterases inhibition by organophosphate and carbamate compounds. ICES Techn Mar Environ Sci 1998;22:1–13.

26. Galgani F, Bocquené G, Cadiou Y. Evidence of variation in cholinesterase activity in fish along a pollution gradient in the North Sea. Mar Ecol Prog Ser 1992;91:77–82.

27. Hogan JW. Water temperature as a source of variation in the specific activity of brain cholinesterase of Bluegills. Bull Environ Contam Toxicol 1970; 5:347–354.

28. OSPAR. Background documents and technical annexes for biological effects monitoring (Update 2013). OSPAR Comm London. Publ 2013;589, 238 pp.

29. See Endnote. 26.

30. Viarengo A, Lowe D, Bolognese C, et al. The use of biomarkers in biomonitoring: A 2-tier approach assessing the level of pollutant-induced stress syndrome in sentinel organisms. Comp Biochem Physiol, Part C 2007;146:281–300.

31. Moore MN. Cytochemical demonstration of latency of lysosomal hydrolases in the digestive cells of the common mussel, *Mytilus edulis,* and changes induced by thermal stress. Cell Tiss Res 1976;175:279–287.

32. For mussels see: Lowe DM, Fossato VU, Depledge MH. Contaminant-induced lysosomal membrane damage in blood cells of mussels *Mytilus galloprovincialis* from the Venice Lagoon: an *in vitro* study. Mar Ecol Prog Ser 1995;129:189–196; for earthworms see: Weeks JM, Svendsen C. Neutral red retention time by lysosomes from earthworm (*Lumbriculus rubellus*) coelomocytes: a simple biomarker of exposure to soil copper. Environ Toxicol Chem 1996;15(10): 1801–1805.

33. ICES. Biological effects of contaminants: Measurement of lysosomal membrane stability. By Moore MN, Lowe D, Köhler A. ICES Techn Mar Environ Sci 2004;36:31 pp. ICES. Lysosomal membrane stability in mussels. Martínez-Gómez C, Bignell J, Lowe D. ICES Techn Mar Environ Sci 2015;56:41 pp.

34. Martínez-Gómez C, Benedicto J, Campillo JA, et al. Application and evaluation of the neutral red retention (NRR) assay for lysosomal stability in mussel populations along the Iberian Mediterranean coast. J Environ Monit 2008;10:490–499.

35. Castaño A, Becerril C, Llorente MT. 11 Fish cells used to detect aquatic carcinogens and genotoxic agents. In: Mothersill C, Austin B, editors. In vitro methods in aquatic toxicology. Berlin:Springer. 2003, pp. 241–278.

36. Hose JE, Cross JN, Smith SG, et al. Elevated circulating erythrocyte micronuclei in fishes from contaminated sites off southern California. Mar Environ Res 1987;22:167–176.

37. Brunetti R, Majone F, Gola I, et al. The micronucleus test: examples of application to marine ecology. Mar Ecol Prog Ser 1988;44:65–68.

38. Brunetti R, Gabriele M, Valerio P, Fumagalli O. The micronucleus test: temporal pattern of base-line frequency in *Mytilus galloprovincialis.* Mar Ecol Prog Ser 1992;83:75–78. Burgeot Burgeot T, Woll S, Galgani F. Evaluation of the micronucleus test in *Mytilus galloprovincialis* for monitoring applications along the French coasts. Mar Pollut Bull 1996;32(1):39–46.

39. Fairbairn DW1, Olive PL, O'Neill KL. The comet assay: a comprehensive review. Mutat Res. 1995 Feb;339(1):37–59.

40. Mitchelmore CL, Chipman JK. DNA strand breakage in aquatic organisms and the potential value of the comet assay in environmental monitoring. Mutat Res. 20, March 1998;399(2): 135—47. Pandrangi R, Petras M, Ralph S, Vrzoc M. Alkaline single cell gel (comet) assay and genotoxicity monitoring using bullheads and carp. Environ Mol Mutagen. 1995;26(4):345—56.

41. Kammann U, Bunke M, Steinhart H, Theobald N. A permanent fish cell line (EPC) for genotoxicity testing of marine sediments with the comet assay. Mutat Res 2001;498(1—2):67—77.

42. Mai H, Cachot J, Brune J, et al. Embryotoxic effects of heavy metals and pesticides on early life stages of Pacific oyster (*Crassotrea gigas*). Mar Pollut Bull 2012;64:2663—2670.

43. Bean T. P., Akcha F. Biological effects of contaminants: Assessing DNA damage in marine species through single-cell alkaline gel electrophoresis (comet) assay. ICES Techn Mar Environ Sci 2016;58. 17 pp.

44. Blaber SJM. The occurrence of a penis-like outgrowth behind the right tentacle in spent females of *Nucella lapillus* (L.). Proc Malacol Soc London 1970;39:231—233.

45. Gibbs P, Bryan GW, Pascoe PL, et al. The use of the dog-whelk (*Nucella lapilus*) as an indicator of tributyl-tin (TBT) contamination. J Mar Boil Ass UK 1987;67:507—523.

46. Huet M. Surveillance du niveau de TBT sur le litoral français de la Manche, par l'utilisation des gastéropodes comme indicateurs biologiques. Université de Bretagne occidentale, Laboratoire de biologie marine U.R.A. C.N.R.S. 1513;1992. Rapport du contrat IFREMER 91 2 43 043 4 DEL.

47. Stroben E, Schulte-Oehlmann U, Fiorini P, Oehlmann J. A comparative method for easy assessment of coastal TBT pollution by the degree of imposex in Prosobranch species. Haliotis 1995;24:1—12.

48. See Endnote. 28.

49. Laranjeiro FMG, Sánchez-Marín P, Galante-Oliveira S, et al. Tributyltin pollution biomonitoring under the WFD: development of a new tool to assess the ecological quality status of European water bodies. Ecological Indicators 2015;57:525—535.

50. See Endnote. 49.

51. Purdom CE, Hardiman PA, Bye VJ, et al. Estrogenic effects of effluents from sewage treatment works. Chem Ecol 1994;8:275—285.

52. Scott AP, Hylland K. Biological effects of contaminants: Radioimmunoassay (RIA) and enzyme-linked immunosorbent assay (ELISA) techniques for the measurement of marine fish vitellogenins. ICES Techniques in Marine Science (TIMES); Nº 31. 2002, 21 pp.

53. Gagné F. Neuroendocrine disruption. Chapter 9. In: Gagné F, editor. Biochemical ecotoxicology London:Academic Press. 2014. pp 145—170.

54. Sanchez-Marín P, Fernandez-Gonzalez LE, Mantilla-Aldana L, et. al. Shotgun proteomics analysis discards alkali labile phosphate as a reliable method to assess vitellogenin levels in *Mytilus galloprovincialis*. Environ Sci Technol 2017;51(13):7572—7580.

55. Beiras R, His E. Toxicity of fresh and freeze-dried hydrocarbon-polluted sediments to *Crassostrea gigas* embryos. Mar Pollut Bull 1995;30(1):47—49. Beiras R, His E, Seaman MNL. Effects of storage temperature and duration on toxicity of sediments assessed by *Crassostrea gigas* oyster embryo bioassay. Environ Toxicol Chem 1998;17:2100—2105.

56. Dillon TM, Moore DW, Jarvis AS. The effects of storage temperature and time on sediment toxicity. Arch Environ Contam Toxicol 1994; 27:51—53. Becker DS, Ginn TC. Effects of storage time on toxicity of sediments from Puget Sound, Washington. Env Toxicol Chem 1995;14(5): 829-35. Beiras R, His E, Seaman MNL. Effects of storage temperature and duration on toxicity of sediments assessed by *Crassostrea gigas* oyster embryo bioassay. Environ Toxicol Chem 1998;17(10):2100-2105. Norton BL, Lewis MA, Mayer FL. Storage duration and temperature and the acute toxicities of estuarine sediments to Mysidopsis bahia and Leptocheirus plumulosus. Bull Environ Contam Toxicol. 1999;63:157—166. Geffard O, His E, Budzinski H, et al. Effects of storage method and duration on the toxicity of marine sediments to embryos of oysters *Crassostrea gigas*. Environ Pollut 2004;129:457—465.

57. American Society for Testing and Materials [ASTM]. Designation E1391-03 Standard guide for collection, storage, characterization, and manipulation of sediments for toxicological testing and for selection of samplers used to collect benthic invertebrates. ASTM Book of Standards 2003;11.06.

58. Watson PG, Frickers PE, Goodchild CM. Spatial and seasonal variations in the chemistry of sediment interstitial waters in the Tamar Estuary. Estuar Coast Shelf Sci 1985; 21:105—119. See also ASTM (2003) op. cit.

59. Nipper M, Qian, Y, Carr, RS et al. Degradation of picric acid and 2,6-DNT in marine sediments and waters: the role of microbial activity and ultra-violet exposure. Chemosphere 2004;56: 519—530.

60. Carr RS. Sediment porewater testing. P. 8(37-41). In: Clesceri LS, Greenberg AE, Eaton AD, editors. Standard methods for the examination of water and wastewater, Section 8080, 20th ed. Washington, DC, USA: American Public Health Association; 1998.

61. Beiras R. Comparison of methods to obtain a liquid phase in marine sediment toxicity bioassays with *Paracentrotus lividus* sea urchin embryos. Arch Environ Contam Toxicol 2001;42: 23—28.

62. ISO 11348—3:2007 Water quality — Determination of the inhibitory effect of water samples on the light emission of *Vibrio fischeri* (Luminescent bacteria test) — Part 3: Method using freeze-dried bacteria.

63. Environment Canada. EPS 1/RM/42. Biological test method for determining toxicity of sediment using luminescent bacteria. 2002.

64. ISO 10253:2006. Water quality. Marine algal growth inhibition test with *Skeletonema costatum* and *Phaeodactylum tricornutum*.

65. US-EPA Short-term Methods for Estimating the Chronic Toxicity of Effluents and Receiving Waters to Freshwater Organisms. EPA-821-R02-013. 2002.

66. Environment Canada Biological test method: growth inhibition test using a freshwater alga. EPS/1-RM/25. 2007.

67. Woelke CE. Bioassay: the bivalve larvae tool. In: Proceedings of the Northwest Symposium on Water Pollution Research, Portland, OR: US Department HEWPHS; 1961. pp. 113—123.

68. ASTM. Standard guide for conducting static acute toxicity tests starting with embryos of four species of saltwater bivalve molluscs. E724-E798 (Reapproved 2004). American Society for Testing and Materials, Philadelphia.

69. Leverett D, Thain J. ICES Techniques in Marine Environmental Sciences No 54, Copenhagen: International Council for the Exploration of the Sea; 2013. 34 pp.

70. Quiniou F, His E, Delesmont R, et al. Bio-indicateur de la toxicité potentielle de milieux aqueux: bio-essai "développement embryo-larvaire de bivalve." Ed. Infremer, Méthodes dánalyse en milieu marin; 2005. 24 pp.

71. OSPAR (2013) op. cit.

72. Wilson DP. A biological difference between natural sea-waters. J Mar Biol Assoc 1951;30: 1—26. See also Bernhard M. Pubbl Stn Zool Napoli 1957;29:80—95.

73. Bougis MP. Sur l'effet biologique du cuivre en eau de mer. Comptes Rendus de l'Academie des Sci 1959;249:326—328.

74. Kobayashi N. Fertilized sea urchin eggs as an indicatory material for marine pollution bioassay, preliminary experiments. Publ Seto Mar Biol Lab 1971;18:379—406.

75. ASTM Standard guide for conducting static acute toxicity tests with echinoid embryos. E 1563-1598 (Reapproved 2004) American Society for Testing and Materials, Philadelphia.

76. Environment Canada. Reference Method for Measuring the Toxicity of Contaminated Sediment to Embryos and Larvae of Echinoids (Sea Urchins or Sand Dollars); 2014. Reference Method 1/RM/58.

77. Beiras R, Durán I, Bellas I, et al. Biological effects of contaminants: *Paracentrotus lividus* sea urchin embryo test with marine sediment elutriates. ICES Techn Mar Environ Sci 2012;51.

78. Saco-Álvarez L, Durán I, Lorenzo JI, Beiras R. Methodological basis for the optimization of a marine sea-urchin embryo test (SET) for the ecological assessment of coastal water quality. Ecotox Environ Saf 2010;73:491−499.

79. Durán I, Beiras R. Assessment criteria for using the sea-urchin embryo test with sediment elutriates as a tool to classify the ecotoxicological status of marine water bodies. Environ Toxicol Chem 2010;29(5):1192−1198.

80. Arizzi Novelli A, Picone M, Losso C, et al. Ammonia as confounding factor in toxicity tests with the sea urchin Paracentrotus lividus (Lmk). Toxicol Environ Chem 2003;85(4/6): 183−190.

81. Losso C, Arizzi Novelli A, Picone M, et al. Sulfide as a confounding factor in toxicity tests with the sea urchin *Paracentrotus lividus*: comparison with chemical analysis data. Environ Toxicol Chem 2004;23(2):396−401.

82. ISO 14699:1999(E) Water quality − Determination of acute lethal toxicity to marine copepods (Copepoda, Crustacea).

83. ASTM Standard Guide for Conducting Renewal Microplate-Based Life-Cycle Toxicity Tests with a Marine Meiobenthic Copepod E2317-04. American Society for Testing and Materials, Philadelphia. 2012.

84. ISO 16778:2015 Water quality − Calanoid copepod early-life stage test with *Acartia tonsa*.

85. ASTM. Standard test method for measuring the toxicity of sediment-associated contaminants with estuarine and marine invertebrates E 1367. American Society for Testing and Materials, Philadelphia. 2003.

86. Roddie B, Thain J. Biological effects of contaminants: Corophium sp. sediment bioassay and toxicity test. ICES Techn Mar Environ Sci 2001;28.

87. See also: U.S. EPA. Methods for Assessing the Toxicity of Sediment-associated Contaminants with Estuarine and Marine Amphipods. EPA/600/R-94/025. Washington, DC 20460: United States Environmental Protection Agency. Office of Research and Development; 1994; ISO 16712:2005. Water quality. Determination of acute toxicity of marine or estuarine sediment to amphipods.

88. USEPA. National Coastal Condition Report IV. EPA-842-R-10−1003. Washington, DC 20460: United States Environmental Protection Agency. Office of Research and Development/Office of Water; 2012.

89. USEPA. Mid-Atlantic Integrated Assessment 1997-98 Summary Report, EPA/620/R-02/003. Narragansett, RI: U.S. Environmental Protection Agency, Atlantic Ecology Division; 2002.

90. Olla BL, Bejda AJ, Pearson WH. Effects of Oiled Sediment on the Burrowing Behaviour of the Hard Clam, *Mercenaria mercenaria*. Mar Enriron Res 1983;9:183193.

91. Pearson WH, Woodruff DL, Sugarman, PC, Olla BL. Effects of oiled sediment on predation on the Littleneck Clam, *Protothaca staminea*, by the Dungeness Crab, *Cancer magister*. Estuar Coast Shelf Sci 1981;13:445−454.

Suggested Further Reading

- Beiras R, Durán I, Bellas I, et al. Biological effects of contaminants: *Paracentrotus lividus* sea urchin embryo test with marine sediment elutriates. ICES Tech Marine Environ Sci 2012;51:13.

- Blaise C, Férard JF, Vasseur P. Chapter 18. Microplate toxicity tests with microalgae: A review. pp. 269−288. In: Wells PG, Lee K, Blaise C, editors. Microscale testing in aquatic toxicology. New York: CRC Press; 1998.

- Bocquené G, Galgani F. Biological effects of contaminants: Cholinesterases inhibition by organophosphate and carbamate compounds. ICES Techn Marine Environ Sci 1998;22:13.

- His E, Beiras R, Seaman MNL. The assessment of marine pollution- Bioassays with bivalve larvae. Advances in marine biology, vol. 37. San Diego: Academic Press; 2000. p. 1–178.
- Martínez-Gómez C, Bignell J, Lowe D. Lysosomal membrane stability in mussels. CES Techn Marine Environ Sci 2015;56:41.
- Teaf CM, Middendorf PJ. Mutagenesis and genetic toxicology. Chapter 12, pp.239–264. In: Williams PL, James RC, Roberts SM, editors. Principles of toxicology. Environmental and industrial applications. 2nd ed. New York: John Wiley & sons, Inc; 2000.
- US EPA. 1994. Metods for assessing the toxicity of sediment-associated contaminants with estuarine and marine amphipods. EPA 600/R-94/025 June 1994.
- Wong CKC, Yeung HY, Cheung RYH, et al. Ecotoxicological assessment of persistent organic and heavy metal contamination in Hong Kong coastal sediment. Arch Environ Contam Toxicol 2000;38:486–93.
- Losso C, Picone M, Arizzi Novelli A, et al. Developing toxicity scores for embryotoxicity tests on elutriates with the sea urchin *Paracentrotus lividus*, the oyster *Crassostrea gigas*, and the mussel *Mytilus galloprovincialis*. Arch Environ Contam Toxicol 53:220–226.

CHAPTER 17

Marine Pollution Monitoring Programs

17.1 INTEGRATED ASSESSMENT OF POLLUTION STATUS; CHEMICAL AND BIOLOGICAL TOOLS

Traditional assessment of environmental pollution consisted of analyzing in environmental samples (normally inert matrices such as water, sediment, or soil), long lists of so-called priority pollutants. This approach was limited by the need to link figures of chemical concentrations with ecologically meaningful assessment criteria to understand and try to predict whether those figures were likely to pose environmental risk or if they were biologically innocuous. This link can only be established by the simultaneous measurement of **pollutant concentrations** and **biological effects**, which is regarded as the integrated approach to environmental monitoring. Currently, institutions in charge of environmental management advocate this integrated approach that combines biological tools with the traditional chemical measurements that provide complementary information regarding the identification of harmful effects on the organisms[1] (Fig. 17.1). For example, imposex, lysosomal membrane stability, and the micronucleus test are proposed by the Helsinki Commission (HELCOM) among the core indicators for monitoring the environmental health of the Baltic sea[2]; whereas ethoxyresosufin-O-deethylase (EROD) activity in benthic fish, scope for growth (SFG) in mussels, and imposex in gastropods are frequently reported by OSPAR,[3] though only the latter is considered mandatory. Chemical and biological data are complementary since the latter informs whether fitness-relevant functions in the organisms are affected, and the former points at potential causal agents of those harmful effects. In addition, biological tools may also provide information useful for environmental management that would be very costly to obtain from analytical chemistry only, considering the mounting number of synthetic chemicals released to the environment.

In addition to the chemical analyses of pollutants and the biological effects measurements, a third component of the integrated monitoring approach is the set of **supporting measurements**. These are environmental (temperature, salinity, food availability, etc.) and biotic (age, condition index, stage in gametogenic cycle) variables typical of each site that interfere with the biological

Biological tools provide ecologically relevant information to assess pollution

Integrated monitoring includes chemical concentrations, biological effects, and supporting measurements

293

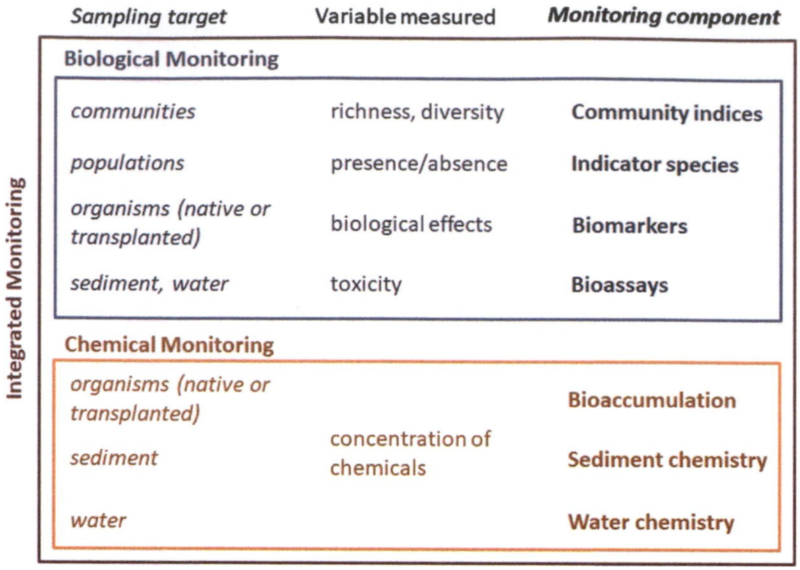

FIGURE 17.1

Chemical and biological components of an integrated marine environmental monitoring strategy. Environmental matrices sampled, variables measured, and components of the monitoring network are indicated. *Own elaboration.*

responses measured helping to understand their results, or are needed to express the chemical concentrations (dry weight, lipid content).

In short, incorporating biological tools to monitoring networks provide the following advantages:

- They allow **detection of** unexpected or **emerging pollutants** not targeted in the chemical monitoring;
- They may be used as a cost-effective **screening tool** where subsequent monitoring efforts may be applied;
- They provide **ecological relevance** to the monitoring, since toxic effects depend on chemical speciation and interactions with environmental factors.

However, lack of consensus on the selection of organisms and biological tools, and a lack of methodological standardization, has hampered the adoption of integrated monitoring schemes for the routine assessment of the marine environment.[4] For example, intercalibration exercises for useful biological tools such as SFG, neutral red retention time (NRRT), AChE, micronuclei, or comet assay are rare or none at present.

Chemical and biological monitoring may disagree; that is why the combination is needed

A frequent misconception in the application of biological monitoring in an integrated monitoring scheme is to expect a perfect match between the chemical and the biological information. First, biological responses are affected by various natural environmental factors and biological traits of the local populations (reproductive stage, food availability, and genetic variability). The wider the geographical extension of the monitoring network, the higher the chance for those natural factors to interfere with the response measured. Second, all chemical monitoring efforts are limited in terms of target analytes. In fact that misconception underrates one of the main strengths of biological tools, i.e., the ability to detect pollutants not targeted in the conventional chemical screenings. For example, J. Widdows and coworkers[5] conducted a comprehensive monitoring of the North Sea coast in the United Kingdom combining SFG measurements and chemical analysis of trace metals, tributyl-tin (TBT), organochlorines, and polycyclic aromatic hydrocarbons (PAHs) in mussels. When the chemical data are combined to calculate for each mussel population the chemical pollution index (CPI, see Section 1.5) using the OSPAR background assessment criteria, the results show a highly significant ($P = 0.007$) negative effect of the bioaccumulation of chemicals on the SFG of the bivalves (Fig. 17.2), but chemical analyses explain 19 percent of the variability in SFG only. In one of the sampling sites (red arrow in Fig 17.2) there was a remarkable disagreement between the low levels of pollutants measured and the poor energy balance of the bivalves. This site would have been classified as nonpolluted according to the chemical dataset, but SFG pointed at unidentified factors decreasing the energy balance of that population. In fact, the site was situated near a sewage outfall and under the influence of the outflow from the Great Ouse, where major sewage inputs had been identified in previous studies. Therefore, this site detected by the biological tool was likely to suffer from high concentrations of unconventional pollutants present in wastewater effluents, which could have been identified by particularly designed further analyses.

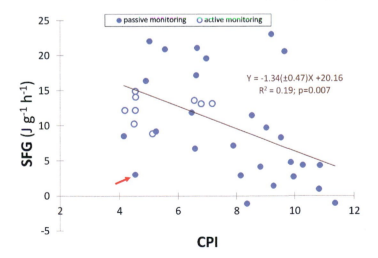

FIGURE 17.2

Inverse relationship between scope for growth (SFG) and chemical pollutants (chemical pollution index, CPI) accumulated in native (passive monitoring) and transplanted (active monitoring) mussels in the North Sea. Transplanted mussels were included in the analysis although their SFG values do not show any relationship with bioaccumulation of chemicals. See text. *Data from Widdows et al. (1995) op. cit.*

Many biological responses measured at different levels of biological organization have been, and continue to be proposed in the scientific literature as useful monitoring tools in aquatic and terrestrial environments. This led to the use, sometimes confusing, of related expressions such as bioindicators, biomonitors, biomarkers, or biosensors.

Indicator species, community indices, biomarkers, and bioassays are all biological indicators

An indicator can be defined as something that gives information and, etymologically, points at certain conclusion. A **biological indicator**, or bioindicator, is thus the most general term encompassing any observation conducted at any level of biological organization (from biomolecules to communities) that provides useful information regarding the pollution status of the study area. Canary birds caged in mining to detect poison gases were definitely replaced by electronic detectors as late as 1986 in the UK. The mouse test is still in use to protect human health in the eventuality of certain algal toxins than can be present in shellfish. None of these examples concern marine pollution, but illustrate the power of extremely simple biological responses to screen for a number of dangerous chemicals in a very cost-effective and yet highly sensitive fashion.

In traditional ecological studies it is well known the more restricted concept of **indicator species**, discussed in Chapter 15, as a species that disappears or is particularly abundant in places with particular environmental conditions, including organic or chemical pollution. For instance, the lichen *Usnea articulate* occurs only where levels of atmospheric sulfur dioxide are low.[7] Studying the distribution of this lichen in trees of a large city, for instance, is much quicker and inexpensive than comprehensive sampling and analysis of SO_2 in the air.

The concept of indicator species, whose absence/presence depends on anthropogenic environmental factors, was further developed up to community level, particularly in benthic communities, originating the **biotic** or **benthic indices**, whose strengths and limitations applied to the study of marine pollution were already discussed in Chapter 15. Community indices such as the saprobic index, based on the classification of the benthic macrofauna species into groups of tolerance to organic enrichment have been successfully applied to evaluate the quality of streams since the beginning of the 20th century. More universal community indices are species richness or **diversity indices**.

Low levels of biological organization give an insight in the mechanisms of effects while higher levels demonstrate ecological relevance

The interpretation of the different biological tools or indicators included in a monitoring network must take into account the advantages and limitations offered by indicators measured at different levels of biological organization. Effects of pollution on biological communities bear the highest ecological relevance, but lack "early warning" value, since significant reductions in species diversity take place quite late in the sequence of events triggered by anthropogenic environmental stress.[8] This limitation may be overcome by studying sensitive biological responses at lower levels of biological organization. Biological indicators at suborganismic level, using molecular, cellular, and physiological responses, are usually termed **biomarkers**, a subject dealt with in Chapter 16. Descending to biochemical pathways pretends, first, to unmask cause-effect relationships, and secondly, to increase sensitivity and thus "early warning" value.

These biological tools are measured in organisms native from or deployed at the locations under study. In contrast, ecotoxicological **bioassays** are conducted using standardized animals from a homogeneous stock, normally in laboratory, which are not native from the study site. This topic is addressed in Sections 16.6–16.8. The advantages of ecotoxicological bioassays are twofold, they detect the presence of harmful chemicals, serving as cost-effective screening tool to identify sites where further analysis may be conducted, and also inform that those chemicals have reached levels that cause toxicity to the marine organisms, which is obviously an ecologically relevant information. Notice that although ecotoxicological bioassays target detection of chemicals, they are in fact a biological tool with the corresponding advantages and limitations, and they are included in the biological monitoring scheme (see Fig. 17.1).

Biomonitors provide biotic matrices for chemical analyses

Chemical analyses on which traditional pollution control is based can be conducted on biological matrices. Although this is not a biological tool, it has generated yet another term, **biomonitor**, applied to the organism (very frequently the marine mussel) used in the surveillance program. Although the term biomonitor has been used in a much broader sense to refer to any biological tool used for environmental monitoring, it is advisable[9] to reserve this term for organisms that are used as bioaccumulators and thus provide a biotic matrix for chemical analysis of pollutants, a topic addressed in Section 17.3. The distinction yet is subtle since biomonitors also frequently provide the matrix for biomarker analysis.

Biosensors continuously record biological endpoints that can trigger real-time alarms

Finally, in the context of biotechnology, biomonitors or **biosensors** are in situ deployed devices connected to a living organism which can measure in continuum a biological response, such as heart beating, respiration rate, or valve closure, useful for the description of the environmental status. The signal obtained from the biosensor is processed by a computer at real time and an alarm may be programmed if the signal falls beyond given values. The main limitation of this technique is to filter the background noise originated by normal responses of the living biosensor to obtain a reliable signal, relevant for environmental monitoring.

In summary, the use of biological tools in environmental pollution studies ideally involves a suite of responses recorded at different levels of biological organization that can gather evidence of the presence of deleterious substances and harmful effects on the exposed organisms. Responses at each level of biological organization provide information that helps us to understand and interpret the relationship between exposure and adverse effects. Variables measured at lower levels of biological organization, such as enzyme activities, are sensitive and occasionally specific[10] for certain kinds of pollutants, unmasking cause-effect relationships between pollutants and biological responses.

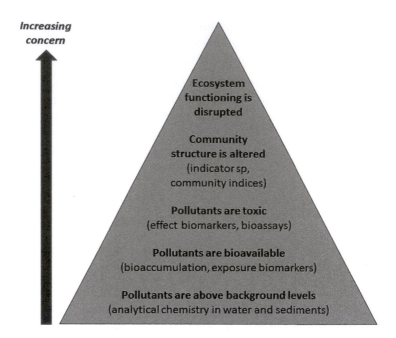

FIGURE 17.3

A hierarchical interpretation of chemical and biological monitoring results in the assessment of environmental pollution. Concern increases in the sequence: pollutants exceed background levels, pollutants bioaccumulate and trigger biological effects, pollutants exert toxicity, the composition of the community is changed, ecosystem functioning is disrupted.

Variables measured at organismic level, such as sediment toxicity, provide a priori evidence of potential risk, while changes in community structure a posteriori certify the effects. While the disappearance of a species from a community takes place only after ecologically harmful levels of pollutants have been reached, the more subtle biochemical effects may hopefully identify incipient pollution and prompt remediation measures.[11]

Therefore, the interpretation of monitoring results must be guided by a hierarchical approach depicted in Fig. 17.3. Increasing concern corresponds to the sequence: pollutants are present above background values, pollutants are bioavailable, accumulate in biota and trigger biological effects, pollutants are toxic, and pollution changes community structure.

17.2   GOALS, DESIGN, AND MANAGEMENT IMPLICATIONS OF MARINE MONITORING PROGRAMS

Monitoring program: scheme of sampling and measurements to assess environmental quality

In the context of environmental pollution, the term **monitoring** can be defined as the direct measurement of a pollutant and/or the indirect measurement of its effects with the aim of evaluating its levels and controlling its impact on the man or on the environment.[12] This task is commonly articulated through monitoring programs undertaken by national or international agencies with

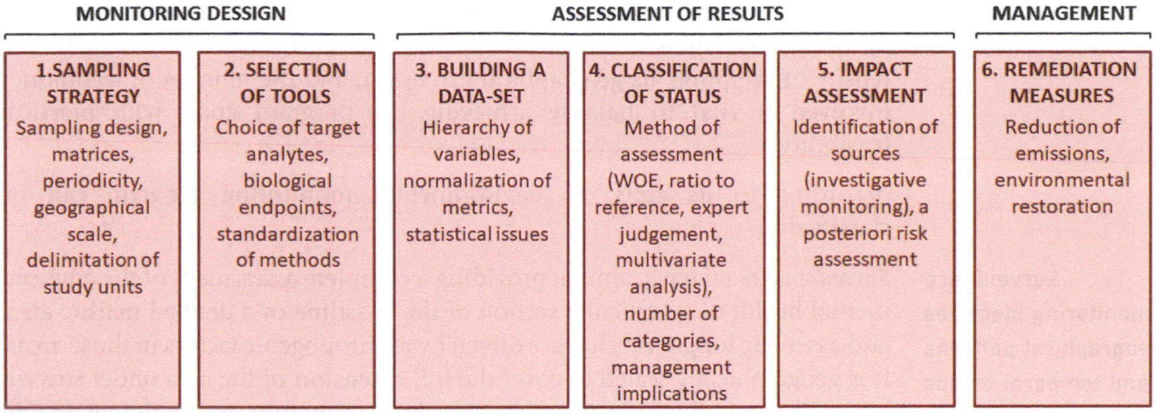

MONITORING DESSIGN		ASSESSMENT OF RESULTS			MANAGEMENT
1.SAMPLING STRATEGY	**2. SELECTION OF TOOLS**	**3. BUILDING A DATA-SET**	**4. CLASSIFICATION OF STATUS**	**5. IMPACT ASSESSMENT**	**6. REMEDIATION MEASURES**
Sampling design, matrices, periodicity, geographical scale, delimitation of study units	Choice of target analytes, biological endpoints, standardization of methods	Hierarchy of variables, normalization of metrics, statistical issues	Method of assessment (WOE, ratio to reference, expert judgement, multivariate analysis), number of categories, management implications	Identification of sources (investigative monitoring), a posteriori risk assessment	Reduction of emissions, environmental restoration

FIGURE 17.4
Steps in the design and implementation of a monitoring strategy to guide environmental management. WOE: Weight-of-evidence.

political competencies for environmental management (See Section 17.4). The steps in the design and implementation of a **monitoring program** are represented in Fig. 17.4. They include a strategy of sampling, and a selection of chemical and biological measurements recorded according to standard procedures (see also Fig. 17.1). A normally complex and heterogeneous environmental dataset is built with the monitoring results. Frequent statistical issues when dealing with these large datasets are choice of metrics, normalization of data, search of reference values, and hierarchical classification of variables as supporting, explicative (or independent), and dependent. The first are used for a posteriori explanations or for normalization of results but they are generally not included in the statistical analyses. Different methods have been used to classify sites into discrete categories of environmental status with correspondingly different management implications according to monitoring results. This overall assessment can follow a weight-of-evidence approach, scoring systems based on ratio-to-reference scores, resort to expert judgment, or rely on more objective multivariate analysis. Some of these alternatives will be further discussed in Section 19.2. In sites or areas classified as polluted, further monitoring activities specifically designed according to the identified pollutants should be conducted to identify the sources of pollution and design remediation measures. Finally, the competent political institutions, which normally are not the same than those conducting the monitoring networks, should implement the remediation measures prescribed, including reduction of emissions and environmental restoration when needed.

Monitoring programs are conducted by agencies owing the ships, personnel, expertise, and facilities that demand a comprehensive monitoring schedule. Coastal monitoring is implemented by national agencies while open sea monitoring programs are normally coordinated by international institutions but dependent on national resources. Marine monitoring networks imply costly sampling and sophisticated laboratory procedures. The decision to scale the intensity of sampling, its geographical extension, and the number of techniques involved is vital to balance achieving the program goals with practical feasibility.

According to its aims, an environmental monitoring program can be classified as:

Surveillance monitoring identifies geographical patterns and temporal trends

Surveillance monitoring: aims at providing a complete assessment of the environmental health of a particular section of the coastline or a defined marine area, and assessing long-term changes driven by anthropogenic factors in those areas. It is geographically scaled to cover the full extension of the area under surveillance. The sampling sites must be sufficiently numerous and representative to identify hot spots for specific pollutants, and at least part of the sites must be sampled with sufficient frequency to elucidate long-term temporal trends. In brief, surveillance monitoring has two aims: description of **geographical patterns** of pollution that allows the assessment of status of each area, and identification of **temporal trends** for those pollutants. This information is vital for the environmental managers to prompt specific remediation actions and examine the efficacy of the actions already taken. For example, in 2014 OSPAR[13] examined by means of the VDSI index the trends in gastropod imposex and identified a significantly decreasing trend in 57.3 percent of the 150 sites monitored along the Atlantic European coast, confirming the efficacy of the TBT ban for antifouling paints implemented by the beginning of the century.

Investigative monitoring seeks the sources of pollution to plan remediation

Investigative monitoring: is specifically designed to identify the source, distribution and effects of certain pollutants in a previously identified hot spot, or as a result of an accidental episode. Thus the intensity of sampling and the number and complexity of techniques applied is much higher, but the geographical range is much more limited. The ultimate aim is to identify the causes to plan remediation actions to ameliorate the environmental status of the impacted area.

The European Water Framework Directive (WFD) defines a third, intermediate, type of monitoring, the operational monitoring, to be undertaken to better establish the status of particular water bodies which are at risk according to the surveillance monitoring, and to assess whether the status changes after remediation measures are taken.

Sediments are the preferred matrix for marine monitoring...

Most pollutants are found in seawater in very small quantities so that their analytical quantification is complex and expensive. Furthermore, concentrations in water, especially in coastal areas, are variable over time and dependent on the tides, currents, winds, or intermittent discharges. Moreover, the information provided by chemical analysis of water reflects only the pollution levels at the time of sample collection, which may change after a few hours or a few days. Therefore, this is not the matrix generally recommended in surveillance programs of marine pollution. Sediments are preferable to water samples as a matrix for monitoring environmental quality, because the concentrations of pollutants in sediments are much higher and less variable in time and space, reflecting in an integrated manner the state of pollution in a certain area. Most anthropogenic chemicals that are introduced into the marine environment are accumulated in the sediment, and the enrichment in polluted sites, relative to reference sites, is even higher for sediments than for mussels, especially for organic chemicals. But sediments do not only act as a reservoir for pollutants, they also serve as a source of toxicants to marine organisms. For these reasons, this matrix is universally used in marine pollution surveillance and control programs.[14]

When a substance cannot be metabolized or is slowly metabolized, tends to accumulate in the tissues of living organisms, or **biota**, reaching higher concentrations than those found in the environment. Many marine organisms accumulate contaminants in their tissues at much higher levels than those present in the surrounding water, with no apparent toxic effects. This feature offers evident advantages for the development of marine pollution monitoring programs: tissue concentrations are commonly several orders of magnitude higher than those in water and thus more likely to be above detection limits of the conventional analytical techniques, are more stable over time (may indicate past pollution events), and reflect only the fraction of a compound present in the environment that may be incorporated by the organism, i.e., the bioavailable fraction.

...but only biota inform about the bioavailable fraction

Therefore, the advantages of biota over inert matrices are the time-integration and the identification of the bioavailable fraction of the chemicals. As mentioned in the previous Section, the organisms used to provide a biotic matrix for chemical analysis of pollutants are termed biomonitors.[15]

According to the origin of the organisms used for the measurement of the biological effects, we can distinguish between **passive monitoring**, when native organisms are used, and **active monitoring**, when organisms from a common origin and with certain characteristics we wish to standardize (size, age, reproductive state, etc.) are transplanted into the different monitoring sites, using cages or other devices, to work with homogeneous biological material and control the exposure time.[16] Surveillance monitoring is normally passive, relying on wild organisms intended to be representative of broad areas. Thus, in surveillance monitoring organisms that live in direct contact with wastewater outfalls or within the mixing zone of effluents are generally avoided. Organisms from extensive aquaculture are also excluded from environmental monitoring networks, although they are surveyed for food safety issues.

Biomonitors can be native or transplanted organisms

On the other hand, for some investigative purposes, active monitoring, though costlier, may be more useful. Transplantation may also allow covering areas or habitats where the monitoring species does not occur naturally, such as the use of caged mussels in open sea. The transplantation locations may be designed to study the effect of distance from a point-source emission or be adapted to any other environmental gradients. Transplantation time in active monitoring campaigns commonly ranges from 1 to 2 months. A 1 month period is enough to reach steady-state concentrations for trace metals and the least hydrophobic organics (e.g., naphthalene, PCB-28), and also, provided deployment date and time was homogeneous, to identify trends and compare sites. However, 1 month is not sufficient to achieve steady state for more hydrophobic PAH and PCB, which may need more than six months to reach values equivalent to native populations.[17] However, longer exposure times are more affected by seasonal fluctuations as confounding factors. In addition, from a practical viewpoint, shorter deployments reduce both field costs and practical problems such as biofouling or the risk of cages loss.

17.3   MARINE MUSSELS, THE UNIVERSAL BIOMONITORS OF POLLUTION

Mussels are the ideal biomonitors…

The ideal characteristics of an organism to be used as a pollution sentinel include: abundant and easy to identify and to catch, ubiquitous along broad geographical areas, long-lived, available all-year round, sessile, or at least sedentary to be representative of the sampling site, large enough for easy handling, dissection, and analysis, amenable for transplantation, tolerant to pollution to be present in polluted sites, and with a moderate to null ability to regulate its internal levels of chemical pollutants to reflect the concentrations in the water. Talking about coastal ecosystems, this is the photo fit of the marine mussel. Mussels (genera *Mytilus*, *Perna*, *Modiolus*) are suspension-feeding bivalves that attach themselves to the substrate using byssus threads, and show a world-wide distribution, including deep ocean (genus *Bathymodiolus*) and freshwater (genus *Dresissena*) ecosystems.

…but other organisms may be better to monitor specific pollutants or habitats

However, in the classical work of G.W. Bryan et al.[18] is already pointed out that, even restricting us to coastal ecosystems, there is no universal indicator organism. This is due to habitat limitations (soft-bottom vs. rocky shores, exposed vs. protected, salinity gradients), but also to differences in the metabolism with regard to different chemical pollutants. Concerning metals, the use of *Fucus*, *Scrobicularia*, and *Mytilus* as more or less universal accumulators is recommended, but certain organisms may be particularly suitable for certain metals, such as *Littorina* for Cd, *Nereis* for Cu, or *Platichthys* for MeHg. The radionuclide ^{137}Cs, used as marker of anthropogenic radioactivity, follows K uptake pathways, and it is commonly measured in algae. Hydrophobic organics preferentially accumulate in lipid tissues. Thus, birds or marine mammals can be suitable bioaccumulators for these chemicals, and they can be sampled from the subcutaneous fat with no need to sacrifice the individual. Also for Hg and halogenated hydrocarbons, due to biomagnification, organisms from higher trophic levels can be more suited as candidate bioaccumulators.

Biomonitoring chemical pollution assumes passive accumulation of pollutants and no ability of the organism to regulate tissue levels. Mussels are normally considered "nonregulators" or "accumulators," in the sense that whole body residues reflect water concentrations. Mussels accumulate waterborne metals between 2 and 10 times more than infaunal bivalves.[19] A relevant exception is the case of copper, whose levels can be regulated by the mussels.[20] Copper content in decapod crustaceans, cephalopods, gastropods, and oysters are normally higher than those of co-occurring mussels.[21] However, even in this case, when environmental concentrations are very high they exceed the regulatory ability of mussels and significant Cu bioaccumulation takes place.[22]

Biomonitoring chemical pollution assumes no ability of the organism to regulate tissue levels

Accumulation does not imply deleterious physiological effects, since metals can be sequestered by metallothioneins or stored in metabolically inert subcellular compartments such as vesicles or granules. When mussels are considered as accumulators it does not mean that they lack regulatory ability at tissular level or regulation of trace metal availability at metabolically active sites.[23] It just means that whole body metal loads increase when concentrations in the water increase. For example, Cd accumulated in mussels at concentrations as high as 150 ppm apparently did not cause detrimental effects on growth and reproduction,[24] and thus viable mussel populations are found in extremely polluted sites.

Biomonitoring also assumes a bioconcentration factor (BCF) dependent on the ratio between uptake rate (determined by k_1) and elimination rate (determined by k_2) only, disregarding the concentration of pollutant in the water (see Eq. 11.5, Section 11.2). Natural biological or environmental factors modifying these rate constants, k_1 and k_2, along a geographical area subjected to biomonitoring will cause deviations from this assumption and thus may interfere with monitoring results. Since mussel size, or rather **growth rate**,[25] affects bioaccumulation, standard sizes must be used in monitoring programs. However, growth rate may be an interfering factor when monitoring spans through areas including populations with different physiological traits, due to either genetic or environmental differences. On the other hand, the rate of water pumped through the gills and the relationship between the **gills area** and the body volume are important factors affecting the uptake rates, which are directly related to the BCF. In addition, the ability of the metabolism to excrete the chemical— that may require previous **biotransformation** induced by local exposure—or the tendency of the organism to store the substances in nonexcretable compartments may also vary among populations. These will affect the elimination rate, inversely related to BCF.

Differences among populations at genetic or environmental level, and seasonal variability, interfere with biomonitoring results

Some pollutants may preferentially accumulate in the digestive gland, and this represents a variable fraction of total body weight. During spawning a drastic reduction in the weight of mussel soft tissues takes place, whereas the weight of the digestive gland remains unchanged. Therefore, sampling during the reproductive season should be avoided. Spawning may also affect physiological and biochemical endpoints. Temperature may also contribute to seasonal differences in metal contents,[26] and background values of physiological rates.

Sampling procedures are also important for quality assured biomonitoring, and technical recommendations are available.[27] For chemical analyses, mussels are commonly left in local seawater to allow elimination of feces and pseudo-feces. For molecular biomarkers, in contrast, mussels must be rapidly immersed in liquid nitrogen. Sample size depends on the variability of the target measurement. For chemical analysis OSPAR recommends sampling 3 pools of 20 individuals of similar size per site. For enzymatic biomarkers minimum sample sizes of 12 to 23, depending on the biomarker, were recommended.[28] For LMS and micronuclei a minimum of 10−20 individuals per site is recommended.

17.4   CASE STUDIES OF NATIONAL AND INTERNATIONAL MONITORING PROGRAMS: US MUSSEL WATCH, FRANCE RNO/ROCCH, OSPAR CEMP

International conventions with regional scope promote monitoring programs for the assessment of the marine environment

Since the sea has no borders, the most important initiatives to control marine pollution and manage marine resources stem from international agreements formalized in conventions voluntarily adopted by signatory countries, that later may evolve into intergovernmental organizations and commissions with regional scope. The oldest one is the **International Council for the Exploration of the Sea (ICES)**, originated in 1906 in Copenhagen and formalized in the 1964 ICES Convention with the aim of promoting international investigation of the sea and particularly the living resources. Concerning marine pollution monitoring, this is currently prompted by the **OSPAR Commission** for the Northeast Atlantic, **HELCOM** for the Baltic Sea, or the **MED POL** program for the Mediterranean, within the umbrella of the UNEP Mediterranean Action Plan that convened all countries with Mediterranean shoreline at the Barcelona Convention in 1976. OSPAR resulted from the merging in 1992 of the 1972 Oslo Convention, dealing with the prevention of pollution by dumping from ships, and the 1974 Paris Convention dealing with discharges from land-based sources. The UN established in 1981 the **Caribbean Environment Programme** (CEP). Countries of the region adopted an action plan that led to the development in 1983 of the Cartagena Convention, the first regionally binding treaty of its kind. It promotes the protection of the marine environment of the Region, provides the legal framework for the CEP, and is supported by technical agreements on oil spills, protected wildlife protection, and land-based sources of marine pollution. However, it has not implemented a monitoring program.

According to the UN Convention on the Law of the Sea (UNCLOS), all states have the right to conduct marine research disregarding their geographical location, but coastal states have the exclusive right to authorize scientific research activities within their exclusive economic zone, and "scientific research activities shall not constitute the legal basis for any claim to any part of the marine environment". Therefore international cooperation is encouraged. Due to political competencies and lack of resources (ships, laboratories, personnel) of those international institutions, in practice most environmental monitoring initiatives are conducted by national agencies. This is particularly evident for coastal monitoring programs, which are always carried out by national authorities. Marine monitoring is costly and labor-intensive, and the degree of engagement of the different contracting parties in the monitoring activities stemming from those international conventions is frequently heterogeneous.[29]

Chemical pollutants measured in the most common monitoring networks (trace metals, PAHs, PCBs, and other organohalogenated hydrocarbons, TBT) are sometimes regarded as "conventional pollutants" or "regulated pollutants," whereas other chemicals such as pharmaceuticals, cosmetic components or plastic additives are frequently referred in the scientific literature as "emerging pollutants." These customary terms lack scientific basis and are not recommended. In addition, new chemicals are successively added to the priority lists of regulated pollutants.

The first comprehensive effort to establish a monitoring program to assess the status of the coast was undertaken in the US by means of the **Mussel Watch Program**, initiated by the US Environmental Protection Agency (US-EPA) in the period 1976−78, and continued by the NOAA from 1986 to date.[30] The program is based on yearly collection of oysters (*Crassostrea virginica*) and mussels (*Mytilus* spp and *Dreissena* spp) in 300 sites from the Atlantic, Pacific, and Great Lakes coasts, and measures more than 140 chemical contaminants, including metals and metalloids, PCB, PAH, DDT, butyltins, and organochlorine pesticides. Sampling and analytical protocols were standardized and submitted to a quality assurance (QA) scheme that includes interlaboratory exercises. This is essential since the geographical scale of the Program requires the involvement of several different laboratories. The Mussel Watch identifies "areas of national concern," and determines the temporal trend for each pollutant at each site. For example, the commercial, military, and recreational boating activity supported made the San Diego Bay an area with elevated butyltin contamination, but several of its sites showed in 2008 a decreasing trend, following the universal ban of TBT at the beginning of the century. In contrast, the elevated levels of cadmium found in the Chesapeake Bay and associated to industrial wastewater discharge and urban stormwater runoff are not decreasing despite years of restoration efforts.

The US Mussel Watch pioneered the use of biota as sentinels of pollution

In Europe, France has been one of the countries that pioneered the routine monitoring of the health status of its coasts, including analysis of pollutants in biota. The national observation network (*Réseau National d'Observation*, **RNO**) run from 1979 to 2006 in over 40 sites along the French coast, and included annual sampling of wild *C. gigas* oysters in the South Atlantic coast and *M. galloprovincialis* mussels in the North Atlantic and Mediterranean coasts. From 1992, RNO actively promoted QA activities such as systematic use of reference materials, participation in international intercalibration exercises,

The French RNO provided on a regular basis monitoring data for the assessment of coastal pollution

and edition of standard method protocols. Within RNO up to 60 different chemicals were measured, including metals, PCB, organochlorine pesticides, and PAH. This allowed geographical identification of hot spots for certain pollutants (Fig. 17.5), and elucidated long-term temporal trends for those pollutants. For example, in contrast with the improvement for organochlorines, PAHs, and most metals, a significant increasing trend in copper pollution was detected in the southwest coast of France,[31] perhaps as a result of the substitution of TBT-based antifouling paints, although the increasing pattern for Cu contamination at national level is previous to the TBT ban (Fig. 17.6).

Several biological effects techniques, such as AChE activity in mussels[34] and EROD activities in benthic fish,[35] were measured in RNO at experimental level from 1991, but they were classified as feasible for investigative monitoring only except for imposex, which was fully included as a routine from 2003 in compliance with OSPAR mandate. Unfortunately, in 2008 RNO was discontinued and a new monitoring scheme adapted to the WFD was imposed where the

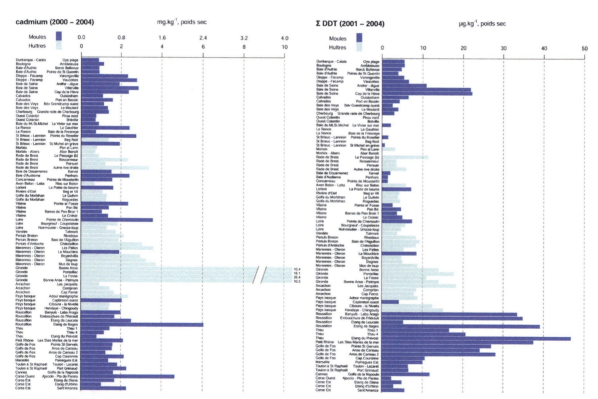

FIGURE 17.5

Median levels of two chemical contaminants, cadmium (left) and DDT (right) in native mussels and oysters along the French coasts. Sampling sites are ordered from the North Atlantic (top) to the East Mediterranean (bottom). Note the different geographical distribution of cadmium, with a hot spot around the La Gironde estuary due to upstream mining activities, compared to the DDT, with maxima in the West Mediterranean, a catchment area for intensive agriculture, and a second lower peak at the Seine river mouth. *From RNO. Surveillance du Milieu Marin. Travaux du RNO. Edition 2006. Ifremer et Ministère de l'Ecologie et du Développement Durable; 2006. ISSN 1620—1124.*

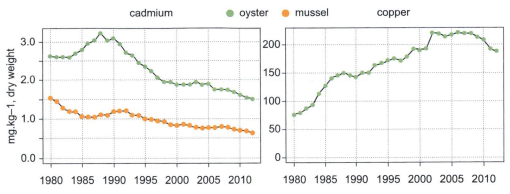

FIGURE 17.6

Five-year median levels of Cd and Cu in native mussels and oysters from the French coasts. The monitoring reveals a decreasing pattern for Cd but an increasing trend for Cu. Note that mussels regulate their Cu levels and they are not used to monitor this metal. *From IFREMER IFREMER. Qualité du milieu marin littoral. Synthèse Nationale de la Surveillance 2012. Nantes: IFREMER; 2013. http://envlit.ifremer.fr/content/download/81717/561807/file/SyntheseNationaleBullSurvED2013.pdf.*

chemical surveillance was downscaled and renamed ROCCH (*Réseau d'Observation de la Contamination Chimique*).

The **Commission for the Protection of the Marine Environment of the North-East Atlantic (OSPAR)** was created in 1992 to prevent pollution originated by dumping from ships, land-based sources, and the offshore industry. The OSPAR Commission later extended the focus to cover also nonpolluting human activities that can adversely affect the sea. The Convention was signed by the governments of Belgium, Denmark, Finland, France, Germany, Iceland, Ireland, Luxembourg, The Netherlands, Norway, Portugal, Spain, Sweden, Switzerland, and United Kingdom.

OSPAR CEMP currently monitors the health status of the Atlantic coast of Europe

OSPAR requires that contracting parties undertake at regular intervals joint assessments to evaluate the quality status of the marine environment, the effectiveness of the measures taken for its protection, and to identify priorities for action. The specific goals are the assessment of the spatial distribution and the changes over time of the concentrations of hazardous substances in the marine environment. The monitoring should identify the sources, levels of discharges, emissions and losses, and to check the effectiveness of measures taken for the reduction of marine pollution within the frame of the current European legislation (WFD 2000/60/EC, MSFD 2008/56/EC and related Directives).

From 2005, OSPAR implemented the **Coordinated Environmental Monitoring Programme (CEMP)**[36] to coordinate the national initiatives and deliver comparable data across the OSPAR maritime area. This monitoring program includes the following components:

- Chemical analyses in biota and sediment of heavy **metals** (Cd, Hg and Pb), **PCBs** (congeners 28, 52, 101, 118, 138, 153, and 180), **PAHs** (anthracene, benzo[*a*]anthrazene, benzo[*g,h,i*]perylene, benzo[*a*]pyrene, chrysene, fluoranthene, indeno[1,2,3-*cd*]pyrene, pyrene, and phenathrene), **brominated flame retardants** (HBCD and BDE congeners 28, 47, 66, 85, 99, 100, 153, 154, 183, and 209[a]), and **TBT** in biota and sediment.

- TBT-specific biological effects (**imposex**).
- Analyses of **nutrients** (ammonia, nitrate, nitrite, phosphate and silicate) in seawater, and direct and indirect eutrophication effects.

An additional set of components is termed pre-CEMP and are to be measured on a voluntary basis, namely PCB congeners 77, 126, and 169, alkylated PAHs, PFOS, dioxins, furans, and biomarkers of biological effects (metallothioneins, ALA-D, oxidative stress, cytochrome p450-1A, DNA adducts, PAH metabolites, liver pathology, vitellogenin induction, intersex, solid- and liquid-phase bioassays, lysosomal stability, neoplasms, fish diseases, and fish reproductive success). Detailed technical guidelines (termed JAMP or CEMP guidelines) are available providing technical guidance for each of the monitoring components.

Not all signatory countries, though, took these endeavors with rigor, and many of these components are monitored for some contracting parties only In fact, monitoring to ensure compliance with the objectives of international treaties remains somewhat controversial, and with only limited exceptions inspection or verification by a third party remains undeveloped in international environmental agreements,[37] including those concerning marine areas. In addition, QA is recognized as an essential component of the monitoring program which is only partially in place, or even absent, for many of the recommended techniques, particularly for the biological effects.[38]

KEY IDEAS

- The integrated approach to environmental monitoring is based on the simultaneous measurement of contaminant concentrations and biological effects to assess whether the pollutant levels measured pose a risk to the organisms.

- Marine pollution monitoring programs have two aims: describe geographical patterns of distribution of specific pollutants that allow assessment of status of each area, and identify temporal trends in the levels of those pollutants. An additional aim of monitoring should be to assess the efficacy of remediation measures.

- Incorporating biological tools to monitoring networks allow detection of nontarget pollutants, serve as screening tool to select sites where further investigation may be applied, and provide ecologically relevant information to assess pollution.

---

[a]This congener in sediment only.

- Passive monitoring relies on autochthonous individuals from wild populations, while active monitoring implies transplantation of standardized individuals to the study sites.

- Mussels are ideal biomonitors because they are abundant and easy to identify and to catch; ubiquitous along broad geographical areas; long-lived; sedentary; large enough for easy handling, dissection, and analysis; amenable for transplantation; tolerant to pollution; and with a moderate-to-null ability to regulate its internal levels of chemicals.

- International conventions with regional scope promote the protection of the marine environment and design monitoring programs undertaken by national agencies from the signatory countries of those conventions.

Endnotes

1. Davies IM, Vethaak AD. Integrated monitoring of chemicals and their effects. ICES Cooperative Research Report No 315, 2012; 277 pp.

2. HELCOM. HELCOM core indicators: Final report of the HELCOM CORESET project. Balt Sea Environ Proc 2013;136.

3. OSPAR Commission 2015. Levels and trends in marine contaminants and their biological effects - CEMP Assessment report 2014.

4. See Endnote. 1.

5. Widdows J, Donkin P, Brinsley MD, et al. Scope for growth and contaminant levels in North Sea mussels *Mytilus edulis*. Mar Ecol Prog Ser 1995;127:131−148.

6. See Endnote. 5.

7. Allaby M. Oxford concise dictionary of ecology; 1994. p. 209.

8. Gray JS. Effects of environmental stress on species rich assemblages. Biol J Linnean Soc 1989; 37:19−32.

9. Wright DA, Welbourn P. Environmental toxicology. Cambridge: Cambridge University Press; 2002.

10. Lam PKS. Use of biomarkers in environmental monitoring. Ocean Coast Manage 2009;52: 348−354.

11. Thain JE, Vethaak AD, Hylland K. Contaminants in marine ecosystems: developing an integrated indicator framework using biological-effect techniques. ICES J Mar Sci 2008;65: 1508−1514.

12. Chapman D, editor. Water quality assessment. A guide to the use of biota, sediments and water in environmental monitoring. London: E&FN Spon; 1996.

13. OSPAR Commission. Levels and trends in marine contaminants and their biological effects - CEMP Assessment report 2014; 2015.

14. Förstner U, Salomons W. Mobilization of metals from sediment. In: Merian E, editor. Metals and their compounds in the environment. Weinheim:VCH; 1991. 379−398. USEPA, United States Environmental Protection Agency. Methods for collection, storage and manipulation of sediments for chemical and toxicological analyses: technical manual. EPA 823-B-01-002. Washington, DC: Office of Water; 2001.

15. Wright DA, Welbourn P. Environmental toxicology. Chapter 3. Routes and kinetics of toxicant uptake. In: Campbell PGC, Harrison RM, De Mora SJ, editors. Cambridge environmental chemistry series; 2002.

16. De Kock WC, Kramer KLM. Active biomonitoring (ABM) by translocation of bivalve mollusks. In: Kramer KLM, editor. Biomonitoring of coastal waters and estuaries. Boca Raton:CRC Press; 1994. p. 51–84.

17. Schøyen M, Allan IJ, Ruus A, et al. Comparison of caged and native blue mussels (*Mytilus edulis* spp.) for environmental monitoring of PAH, PCB and trace metals. Mar Environ Res 2017; 130:221–232.

18. Bryan GW, Langston WJ, Hummerstone LG, et al. A guide to the assessment of heavy-metal contamination in estuaries using biological indicators. MBA UK Occasional Publication No. 4; 1985. 92 pp.

19. Jackim E, Monrison G, Steele R. Effects of environmental factors on radiocadmium uptake by four species of marine bivalves. Mar Biol 1977;40:303–308.

20. Lorenzo JI, Aierbe E, Mubiana VK, et al. Indications of regulation on copper accumulation in the blue mussel *Mytilus edulis*. In: Villalba A, Reguera B, Romalde JL, Beiras R, editors. Molluscan shellfish safety. 2003; pp. 533–544.

21. Reviewed by Bryan GW. 3. Heavy metal contamination in the sea. In: Johnston R. editor. Marine Pollution. London: Academic Press; 1976. pp. 185–302.

22. Amiard JC, Amiard-Triquet C, Berthet B, et al. Comparative study of the patterns of bio-accumulation of essential (Cu, Zn) and non-essential (Cd, Pb) trace metals in various estuarine and coastal organisms. J Exp Mar Biol Ecol 1987; 87(106):73–89.

23. Depledge MH, Rainbow P. Models of regulation and accumulation of trace metals in marine invertebrates. Comp. Biochem. Physiol. 1990;97C (1):1–7.

24. Poulsen E, Riisgard HU, Møhlenberg F. Accumulation of cadmium and bioenergetics in the mussel *Mytilus edulis*. Mar Biol 1982;68:25–29.

25. Wang W-X, Fisher N, Modeling the influence of body size on trace element accumulation in the mussel *Mytilus edulis*. Mar Ecol Prog Ser 1997; 161:103–115.

26. Zaroogian GE. *Crassotrea virginica* as an indicator of cadmium pollution. Mar Biol 1980;58: 275–284.

27. OSPAR Comission. JAMP Guidelines for Monitoring Contaminants in Biota. Review 2012. Available at: https://www.ospar.org/work-areas/cross-cutting-issues/cemp.

28. Vidal-Liñán L, Bellas J. Practical procedures for selected biomarkers in mussels, Mytilus gallo-provincialis — Implications for marine pollution monitoring. Sci Total Environ 2013; 461–462:56–64.

29. Grip K. International marine environmental governance: A review. Ambio 2017;46:413–427.

30. Kimbrough KL, Johnson WE, Lauenstein GG, et al. An assessment of two decades of contaminant monitoring in the nation's coastal zone. SilverSpring, MD. NOAA Technical Memorandum NOS NCCOS 2008;74, 105 pp.

31. RNO. Surveillance du Milieu Marin. Travaux du RNO. Edition 2000. Ifremer et Ministère de l'Aménegement du Territoire et de l'Environnement; 2000.

32. RNO. Surveillance du Milieu Marin. Travaux du RNO. Edition 2006. Ifremer et Ministère de l'Ecologie et du Développement Durable; 2006. ISSN 1620–1124.

33. IFREMER. Qualité du milieu marin littoral. Synthèse Nationale de la Surveillance 2012. Nantes: IFREMER; 2013. http://envlit.ifremer.fr/content/download/81717/561807/file/SyntheseNationaleBullSurvED2013.pdf.

34. Bocquené G, Galgani F, Burgeot T, et al. Acetylcholinesterase levels in marine organisms along French coasts. Mar Pollut Bull 1993;26(2):101–106.

35. Burgeot T, Bocquené T, Pingray G, et al. Monitoring Biological Effects of Contamination in Marine Fish along French Coasts by Measurement of Ethoxyresorufin-0-deethylase Activity. Ecotox Environ Safety 1994;29:131–147.

36. http://www.ospar.org/work-areas/cross-cutting-issues/cemp.

37. Sands P. Principles of international environmental law. 2_{nd} ed. Cambridge. 2003. (p. 848).

38. JAMP Guidelines for General Biological Effects Monitoring (OSPAR Agreement 1997-7) Technical annexes revised in 2007.

Suggested Further Reading

- Davies IM, Vethaak AD. Integrated monitoring of chemicals and their effects. In: ICES Cooperative Research Report No. 315, 2012; 277 pp.

- IFREMER chemical monitoring results: http://envlit.ifremer.fr/var/envlit/storage/documents/parammaps/contaminants-chimiques/.

- Kimbrough KL, Johnson WE, Lauenstein GG, et al. An assessment of two decades of contaminant monitoring in the nation's coastal zone. SilverSpring, MD. NOAA Technical Memorandum NOS NCCOS 2008;74. 105 pp.

- O'Connor TP, Paul JF. Misfit between sediment toxicity and chemistry. Mar Pollut Bull 2000; 40(1):59—64.

- OSPAR monitoring results: https://odims.ospar.org/.

- Schmitt CJ, Dethloff GM. Biomonitoring of Environmental Status and Trends (BEST) Program: selected methods for monitoring chemical contaminants and their effects in aquatic ecosystems. U.S. Geological Survey, Biological Resources Division, Columbia, MO: Information and Technology Report USGS/BRD-2000—0005, 2000; 81 pp.

CHAPTER 18

Pollution Control: Focus on Emissions

18.1 IDENTIFICATION OF PRIORITY POLLUTANTS

Environmental law addresses the issue of environmental protection with different tools (registration of chemicals, limits of emission, discharge permits, fines, environmental standards) that target both anthropogenic inputs and receiving environmental compartments. Both sets of tools are complementary. **Control of emissions** allows identification of sources and makes easier the implementation of effective remediation measures such as improved depuration, economic incentives to reduce input of pollutants, or direct identification and prosecution of offenders. On the other hand, **control of receiving waters** allows an ecologically relevant assessment of the pollutant effects and protects human health by preventing certain uses of polluted waters (fishing, swimming, aquaculture, etc.). Standards for drinking waters and foodstuffs may be also considered among the latter.

The legislation instruments mentioned in this section are derived for those substances most likely to cause environmental problems. These are termed **priority substances** or **priority pollutants**, and they are identified in lists issued by the various national and international entities concerned with environmental protection. One of the earliest attempts to protect the sea against pollution was the production of two lists of substances so hazardous that they should not be discharged to the environment. In 1972, the International Maritime Organization (IMO) hosted in London the "Convention on the prevention of marine pollution by dumping of wastes and other matter" that issued a **"black list"** of substances whose dumping was prohibited, and a **"gray list"** of substances whose dumping was restricted and required a special permit (Table 18.1).

> Lists of priority pollutants identify chemicals of environmental concern whose release should be prohibited or restricted.

In 1996, the new London Protocol reversed the procedure, i.e., prohibiting all dumping except for a list of substances that include dredged material, sewage sludge, fish wastes, vessels and platforms, mining wastes and other inert materials, organic materials of natural origin, bulky items made of carbon, steel, or concrete, and CO_2 streams. However, the idea of listing priority pollutants has

Marine Pollution. https://doi.org/10.1016/B978-0-12-813736-9.00018-0

Table 18.1 Black and Gray Lists From the 1972 IMO London Convention, Prohibiting and Restricting Respectively Dumping of Harmful Substances Into the Sea

Black List	Gray List
Organohalogens Hg, Cd, and their compounds Plastics Crude and refined oil Radioactive wastes Biological and chemical weapons	As, Be, Cr, Cu, Pb, Ni, Va, Zn, and their compounds Organosilicon compounds Cyanides and fluorides Pesticides not included in the black list and their byproducts Obstacles to fishing or navigation Materials that due to the quantities dumped may become hazardous or reduce amenities

continued, and thousands of chemicals have been identified as posing environmental risks.[1]

Priority pollutants are selected according to their persistence, bioaccumulation, and toxicity.

Persistence, bioaccumulation, toxicity (including carcinogenicity, mutagenicity, teratogenicity, and reproductive toxicity), and amount released are the main properties considered for a chemical to be included in these lists. Prioritization methods may assign a score to each individual property and combine them in an overall score, or assess the importance of each property in a pass or fail decision scheme.[2] Thresholds used to classify a substance as toxic at short (acute toxicity) and long (chronic toxicity) term, bioaccumulative, and persistent are summarized in Table 18.2. The outcome of the assessment typically classifies a chemical into one of the following groups: "of concern," "not of concern," or "uncertain," requiring further information. However, the final composition of priority lists is not defined purely by science; they are also subject to the aims of policy.

Persistence and bioaccumulation are the basis of the priority pollutants targeted by the Stockholm Convention (see Box 8.1). The US Environmental Protection Agency (US-EPA) has identified a list of 129 priority chemical compounds regulated in Clean Water Act programs.[3] The Canadian Environmental Protection Act (1989) originally identified 44 priority substances, plus 26 substances added in 1999, whereas the EU has issued a list of priority substances or groups of substances in the field of water policy, currently 45,[4] reexamined periodically.

The main criticism to the priority lists is that they are made on the basis of the potential **hazard** of the substances per se, rather than on the actual risk posed by them under specific environmental scenarios. This may be related to the **precautionary principle**, supporting that the release of a chemical into the environment must be avoided when there is reason to assume that harmful effects on organisms are likely to be caused, even when there is no scientific evidence to prove a causal link between emission and effects. The precautionary

Table 18.2 Threshold Values Used in the Assessment of Properties of Substances Considered in Priority Lists From Europe and North America and Proposed Common Criteria.

Property	Parameter	Unit	Categories	Proposed Criteria
Persistence in water	Half-life ($T_{1/2}$)	days	L/M/H: <10 / 10–100 / >100[a] P: ≥50[d] M/H: >2 months / >6 months[b] P and vP: >60 (marine water)[e]	$T_{1/2}$>60 d
Persistence in sediment			P and vP: >180 (marine sediments)[e]	$T_{1/2}$>180 d
Bioaccumulation	BCF	L/Kg ww	L/M/H: <100 / 100–1000 / >1,000[a] B: ≥500[d] M/H: ≥1000 / ≥5000[b] B/vB: >2000 / >5,000[e]	BCF > 1000 L/Kg ww
Toxicity (acute aquatic)	LC_{50}/EC_{50} (48h invert, 96h fish)	mg/L	L/M/H: >100 / >1–100 / ≤1[a] T: ≤1[c,d]	EC_{50} < 1 mg/L
Toxicity (chronic aquatic)	NOEC/EC_{10}	mg/L	L/M/H: >10 / >0.1–10 / ≤0.1[a] T: ≤0.1[d] T: <0.01[e]	EC_{10} < 0.1 mg/L
Mammalian toxicity			T: carcinogenic (category 1A or 1B), germ cell mutagenic (category 1A or 1B), or toxic for reproduction (category 1A, 1B or 2)[e]	—

B, bioaccumulative; L/M/H, low/medium/high; P, persistent; T, toxic (own elaboration); vB, very bioaccumulative; vP, very persistent.
[a]Smrchek and Zeeman (1998) op. cit.
[b]op. cit.
[c]op. cit.
[d]op. cit.
[e]op. cit.

principle allowed reversing the burden of the proof, since previous traditional approaches restricted action to situations where there was scientific evidence that significant environmental damage was already occurring.[5] Although this principle is central in environmental protection its implementation poses a number of technical and legal problems. Harmful effects must be technically defined and numerical criteria to consider a substance as persistent, bioaccumulative, and toxic must be established.

18.2   THE RATIONALE OF ECOLOGICAL RISK ASSESSMENT

Ecological risk assessment (ERA) is a tool to make decisions on use or release of new chemicals or planned activities that may threaten the ecosystems

Once the priority pollutants have been identified, effective prevention of environmental pollution implies a good knowledge of the maximum tolerable concentrations of those pollutants in the environment to keep them below the threshold of deleterious effects on the organisms. Health risk assessment (HRA) focuses on prevention of threats to human health normally caused by occupational exposure or dietary intake of chemicals. Ecological risk assessment (ERA) is concerned with the prevention of significant alterations of the community structure, conservation of biodiversity, and ecosystem services.

Traditionally, ERA was divided in a priori or prospective and a posteriori or retrospective. The use of the expression retrospective risk assessment has been criticized, since by definition risk refers to the possibility of something to happen in the future. While the rationale under ERA is to evaluate alternatives and make decisions before effects take place, ERA methods are also applicable to scenarios where a substance is already present in the environment, although in this case "risk" might not be the more correct term and "impact" should rather be used. A priori ERA is useful to test a request for introducing a new chemical in the environment, to choose between management options (e.g., alternative cleanup methods, dredging), or alternative uses of an aquatic ecosystem. A posteriori ERA is useful to identify ecosystems at risk after waste disposal or accidental spillages, rank sites according to the priority for remediation actions, rank chemicals according to current threat posed by them to the environment, or test the efficacy of already implemented restoration measures.

ERA quantifies the risk by analyzing how close to toxicity thresholds predicted or actual environmental concentrations are

Fig. 18.1 shows the main steps in an ERA.[6] This relatively simple scheme derives from HRA, where the likelihood of a deleterious effect to human health is estimated by comparing available acute and chronic mammalian toxicity data (effects assessment) with levels of exposure of humans to a given substance (exposure assessment). Since both toxicity and exposure are estimates, risk is expressed in terms of probabilities. That simple scheme is considerably complicated when applied to environmental studies because of the many factors that determine the environmental fate of a substance in the different environmental compartments, and the multiple organisms with complex interactions, very

ECOLOGICAL RISK ASSESSMENT (ERA)

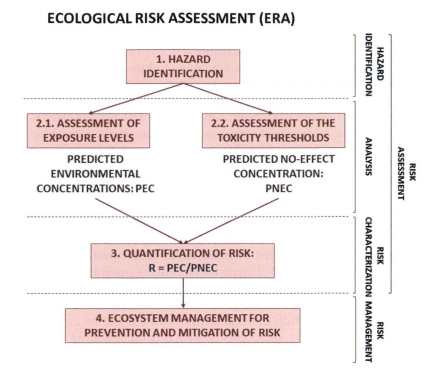

FIGURE 18.1

Conceptual framework for an a priori Ecological Risk Assessment (ERA) study. First step includes identification of hazardous chemicals on the basis of properties (persistence, bioaccumulation, toxicity), amount released, and exposure pathways. Second step includes the assessment of both environmental concentration (PEC) and thresholds for toxic effects (PNEC). Third step quantifies risk, frequently by means of the Risk Quotient (R = PEC/PNEC). After this assessment environmental managers should make decisions to prevent or mitigate the risk. The process may be repeated in an iterative way. *Modified from NRC (1983) op. cit.*

different life histories, sensitivities, and exposure pathways that comprise the biological communities. Thus the hazard identification in an environmental assessment is sometimes referred to as **problem formulation**. The problem formulation is the most critical step in ERA since it involves not only the choice of substance, but also the environmental compartments, conceptual model, test species, and biological endpoints to be used in the study. To illustrate this complexity with just an example, we can be interested in assessing the risk posed by a chemical to a hypothetical commercial fish species. The problem may seem simpler than a conventional ERA since we target a single organism. However, let us assume the fish eggs inhabit the sea surface microlayer, the juveniles swim in the water column, and the adults are top predators living in benthic habitats. The problem formulation will have to take into account that these three life stages show completely different properties in terms of both exposure pathways and sensitive endpoints to the substance. Spatial considerations are rarely introduced in aquatic ERA, but they may contribute to solve these kinds of problems.[7]

18.2.1   Hazard Identification

This is the first step that includes the identification of a given substance potentially hazardous to representative species of the ecosystem according to the available information on the amounts produced, the release into a given type of ecosystem, and its inherent properties, which are mainly environmental persistence, bioaccumulation potential, and toxicity. Typical examples are the commercialization of a new chemical that will be produced at a given scale and applied in a given manner to the environment, or an activity planned in the environment that may imply habitat alteration or destruction. Occasionally we may be interested in particular species for economic or ecological conservation reasons.

18.2.2   Analysis

Second in the ERA framework is the step of **analysis**. This includes the assessment of both exposure and toxicity, intended to provide two quantitative estimates, the **Predicted Environmental Concentration (PEC)**, and the toxicity threshold, here called **Predicted No-Effect Concentration (PNEC)**, respectively.[8]

18.2.2.1   Exposure Assessment

The PEC at which the organisms of concern will be exposed is obtained either from actual (reported environmental concentrations) or estimated maximum environmental concentrations in the compartment of interest, e.g., the water column. In retrospective ERA, actual concentrations may be obtained from a literature survey or sampling campaigns. In a priori ERA, exposure concentrations are estimated by models of environmental distribution, such as the fugacity models described in Section 10.2, fed with data on amounts released, pathways, and properties (solubility, volatility, partition coefficients, persistence, bioaccumulation, etc.) determining the environmental fate of the substance.

18.2.2.2   Toxicity Assessment

In this step, laboratory toxicity tests are conducted with test species representative of the main taxa present in the ecosystem to estimate the acute and chronic toxicity thresholds of the substance (NOEC/LOEC, EC_{10}; see Section 13.2), and the lowest value is used as PNEC in the subsequent risk characterization step. For environmental protection value, endpoints selected and exposure conditions should maximize sensitivity. Choice of early life stages, sublethal endpoints, and chronic or subchronic exposures allow achieving this sensitivity in a more objective and environmentally relevant manner than the application of high and arbitrary assessment factors.

18.2.3   Risk Characterization

Once PEC and PNEC have been obtained, a quantitative estimation of risk can be obtained by comparing both estimates. This is frequently done by using the **risk quotient** (R), where

R = PEC/PNEC   (Eq. 18.1)

Notice that an **assessment factor** (**AF**), also called uncertainty factor, intended to make conservative allowances for the various sources of uncertainty[a] associated to both PEC and PNEC, is frequently used to calculate this quotient. Typical sources of uncertainty are extrapolation of chronic effects from acute toxicity data, laboratory to field extrapolations, or allowances for the representativeness of the chosen test species. Uncertainty concerning PNEC can be reduced through additional toxicity testing resorting to experimental designs beyond standardized methods, which take into account relevant abiotic and biotic sources of variability in the field. Whereas standard methods are essential to allow comparisons of toxicity among chemicals, additional testing in more environmentally relevant conditions is essential to reduce uncertainty in the laboratory to field extrapolation inherent to ERA.

The AF values generally range from 1000 to 10, decreasing as the amount and quality of ecotoxicological information increases. Obviously, high AF values increase the risk of too conservative—and thus in practice hardly applicable—estimates of risk. This AF may be already included in the calculation of PNEC estimates,[9] or be explicitly indicated in the value of the quotient that then takes the form:

$$R = PEC \cdot AF / PNEC \qquad \text{(Eq. 18.2)}$$

Thus formulated, PEC and PNEC are fixed values, and risk is characterized in a deterministic way. Alternatively, more complex probabilistic approaches can produce distributions of PEC and PNEC values, and quantify the likelihood of adverse effects in terms of the probability of PEC > PNEC.[10] Probabilities may be introduced into the step of analysis at several stages. Probability of exposure may be distributed with respect to the space occupied by a species, the time the species is exposed, the food preferences, etc. Probabilities can also be introduced in the effects assessment, first, by taking into account the confidence limits of the estimates of PNEC obtained from concentration: response relationships (e.g., EC_{10}) on a given species. Second, sensitivity of the different species in the ecosystem varies, and the species sensitivity distribution approach allows the choice of PNEC values protective for a certain percentage of the species at a given confidence level.

Probabilistic risk assessment works with distributions of PEC and PNEC values while deterministic approach uses the most conservative values and an AF

Risk may thus be expressed in terms of probabilities when PEC, PNEC, or both are distributed values, rather than the fixed values reflected in Eqs. (18.1) and (18.2). If only either PEC or PNEC is a distributed value, then expressing risk in terms of probabilities is straightforward. If both PEC and PNEC are distributed

---

[a] Uncertainty must not be confused with variability. The first is due to lack of knowledge and can be reduced by obtaining additional information. The latter is an inherent property of any event; it can be estimated but not reduced.

values then risk is expressed as joint probability, but concordance of both distributions must be previously ensured.

If PEC and PNEC are distributed values with a known variance it can be assumed that R follows a log-normal distribution. Therefore, taking logarithms in Eq. (18.1),

$$\ln R = \ln PEC - \ln PNEC \qquad (Eq.\ 18.3)$$

and thus the variance of R can be obtained from the antilogarithm of the sum of the variances of the logarithms of PEC and PNEC.

Use of AF is viewed by some authors as too simple, with little or no theoretical foundation, whereas for others remains a useful and effective tool as compared with alternative probabilistic approaches that have not proved yet more accuracy or more conservative value.[11] Whatever the approach used, probabilistic or deterministic, it must be considered that ERA is a **theoretical exercise** whose predictive value depends on the amount and quality of data available and models chosen. It is not feasible to produce accurate estimates of risk from incomplete pieces of information or by using theoretical models that estimate rather than measure toxic effects.

The task of scientists in an ERA study ends up here, but the process should be continued by the environmental managers. The following section discusses the next step.

18.2.4 Risk Management

At this step the competent authorities must classify the risk on the basis of the assessment made as acceptable, if R << 1 (or probability of PEC > PNEC is very low), or unacceptable if it approaches 1 (or probability of PEC > PNEC is not negligible). In the latter case take the preventive or corrective measures needed to reduce the risk, such as restrictions to amounts produced or application scenarios to achieve acceptable R values.

The scheme may work as an **iterative process** in cycles where the feedback provided by an initial assessment refine the problem formulation and help to design a new analysis to produce more accurate characterization of risk, or test the efficacy of implemented remediation measures until the goal of prevention/remediation of deleterious effects on the environment was achieved.

Finally, we should pay attention to the practical differences between hazard identification and risk assessment as tools for the management of chemicals. A chemical may pose potential threat to the environment because of its inherent properties, such as persistence or toxicity, which may well support the classification of this chemical as a priority pollutant according to hazard identification studies, and thus the precautionary principle will prescribe prohibition of its use. However, if we wish to assess the likelihood of deleterious effects of this chemical when used under well-defined conditions in real world scenarios then we must conduct a risk assessment study, compare estimated environmental concentrations with toxicity thresholds, and characterize the risk. Frequently this produces more realistic, scientifically sound, and valuable information than the mere application of the precautionary principle.[12]

18.3 REGULATIONS FOR SINGLE CHEMICALS: THE REACH PARADIGM

In the United States, the main regulations for new commercial chemicals derive from the **Toxic Substances Control Act (TSCA)**, originally enacted in 1976 and last amended in 2016. TSCA created an inventory of existing marketable chemicals, updated by the US Environmental Protection Agency (US-EPA) to the current figure of 85,000. Similar inventories of marketable chemicals are enforced by analog pieces of legislation in Canada,[13] Korea,[14] Japan,[15] and Australia,[16] among other countries. If a chemical is not in the inventory it is considered as "new" and a notification to EPA must be submitted prior to manufacturing or importing for commerce. Exemptions include drugs, cosmetics, and pesticides, addressed by other pieces of legislation.[17] The information to be submitted includes chemical identity, production volume, byproducts, use, environmental release, disposal practices, human exposure, and existing available test data. Risk assessment under TSCA consists of the integration of the hazard assessment for a chemical with the chemical's exposure assessment.[18] However, toxicity testing is not compulsory, and frequently replaced by predictions made from QSAR models, submitted to broad margins of error (See Section 13.2). Even when actual experimental data are provided, in almost all cases hazard profiles for new chemicals are incomplete. If the submitted information is considered sufficient, EPA conducts a human health and environmental risk assessment and classifies the substance as "not likely to present unreasonable risk" or "likely to present unreasonable risk." In the latter case, EPA may, (1) limit the amount distributed in commerce or impose other restrictions on the substance, or (2) issue an order to prohibit or limit the manufacture, processing, or distribution.

Chemicals must be submitted to official inventories for commercialization

In Europe, the commercialization of potentially hazardous chemical substances is addressed by the comprehensive Regulation (EC) No 1907/2006 (Registration, Evaluation, Authorization, and Restriction of Chemicals, REACH), considered as one of the most complex pieces of legislation in the EU history. However, similarly to TSCA, pharmaceuticals and cosmetics are partially exempted since they are covered by specific pieces of legislation.[19] According to REACH, substances manufactured or imported into the EU in quantities above 1 ton per year need to be registered with the European Chemical Agency (ECHA). With that aim, companies must submit a registration dossier to ECHA, which may impose restrictions on certain uses or even a ban, based on the available information. Commercialization in the EU of substances not following these procedures is illegal, a philosophy known as "no data, no market." This paradigm has been successful and internationally spread. Regulations with similar philosophy to REACH have been enforced in China since 2010.

The EU Regulation Registration, Evaluation, Authorization, and Restriction of Chemicals (REACH) imposes an a priori ERA before a chemical is placed in the market

Information required for registration includes identity of the substance, physicochemical properties, environmental fate, and toxicological characterization, so that a complete a priori **ERA** can be conducted. The amount of information required depends on the amount of substance to be placed in the market. Above 10 tons per year human health and environmental hazard, persistence, bioaccumulation, and toxicity must be reported, and if the substance is considered as PBT (see Table 18.1) an exposure assessment and risk characterization must be conducted (see Section 18.2).

Regarding the toxicological information, it is mainly based on in vivo rodent tests and in vitro mammal cell lines tests, including lethal and sublethal (carcinogenesis, mutagenesis, and toxicity on reproduction) endpoints. Chemicals are considered toxic for reproduction whether they cause abnormal embryo development (teratogenicity), interference with normal functioning of hormones (endocrine disruption), or reduction of fertility or offspring performance by any other mechanism. Aquatic toxicity is also represented, but assessed with **freshwater species only**; including microalgae, *Daphnia*, and fish adult, juveniles, and early life stages tests. The LC_{50} or EC_{50} is used as endpoint in acute (48 h for invertebrates, 96 h for fish) tests whereas the LOEC/NOEC or EC_{10} is used in chronic or subchronic tests.[20]

Again, the information on the different kinds of toxicity required depends on the amount of chemical to be produced. For example, at the lowest tonnage level (1–10 T/y), only acute toxicity to *Daphnia*, algal growth and mutagenic tests with bacteria are required. For the next tonnage band (10–100 T/y) additional information required includes acute lethal toxicity to fish and in vitro tests for mutagenicity and chromosomal aberrations, and in case a mutagenic effect is seen in the in vitro studies, information from an appropriate in vivo somatic cell genotoxicity study is required. Between 100 and 1000 T/y, aquatic toxicity requirements include long-term invertebrate toxicity and fish early life stage development or juvenile growth tests. Finally, standard information requirements on carcinogenicity are set only at the highest tonnage level (above 1000 T/y), and whether the substance is classified as mutagen category 3 or is able to induce hyperplasia and/or preneoplastic lesions in repeated dose studies.

18.4   REGULATIONS FOR COMPLEX EFFLUENTS AND DEPURATION OF WASTE WATERS

Complex effluents are characterized on the basis of general parameters (Biological Oxygen Demand [BOD], suspended solids [SS], nutrients) and presence of priority substances

Point-source control of pollution is based on **emission standards** the discharges must meet to be authorized. These standards address general water quality parameters such as Biological Oxygen Demand (BOD), suspended solids (SS), or nutrients. In addition, specific legislation may target priority substances, such as trace metals or persistent organics (POPs), whose presence in the discharges may be restricted to maximum levels or prohibited. In Canada, current regulations include testing acute lethality of the effluent to rainbow trout.[21]

The US **Clean Water Act**, derived from the 1948 Water Pollution Control Act, was designed to control the point-source discharge of effluents into surface waters. Effluent standards are set for categories of existing sources, including WWTP, and **permits** of discharge are issued for emissions provided the discharges meet those performance standards. The standards are issued by EPA on the basis, not of ecotoxicological impact or risk assessment, but on the application of the best available technology each type of source can achieve.

In Europe, the protection of the environment from the adverse effects of urban waste water discharges is addressed in Directive 91/1971, (later amended by Directive 98/15/EC) which enforces appropriate collection, secondary treatment for all emissions above 2000 inhabitant equivalents, and additional treatment for nutrient removal for emissions to areas sensitive to hypereutrophication. Also, compliance with general emission standards for BOD (See Table 2.2), Chemical Oxygen Demand (COD), SS, and for sensitive areas, nutrients (see Table 3.1) is required. In addition, Directive 2006/11 issues two lists of priority substances (see Section 18.1) termed dangerous substances, whose emission must be subjected to a previous permit and according to substance-specific emission standards. List I, substances whose pollution in the water should be eliminated, include organohalogens and organophosphates, organotin compounds, Hg, Cd, mineral oil, and petroleum hydrocarbons, carcinogenic and persistent floating substances interfering with navigation, leisure, or other uses of water. List II, whose pollution in the water should be reduced, includes a wide array of trace elements and organic chemicals. The lists are similar to the black and gray lists from the 1972 London Convention (see Table 18.1).

These direct regulations based on enforced standards of compulsory compliance are combined in some national legislations like in Spain[22] with economic **incentives for depuration** or emission to less sensitive waters, based on modulating the amount of the fee paid by emitters as a function of the degree of depuration, or use of receiving waters (Table 18.3). The weakness of this system is twofold. First, it follows the "polluter pays" principle which can be interpreted as polluting is allowed if polluter can afford it. On the other hand, the amount of the economic incentive is quite moderate. For example, for a class 2 (medium impact) industrial effluent, elimination of dangerous substances following Directive 2006/11/EC from the wastes represents saving only around 15 percent of the fee value. Moreover, except in the unlikely event of an inspection posterior to the concession of the permit establishing otherwise, the C_2 value used for all legal discharges is always the corresponding to adequate depuration. Once again, the implementation phase of an environmental law prevents the full benefits that inspired the creation of that law.

Depuration may be encouraged in regulations by modulating the emission fees as a function of the contaminant loads in the effluent

Table 18.3 Basic Prices and Coefficients (C_1 to C_3) Used for Calculating the Fee Applied to Wastewater Effluents According to the Spanish Legislation. The Fee Amount Is Calculated by Multiplying the Volume of the Effluent (V, m^3) Times the Final Price Resulting From the Characteristics of the Effluent and Receiving Waters: $V \times P \times C_1 \times C_2 \times C_3$. Notice that the Final Price Coefficient ($C_1 \times C_2 \times C_3$) Ranges from 0.5 to 4. Special Cases Are Fish Farming Effluents and Cooling Waters for Thermic Plants, which Receive Special Discounts

Basic Price (P)	Coefficient C_1: Nature of Effluent	Coefficient C_2: Degree of Depuration	Coefficient C_3: Quality of Receiving Waters
Urban: **0.012 €/m³**	≥10000 i.e.,: **1.28** 2000–9999 i.e.,: **1.14** <2000 i.e.,: **1.00**	Not adequate: **2.5** Adequate: **0.5**	Drinking, bathing, salmonids, underground, sensitive, or special: **1.25**
Industrial: **0.030 €/m³**	Presence of dangerous substances: **1.28** Class 3 (leathering, surface treatment, farms): **1.18** Class 2 (mining, chemical, building, drinks, tobacco, vegetal oil, meat, dairy, textile, paper): **1.09** Class 1 (others): **1.00**	Not adequate: **2.5** Adequate: **0.5**	Shellfish, cyprinids, leisure: **1.12** Others: **1.00**

18.5   REGULATIONS FOR THE PREVENTION OF OIL SPILLS

Accidental oil spills triggered legislative changes for improved tanker navigation safety

Concerning marine pollution, accidental oil spills are by far the most high-profile media events, and not surprisingly some of them gave origin to, on the one hand, thorough scientific studies on the ecological impact, and, on the other hand, triggered legislative changes intended to improve prevention of their occurrence and remediation of their effects.

As mentioned in Section 7.5, the case of the *Exxon Valdez* in Mar 1989, although minor in terms of amount spilled (38,000 T of crude oil), raised unprecedented public awareness due to the ecological value of the affected area in Alaska, and prompted legislative initiatives intended to enhance maritime safety, prevent tanker accidents, and improve liability and economic compensations to affected stakeholders. In 1990, the United States passed the **Oil Pollution Act (OPA)** meant to be the primary federal legislation addressing oil spills, with the purpose to set practical tools for the prevention of spills and abatement of their consequences.[23] OPA provided a rapid legislative tool to ensure effective cleanup activities, required all new built tankers to have a double hull, and increased the liability in case of pollution to unlimited damages for the cases of negligence, federal law infringement, or failure to cooperate with the removal activities. OPA also created a fund, the Oil Spill Liability Trust Fund, covered by taxes on both domestic and imported oil, to address recovery activities in case compensation is not available from the responsible party within 90 days.

At international level, the 1992 **Civil Liability Convention (CLC)** updated the original 1969 CLC adopted following the *Torrey Canyon* accident in the south-west coast of Britain. It establishes the liability of the owner of a ship causing pollution due to incidental oil spilling that damages the territorial waters of a party, including the costs of preventive measures, and sets an economic limit to that liability of up to 59.7 million SDR, an international asset created by the IMF with a current equivalence of around 1.5 US dollars. In addition, the 1992 Fund Convention established an International Oil Pollution Compensation (IOPC) fund intended to provide economic compensation for pollution dam-age inadequately covered by the ship owner, either because no liability is estab-lished by CLC or the owner cannot meet the obligations of the CLC. The IOPC fund in each party receives annual contributions from entities receiving oil car-ried by sea.[24]

As a result of the measures taken worldwide the amount of oil spillages due to tanker accidents showed a significantly decreasing trend, particularly remark-able from mid 1990s on (Fig. 18.2).

In Europe, the *Erika* accident (December 1999) prompted some initiatives such as the adoption of two legislative packages (*Erika I* and *II*) that included the cre-ation of the European Maritime Safety Agency (EMSA) in 2002 with the aim,

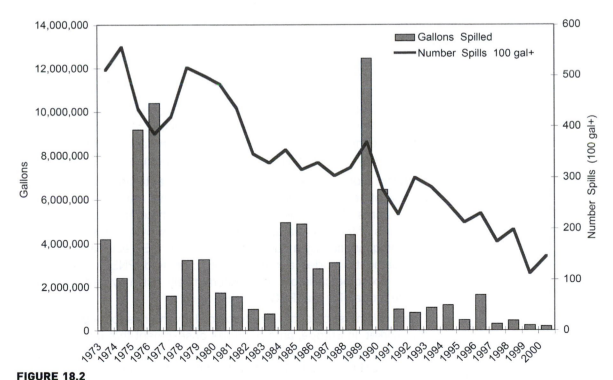

FIGURE 18.2

Historical trend (1973–1999) in oil spills from vessels into U.S. marine waters. Notice the substantial reduction from 1991 on. *From NRC (2002) op. cit.*

among other purposes, to ensure prevention of pollution caused by ships, and response to marine pollution caused by ships and oil and gas installations. The *Erika* packages also dealt with improved controls to ensure that vessels meet safety standards, rules for monitoring traffic in EU waters, identifying ports of refuge, as well as compensation of payment to victims of oil spills.

The subsequent *Prestige* accident (November 2002) evidenced that there were not enough oil combating ships in Europe, and somewhat accelerated the *Erika* legislative packages, including publication of a black list of banned ships identified through a more strict system of inspections, accelerated the phasing-out of single-hull tankers, and set the basis for a better identification and monitoring of tankers.[26] With that aim, automatic identification systems for vessels coming into EU waters, which can provide information about ships to coastal authorities, and voyage data recorders, which help investigations after incidents, have been introduced. However, no analog process to the enforcement of OPA in the United States took place in Europe. The *Prestige* accident sadly illustrates that adopting environmental protection measures is not enough; they have to be followed. The accident could have been prevented if the *Erika* legislation had been in force at the time, since the vessel would have been taken out of service two months prior to the incident.[27] No major tanker accidents have occurred in European waters after the *Prestige*, and no new pieces of legislation for their prevention were produced.

KEY IDEAS

- Lists of **priority pollutants** identify hazardous chemicals of environmental concern on the basis of anthropogenic enrichment, environmental persistence, bioaccumulation potential, and toxicity. Identification of hazardous chemicals is the first step in many legislative instruments intended to prevent and abate pollution, such as environmental standards or limits of emission.

- **ERA** is a tool to make decisions on planned activities such as release of new chemicals or developments that may threat ecosystems. The decision is made by environmental managers on the basis of the risk characterization provided by scientists. **Risk** may be quantified by a quotient between the PEC estimated by environmental models and the toxicity threshold (PNEC) estimated by the toxicological information. Alternatively, risk may be assessed as the probability that PEC > PNEC.

- Chemicals must be submitted to official inventories to be placed in the market. In Europe, an a priori risk assessment that takes into account effects on human health and aquatic toxicity is imposed by the **REACH** regulation. This paradigm has been adopted by China.

- Complex effluents are characterized on the basis of general parameters (BOD, SS, nutrients) and loads of priority substances. These parameters must meet **emission standards** to obtain a **permit** of discharge into natural waters.

- Depuration may be promoted through legislation by implementing **economic incentives** based on the modulation of the discharge fee on the basis of contaminant loads in the effluent.

- Accidental oil spills triggered legislative changes for improved tanker navigation safety. A case of effective enforcement of improved regulations is the **OPA**, which passed in the United States in the aftermath of the *Exxon Valdez* oil spill and was instrumental in reducing the number of tanker accidents.

Endnotes

1. Gray JS. Risk assessment and management in the exploitation of the seas. In: Calow P, editor. Handbook of environmental risk assessment and management. Oxford: Blackwell, 1998; pp. 453−474.

2. Hedgecott S. 16: Priorization and standards for hazardous chemicals. In: Calow P, editors. Handbook of ecotoxicology. vol. 2. Oxford: Blackwel, 1993; pp. 368−393.

3. https://www.epa.gov/sites/production/files/2015-09/documents/priority-pollutant-list-epa.pdf.

4. Directive 2013/39/EU of the European Parliament and of the Council of 12 August 2013 amending Directives 2000/60/EC and 2008/105/EC as regards priority substances in the field of water policy.

5. P. 268 in: Sands P. Principles of international environmental law. 2nd ed. Cambridge; Cambridge University Press; 2003. (p. 268).

6. NRC. Risk assessment in the Federal Government: managing the process. National Research Council. Washington DC: National Academy Press; 1983.

7. See Chapter 11 In: Landis WG, Yu M.-H. Introduction to environmental toxicology. Boca Raton: Lewis Publishers; 1999.

8. Ahlers J, Diderich R. Legislative perspective in ecological risk assessment. Pp. 841−868. In: G. Schüürmann, B. Markert, editors Ecotoxicology. New York:Wiley; 1998.

9. EC. Technical guidance document for risk assessment. Part 2. European Commission Joint Research Center. EUR 20418 EN/2. 2003. Nabholz JV, Environmental hazard and risk assessment under the United States Toxic Substances Control Act. Sci Total Environ 1991;109/110: 649−665.

10. Suter II, GW, Ecological risk assessment. 2nd ed. Chapter 30. Boca Raton:CRC Press; 2006.

11. Smrchek JC, Zeeman MG. Chapter 3. Assessing risks to ecological systems from chemicals. In: Calow P, editor. Handbook of ecological risk assessment and management. Oxford: Blackwel; 1998. pp. 24−90.

12. See GW Sutter II, Vignette 13.1, pp. 399−401. In: Newman MC, editor. Fundamentals of ecotoxicology. The science of pollution. Boca Raton: CRC Press; 2015.

13. https://www.canada.ca/en/environment-climate-change/services/canadian-environmental-protection-act-registry/substances-list.html.

14. http://ncis.nier.go.kr/en/main.do.

15. http://www.nite.go.jp/en/chem/chrip/chrip_search/systemTop.

16. https://www.nicnas.gov.au/chemical-inventory-AICS.

17. For pesticides see: Federal Insecticide, Fungicide, and Rodenticide Act (FIFRA) and subsequent amendments. For drugs see: Federal Food, Drug, and Cosmetic Act (FD&C Act) and subsequent amendments.

18. Nabholz JV, Miller P, Zeeman M. Environmental risk assessment of new chemicals under the toxic substances control act TSCA section five. In: Landis WG, et al., editor. Environmental toxicology and risk assessment. ASTM special technical publication 1179; 1993 Nabholz (1991) op. cit.

19. For pharmaceuticals see: Regulation (EC) No 726/2004 and Directive 2001/83/EC. For cosmetics see: Regulation (EC) No 1223/2009.

20. ECHA. Guidance on information requirements and chemical safety assessment. Part B: hazard assessment. Helsinki: European Chemicals Agency; 2011. 59 pp.

21. http://laws-lois.justice.gc.ca/eng/regulations/SOR-2012-139/FullText.html.

22. Ministerio de Medio Ambiente. Manual para la gestión de vertidos. Autorización de vertido. Centro de Publicaciones, Secretaria General Técnica. Madrid: Ministerio de Medio Ambiente; 2007. 270 pp.

23. See: NRC. Oil in the sea III. National Research Council. Washington DC: National Academy Press; 2002. pp. 255–257.

24. Sands (2003) op. cit.

25. National Research Council. Oil in the sea III; 2002.

26. Luoma E. Oil spills and safety legislation. Publications from the Centre for Maritime Studies. University of Turku; 2009. http://www.merikotka.fi/safgof/Oil%20spills_luoma_2009.pdf.

27. Wene J. European and international regulatory initiatives due to the Erika and Prestige Incidents. MLAANZ J 2005;19:56–73.

Suggested Further Reading

• Ornitz BE, Champ MA. Oil spills first principles. Prevention and best response. Amsterdam: Elsevier, 2002; 678 pp.

• Petry T, Knowles R, Meades R. An analysis of the proposed REACH regulation. Regul Toxicol Pharmacol 2006;44:24–32.

• Suter II GW, Ecological risk assessment. 2nd ed. Chapter 30. CRC Press, Boca Raton.

• US EPA. Priority pollutants under the Clean Water Act: https://www.epa.gov/eg/toxic-and-priority-pollutants-under-clean-water-act.

CHAPTER 19

Pollution Control: Focus on Receiving Waters

19.1 MANAGEMENT OF RECEIVING WATERS UNDER THE ECOSYSTEM APPROACH

When in 1935 the English botanist Sir Arthur Tansley coined the term "ecosystem" to refer to the assembly of biological and physical elements of a living system, within a similar conceptual framework that would inspire von Bertalanffy's general systems theory, he could hardly imagine that 70 years later his highly theoretical concept would be reflected in important pieces of legislation and enforcing laws. The so-called **ecosystem approach** for environmental management was defined by the UN Convention on Biological Diversity as "a strategy for the integrated management of land, water and living resources [...] based on the application of appropriate scientific methodologies focused on levels of biological organization which encompass the essential processes, functions and interactions among organisms and their environment."[1] In practice, key aspects of the ecosystem approach applied to environmental monitoring networks are the following (Fig. 19.1). (1) Assessment is conducted within an integrative framework that encompasses measurements of **chemical** and **biological** variables, the latter at different levels of organization. The synergism gained by combining biological tools that unveil effects and chemical tools that identify potential causal agents has already been discussed (Section 17.1). (2) Effects of human activities on natural **biogeochemical cycles**, **ecosystem functional traits**, and **human health** are all included in the overall assessment, and remediation actions should be planned bearing in mind the three types of effects. (3) **Conservation and sustainability** issues, related to rates of consumption of natural resources, and monitored through stock assessments, must be included in the overall assessment along with pollution issues. Excessive consumption through water extraction, occupation of coastal habitats for commercial, aquaculture and touristic activities, and overfishing are in fact among the main threats for aquatic ecosystems, and remedial actions cannot ignore these standpoints. Sustainable exploitation of a natural resource means

Ecosystem: From a theoretical ecology concept to the object of regulations

329

Marine Pollution. https://doi.org/10.1016/B978-0-12-813736-9.00019-2

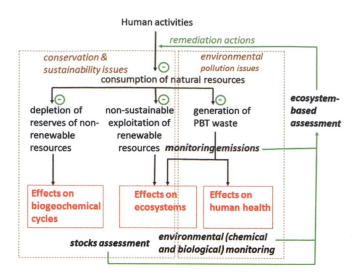

FIGURE 19.1

Scheme of environmental monitoring and management based on the ecosystem approach. The ecosystem-based assessment should provide solid grounds for planning remediation actions to prevent depletion of nonrenewable natural resources, nonsustainable exploitation of renewable natural resources, and pollution. *Modified from Beiras (2016) op. cit.*

that consumption rate does not exceed natural renewal rate, and thus current exploitation does not compromise the availability of the resource for future generations.

The ecosystem approach has been internationally embraced in environmental regulations... The ecosystem approach was embraced as the way forward to tackle different environmental issues, from conservation of the biodiversity[2] to management of the marine resources.[3] Regarding the latter, this approach was invoked by Canada's Oceans Act in 1997, by Australia's Ocean Policy in 1998, and by the two USA Commissions for marine policies (the nongovernmental Pew Oceans Commission and the congressionally mandated US Commission on Ocean Policy), and the resulting Oceans 21 bill in the early 2000s.[4] In Europe, with the turning of the century the Water Framework Directive (WFD) (Directive 2000/60/EC) was adopted, and that legislative initiative triggered a cascade of environmental regulations, including the marine environmental policies (Directive 2008/56/EC) and the environmental quality standards (EQS) for priority pollutants in continental and marine waters (Directive 2008/105/EC and Directive 2013/39/EU). The first purpose stated in the WFD was "to establish a framework which [...] prevents further deterioration and protects and enhances the status of aquatic ecosystems." With that aim, acute (termed maximum admissible concentration) and chronic (termed annual average) standards were adopted as safe levels that must not be exceeded in the short and long term, respectively. Similarly, acute and chronic water quality criteria are provided in North America by US-EPA[5] and the Canadian Government,[6] and in Oceania (chronic only) by ANZECC.[7]

While the ecosystem approach permeated the well-meaning environmental regulations, it must be born in mind that assessment is not an objective per se, but a means to design and implement **remediation actions**—the negative feedback depicted in green in Fig. 19.1, which must be instrumental in reducing those anthropogenic pressures identified in the assessment as most threatening for the preservation of the natural ecosystems and the services they provide to human kind. Throughout the previous chapters we have studied some stories of success, such as the reduction of Pb pollution at global scale derived from the use of unleaded petrol, or the recovery of snail populations affected by imposex after phasing out TBT-based antifouling paints. None of those cases were identified within the framework of a holistic ecosystem-based monitoring scheme. In fact the comprehensive monitoring networks implemented in many countries largely focus on the identification of pressures and impacts at regional scale, but there is limited evidence that the findings of these assessments have ever been used in environmental policy making.[8]

...**but there is limited evidence that the findings of ecosystem assessments have ever been used in environmental policy making**

The most common legal tools to protect receiving waters are the EQS; maximum admissible levels that concentrations measured in receiving waters should not exceed nor tend to approach, in the light of scientific knowledge on deleterious ecological effects. The rationale underneath the implementation of environmental standards was introduced in Section 1.4. Basically, for each chemical pollutant, the safe level for an environmental compartment is extrapolated from laboratory toxicity tests with representative species of the environmental compartment concerned. The most scientifically sound approach for this extrapolation is based on the concept of a distribution of tolerances, frequently termed **Species Sensitivity Distribution (SSD)**, for the different species to the chemical under evaluation. Obviously, not all species in an ecosystem show the same sensitivity to a given pollutant. It has been empirically found that for a given chemical, the distribution of toxicity thresholds—estimated as EC_{10} or NOEC/LOEC—on a set of species fit well to a sigmoid curve, once concentrations are transformed into logarithms.[9] Thus, distribution of tolerances among species seems to follow a pattern analogous to distribution of tolerances among individuals within a population (see Section 13.3). Once the distribution is fitted by nonlinear regression (see Section 19.4), a value can be obtained from the equation of the SSD curve that protects a given percentage of the species (e.g., 95%, i.e., the 5% percentile) with a given certainty (the lower limit of the confidence interval of the percentile).[10] The degree of protection may be selected by the political authorities by choosing a certain percentile (e.g., 90, 95, 99%) and/or confidence level (e.g., 50, 90, 95%), for example on the basis of the ecological or economic value of the ecosystem to be protected (see ANZEC, 2000[11]).

Environmental Quality Standards (EQS), the legal tools to protect receiving waters

19.2   EUROPEAN REGULATIONS FOR THE PROTECTION OF AQUATIC ECOSYSTEMS: THE WATER FRAMEWORK DIRECTIVE AND THE MARINE STRATEGY FRAMEWORK DIRECTIVE

Current European legislation prompts cycles of assessment intended to implement remediation measures that warrant good status of all surface water bodies or marine areas

In Europe, environmental quality assessment in marine waters is addressed by two quite different pieces of legislation, the WFD, which covers all kind of underground and surface waters, and the specific **Marine Strategy Framework Directive (MSFD)**. A comparative overview is reflected in Table 19.1. Common to both Directives is the task to classify aquatic areas into discrete categories of ecological status (in the WFD), termed environmental status in the MSFD. The unit of assessment is called the water body in the WFD and the subdivision of broad marine regions and subregions in the MSFD. The WFD establishes typologies for the surface waters that include transitional (estuaries) and coastal water bodies, and a complex taxonomy of subtypes with subtype-specific reference conditions for each indicator of status.

The classification of status of each water body takes into account an arbitrary set of quality elements (WFD) or descriptors (MSFD), for which quantitative indicators must be developed. In the WFD, the environmental quality elements are related to **biological** (phytoplankton abundance, macroalgae, and benthic invertebrate community composition), **physicochemical** (temperature, oxygen, nutrients, etc.), and **hydromorphological** traits. The MSFD uses 11 descriptors for the same purpose. As reflected in Table 19.1, both sets of indicators have little in common, with a more explicit focus toward preservation of ecosystem services in the case of the MSFD.

WFD classification of status is based on type-specific reference conditions and ratio-to-reference (RTR) values of indicators

The classification of ecological status (sensu WFD) relies on a quite complex framework in continuous technical development since the approval of the Directive.

1. All surface waters are divided into water bodies, each classified into a type (lake, river, transitional or coastal), and a **subtype** according to geomorphological and hydrological conditions (degree of exposure, bottom texture, tidal range, salinity, etc). Special types are artificial and heavily modified surface water bodies (channels, harbors, etc.) for which special allowances are made concerning quality objectives.
2. Arbitrary **reference conditions**, specific for each subtype of water body, are established for each quantitative indicator of quality.
3. The RTR (**Ecological Quality Ratio**, EQR, ranging from 0 to 1) is calculated for the value of each indicator.
4. Four arbitrary **EQR boundary values** should be adopted for each indicator to obtain five discrete categories of status; high, good, moderate, poor, and bad. Fig. 19.2 illustrates these categories for the case of the benthic quality index (BQI). In practice, often only the

Table 19.1 Comparative Overview of the Water Framework Directive (WFD) and Marine Strategy Framework Directive (MSFD) European Directives, With Emphasis in the Variables Considered for the Classification of Status. Coastal Waters Sensu WFD Extend One Nautical Mile Seawards From the Baseline From Which Territorial Waters Are Measured. The Scope of the MSFD is the Jurisdictional Waters of EU Countries From the Baseline From Which Territorial Waters are Measured. In that One Mile Wide Strip Where Both Directives Overlap, WFD Dispositions Prevail Except for Those Aspects of Environmental Status Not Addressed by WFD

WFD

Unit of study: **Water body**

Unit of management: **River basin district**

Objective: **Good ecological status by year 2015**

Variables considered for the classification of status:

Quality elements	Examples of indicators
Composition and abundance of macroalgae-angiosperms	Macroalgae community indices
Composition and abundance of benthic invertebrate communities	Benthic invertebrate community indices
Composition and abundance of fish fauna (transitional waters only)	
Composition, abundance, and biomass of phytoplankton	Chlorophyll, frequency of blooms
General physico-chemical conditions.	Transparency, temperature, dissolved oxygen, salinity, nutrients
Hydromorphological elements	Depth variation, structure of substrate and intertidal zone, freshwater flow, wave exposure

MSFD

Unit of study: **Subdivision**

Unit of management: **Regions and subregions**

Objective: **Good environmental status by year 2020**

Variables considered for the classification of status:

Descriptors (D)	Examples of indicators [12]
Biological diversity is maintained (D1)	Population abundance, structure, and distribution of representative vertebrates and cephalopod species
Nonindigenous species do not alter ecosystems (D2)	Number of nonindigenous species newly introduced via human activity per assessment period (6 years)
Stocks of commercial fish and shellfish are healthy (D3)	Catches below Maximum Sustainable Yield; High relative abundance of large fish
Marine food webs preserve their integrity (D4)	Balance of total abundance between the trophic guilds
Human-induced eutrophication is minimized (D5)	Chlorophyll, nutrients, dissolved oxygen, transparency, frequency of harmful algal blooms
Seafloor integrity is not adversely affected (D6)	Physical loss or physical disturbance of natural seabed
Hydrographical conditions are not altered (D7)	Permanent alterations in the wave action, currents, salinity, temperature.

Continued

Table 19.1 Comparative Overview of the Water Framework Directive (WFD) and Marine Strategy Framework Directive (MSFD) European Directives, With Emphasis in the Variables Considered for the Classification of Status. Coastal Waters Sensu WFD Extend One Nautical Mile Seawards From the Baseline From Which Territorial Waters Are Measured. The Scope of the MSFD is the Jurisdictional Waters of EU Countries From the Baseline From Which Territorial Waters are Measured. In that One Mile Wide Strip Where Both Directives Overlap, WFD Dispositions Prevail Except for Those Aspects of Environmental Status Not Addressed by WFD *Continued*

WFD	MSFD
Concentrations of priority and nonpriority substances in water and biota	
Concentrations below environmental quality standards for priority substances	Concentrations of contaminants do not cause effects (D8)
	Concentrations of contaminants in water, sediment, and biota. Biomarkers of effects
	Proportion of contaminants exceeding EU standards
	Contaminants in seafood meet standards for human consumption (D9)
	Marine litter do not cause harm (D10)
	Amount of litter on the coastline
	Amount of microlitter (<5 mm) on the surface layer
	Litter and microlitter ingested by seabirds
	Underwater noise does not cause harmful effects (D11)
	Annual average of the squared sound pressure in each of two '1/3-octave bands', one centered at 63 Hz and the other at 125 Hz, in decibels

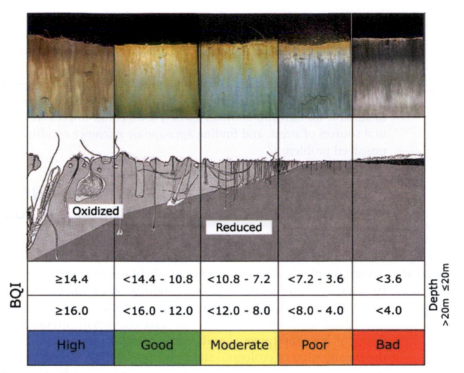

FIGURE 19.2

Classification of the successional stages of a soft bottom benthic community along a left to right increasing gradient of organic enrichment. From top to bottom: sediment profile images where brownish color indicates oxidized conditions and black reduced conditions; diagram of changes in fauna and benthic structure; benchmark values of the benthic quality index (BQI) for the different environmental status according to the Water Framework Directive (WFD), for depths >20 and ≤ 20 m. *Modified from Rosenberg et al. (2004) op. cit.*[13]

most relevant boundary between good and less than good (which implies failure in the assessment) status is established.

5. The overall **ecological status** of each water body is established according to a decision tree where all indicators must show at least good status to the water body to be classified as in **good status.**

6. **Chemical status** is also independently assessed by comparing concentrations of contaminants with EU environmental standards. In this case only two categories, good or less than good—the latter if any concentration exceeds the corresponding standard—are established.

Within this framework, environmental management should progress through pluriannual cycles of monitoring, assessment of status, and implementation of remediation measures in areas classified in nonacceptable categories (less than good status). The procedure was ambitiously intended to achieve at least good status in all areas, according to a timescale already expired for the case of the WFD, and close to expire for the MSFD (see Table 19.1).

The approval of those Directives prompted a massive production of technical documentation intended to standardize the assessment methods. The task proved to be particularly complex in the case of marine water body evaluations.[14] For example, in the Southern Rias of Galicia (Northwest Iberian Peninsula) five subtypes of coastal and transitional waters have been defined. Eleven indicators applied to the five subtypes make 55 reference condition values. This makes a total of arbitrarily (expert judgment) set 220 benchmark values. The procedure of comparison with fixed reference condition values seems especially unsuitable for transitional waters, where local communities are adapted to natural sources of stress, and finding appropriate reference conditions remains an unsolved problem.[15]

19.3   TOWARD SIMPLE, OBJECTIVE, AND UNIVERSAL TOOLS TO CLASSIFY ECOLOGICAL STATUS

Multimetric indices summarize a high amount of information in a single figure

Quantitative indicators of environmental quality currently used or proposed for the classification of status are simple variables (concentrations of dissolved oxygen, nutrients, chemicals, amount of chlorophyll, etc.), univariate indices (species richness, diversity, benthic indices based on the indicator species concept), or multimetric indices resulting from combination of univariate ones. WFD states that the presence of "taxa associated with undisturbed conditions" or "taxa indicative of pollution" are biological quality elements used in the classification of ecological status. Despite the controversy around the static concept of indicator species with a fixed sensitivity to anthropogenic stress, that statement triggered the development of a myriad biotic indices based on the abundance of species a priori classified in groups of tolerance (see Section 15.2), later combined with classical community indices to produce numerous multimetric indices (e.g., B-IBI,[16] RBI,[17] BQI,[18] M-AMBI[19]), with different countries choosing different indices. **Multimetric indices** have the appeal to collapse a high amount of information into a single figure easy to handle by environmental managers and readily comparable to a target value, and they have flourished with the implementation of WFD and derived environmental regulations enforcing assessment of ecological state.

classification of sampling sites into classes of ecological status is index dependent...

However, the classification of sampling sites into different classes of ecological status is index-dependent and does not necessarily reflect the pollution status of the study areas for sources of pollution other than organic matter.[20] An inter-comparison study found full agreement among indices in the split between acceptable and nonacceptable sites in less than 2% of the cases.[21] Furthermore, many multimetric indices are highly redundant. M-AMBI for example is a trivariate index that includes two classical parameters: Shannon diversity and species richness. Since the first is a function of the latter high correlations are generally found between both,[22] and the index is reducible to a bivariate version.[23]

...so international consensus is currently needed to choose common assessment tools

In other cases international consensus on standard tools has been adopted. For example, regarding the marine litter descriptor of the MSFD, it has been noticed that monitoring the accumulation of stranded plastic in beaches is biased by natural (beach dynamics) and anthropogenic (clean ups) factors.

In contrast, monitoring of plastic loads in seabirds was effective in detecting spatial differences, temporal trends, and even changes in the composition of environmental plastics.[24] Close cooperation between researchers around the North Sea permitted the development of a common monitoring tool and associated metrics, based on the plastic content in stomachs of fulmars, already adopted to test the OSPAR Ecological Quality Objective for marine litter.[25] The preliminary target for acceptable ecological conditions, applicable to the MSFD, is defined as "less than 10% of northern fulmars having 0.1 g or more plastic in the stomach in samples of 50−100 beached fulmars from each of 5 different subregions of the North Sea over a period of at least 5 years." The tool seems useful for both identification of geographical patterns—with average plastic loads showing a fourfold increase from the Faroe Islands to the southern North Sea—and temporal trends—with plastic ingestion increasing from the 1960s to the 1980s but stabilized or decreasing more recently. The composition of ingested plastic also changed, with a decrease in industrial plastic (virgin pellets) and an increase in the percentage of consumer plastics fragments.

The assessment of status may be based, instead of on a single figure, on the combination of different types of measurements that contribute with complementary pieces of information. This strategy is commonly termed **weight-of-evidence approach**. A classic example of this approach integrating different lines of evidence for the assessment of sediment pollution is the **sediment quality triad**, proposed by P. Chapman and coworkers, which combines measures of chemical contamination, sediment toxicity using ecotoxicological bioassays, and benthic community composition.[26] Within each line of evidence, data can be aggregated by calculating for each variable the RTR and taking the mean of the RTR values, which can be depicted in a three-axis star plot.[27] The triad places emphasis on the need to take into account the three sources of information at chemical, toxicological, and faunistic levels to produce correct assessments of sediment pollution, and it has been successfully applied in the evaluation of marine pollution world-wide.[28] However, this approach underexploits the information gathered by not attempting exploration of the interrelations between the different lines of evidence. In the same vein, F. Regoli and coworkers proposed a sediment hazard assessment strategy based on four lines of evidence: sediment **chemistry**, **bioaccumulation**, molecular **biomarkers**, and ecotoxicological **bioassays**, and developed a computer assisted method to provide synthetic indices suitable for classification of the sediments in categories of status.[29]

Weight-of-evidence approaches combine for the assessment different types of information, called lines of evidence

Multivariate methods are powerful statistical tools suitable to ordinate and classify sampling sites from monitoring networks (see Box 19.1). When multivariate methods are compared to univariate or multimetric indices using the same data sets, the first result in a greater ability to differentiate spatial and temporal trends.[30] Only multivariate analysis allows the identification of the key environmental variables responsible for community change.[31] In addition,

Multivariate methods are more universal, objective, and discriminant but less user-friendly

BOX 19.1 USING MULTIVARIATE ANALYSIS FOR THE OBJECTIVE CLASSIFICATION OF STATUS IN AQUATIC ECOSYSTEMS.

Ecological studies frequently combine observations of the physical environment and the organisms, the latter at different levels of biological organization. Current environmental monitoring programs in particular integrate chemical and biological information, generating complex and heterogeneous data sets. Multivariate analyses are specially indicated for the treatment of those heterogeneous—in terms of metrics, scales, and sampling design-data sets.[34] Among multivariate methods, **classification** (clustering) and **ordination** methods normally deal with comparisons among objects, which can be the sampling sites in a monitoring network. Both classification and ordination methods are complementary, since the latter are useful to visualize the similarities between sites, summarizing the multidimensional information in a 2-D plot, while the earlier reveal discontinuities that allow the partition of the sampling sites into discrete groups. Thus ordination and classification should be applied jointly for a better understanding of an environmental data set.

Principal Component Analysis (PCA) is the most common ordination method. It allows the identification of weighed combinations of the original variables (components) than better explain the total variability of the data. Sites can later be plotted as a function of the orthogonal principal components explaining higher percentages of variability in two-dimensional (2D) graphics. However, PCA assumes multinormal distribution of the data set, and previous data transformation or standardization is required. Moreover, metrics also affect the PCA results, and if variables are not standardized to relative scales, the PCA may be dominated by few variables with large units of measurement. **Multidimensional scaling (MDS)**[35] is an alternative nonmetric (although metric versions exist) ordination method particularly useful when assumptions to conduct metric techniques are unwarranted.[36] Incidentally, this technique was initially developed in the field of psychology, where similarities between the not metrically measurable perceptions of stimuli (objects) among persons (variables) were targeted. Nonmetric techniques such as MDS, using ranks rather than metrics, are robust against background noise from random variables, work equally well with categorical, ordinal, and numeric attributes, and are applicable to "shallow" matrices, such as those obtained in monitoring campaigns, where a large number of variables are measured in a relatively small number of sites.[37] In addition, MDS lacks the assumption of linear relationships between variables typical of PCA and other classical ordination methods, and thus is more effective in reducing dimensionality when variables involved present nonlinear responses, such as many environmental processes.

The objective of MDS is to construct a bidimensional (or more rarely tridimensional) configuration of the objects such that the agreement between the distances in the configuration and the dissimilarity values between all possible pairs of objects is maximized. Whereas in the metric version of MDS this agreement is quantified through Euclidean distances or similar equations, in nonmetric MDS the agreement is assessed by comparing the rank order of distances to the rank order of proximity measures.[38] The best configuration is iteratively achieved, and the departure from the optimal solution is quantified by a measure called "stress."

Clustering analysis allows the partition of the sampling sites into discrete groups with a given degree of dissimilarity. Both MDS and clustering analysis use a symmetrical matrix of dissimilarities obtained by comparing all possible pairs of sites. This matrix is called the association matrix, and it has $n \times n$ dimensions, where n is the number of sites in the data set. Dissimilarity between each pair of sites, j, k, may be calculated through several expressions, including the Bray–Curtis index:

$$\delta_{jk} = \left(\sum\nolimits^m |Y_{ij} - Y_{ik}| \right) \Big/ \left(\sum\nolimits^m (Y_{ij} + Y_{ik}) \right) \qquad \text{(Eq. 19.1)}$$

where Y_{ij} and Y_{ik} are the scores of variable i, for a total of m variables, at sites j and k respectively. The matrix is symmetrical since $\delta_{jk} = \delta_{kj}$, and square (number of rows = number of columns = n). The diagonal of the matrix is made of 0, the dissimilarity of each object with itself. An obvious requirement not always fulfilled in overambitious monitoring designs is that all variables are to be measured at all sites in the monitoring network.

MDS in combination with clustering analysis has been successfully applied to monitor the spatial homogeneity and seasonal dynamics of phytoplankton communities,[39] to classify coastal sites according to the composition of their benthic fauna,[40] or to track the effects on marine mussels of dredging and dumping of dredged materials.[41] Superimposing the results of

BOX 19.1 USING MULTIVARIATE ANALYSIS FOR THE OBJECTIVE CLASSIFICATION OF STATUS IN AQUATIC ECOSYSTEMS.—cont'd

clustering analysis on the MDS configuration provides also an easy way to visualize groups of sites that can be interpreted as with similar status. In Fig. 19.3, using data from R. Beiras & I. Durán,[42] coastal sites sampled three successive years are classified according to a comprehensive data set that includes benthic community indices, sediment chemistry, and ecotoxicological bioassays.

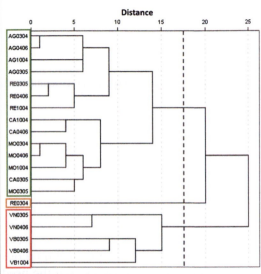
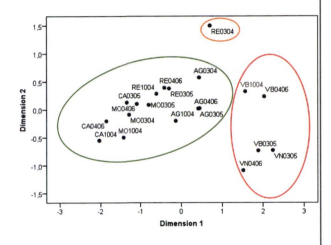

FIGURE 19.3

Clustering analysis (left) and MDS configuration (right; stress = 0.086) of sites in Ría de Vigo (Northwest Iberian Peninsula) sampled three consecutive years to study the benthic fauna composition, sediment chemistry, and sediment toxicity. AG0304, for example, indicates site AG sampled on month 3 (March), year 2004. An arbitrary distance between 15 and 20 classifies site x cruise combinations in three groups. According to a traffic-light color system, green was used for the cluster that includes the reference site (CA). Notice that sites from the Port of Vigo (VN and VB) were always classified in the red group.

By choosing the appropriate similarity thresholds, one can obtain the wished number of groups of status. An approach sometimes regarded as "**traffic light**" **system**, very intuitive for presentation, is advocated by ICES to assess pollution related variables, and can be also used for representing sites. This approach considers three different categories: background or reference, represented in green, elevated values compared to background, depicted in orange, and high values that are cause for concern, depicted in red. The WFD though specifies five different categories of ecological status with their correspondent colors: **high** (blue), **good** (green), **moderate** (yellow), **poor** (orange), and **bad** (red). K-means clustering is a method that allows you to specify a priori how many clusters you wish, avoiding the arbitrary choice of the splitting distance. However, the split is dependent on the set of monitored sites. If we wish to fix that criterion then a fixed set of well-known reference sites and highly polluted sites must be included in all analyses.

The multivariate methods so far described use a single data matrix with no assumptions underneath. We can further sophisticate the analyses by making a distinction between **explanatory** and **dependent variables**, assuming that the later (frequently biological) depend on the first (frequently physicochemical and geomorphological). This approach, pioneered by plant ecologists and known as gradient analysis, seeks the combination of explanatory variables that best explain the

Continued

BOX 19.1 USING MULTIVARIATE ANALYSIS FOR THE OBJECTIVE CLASSIFICATION OF STATUS IN AQUATIC ECOSYSTEMS.—cont'd

variation of the dependent matrix. It is therefore a constrained ordination process; we constrain the axes of the new configuration (now called canonical axes) to be linear combinations of the explanatory variables. The canonical axes are again orthogonal (uncorrelated) linear combinations (multiple regression models) of all explanatory variables. The most used canonical analysis is **redundancy analysis (RDA)**, the canonical version of PCA.[43]

Despite their higher objectivity, less redundancy, and higher discriminant power, multivariate methods are complex, require some statistical training, and are difficult to convey to managers and the public. This has precluded their generalized use in environmental assessment, particularly when time is present as a factor and temporal changes are to be studied. The increase or decrease in a multimetric index value is much easier to follow than the trajectories in a multivariate configuration, particularly because those trajectories are often nonlinear in such plots. The **Principal Response Curves (PRC)** is a method derived from RDA that produces a simple and user-friendly graphical representation of the temporal evolution of studied sites compared to a reference. The reference is constrained to a horizontal line, buffering any seasonal or successional change, whereas the curves corresponding to the studied sites may evolve showing decline or recovery over time.[44]

---

multivariate approaches are not dependent on the type of pollution,[32] in contrast to the biotic indices developed on the basis of the differential sensitivity of benthic species to organic enrichment (see Section 15.2). Also, multivariate methods are the least subjective within the context of integration of information and selection of reference conditions.[33] Among their limitations are the training required to deal with complex statistical issues, and the need to develop more simple and user-friendly ways to convey the information extracted from these statistical methods.

19.4 TOWARD ECOLOGICALLY MEANINGFUL ENVIRONMENTAL QUALITY STANDARDS

The strategies of environmental protection based on the surveillance of the receiving waters demand the knowledge of the ranges of anthropogenically disturbed physicochemical variables and maximum concentrations of chemicals that are safe and should not be exceeded to avoid harmful effects on the organisms and risk to human health. Those figures are known as environmental quality criteria (EQC) or guidelines, as long as they are recommendations, and EQS, when they are enforced by legislation. Section 1.4 introduced the rationale for deriving scientifically sound EQC and EQS (hereafter EQC/S), and the technical aspects will be developed here.

EQC/S have been developed for the three main matrices used in marine pollution monitoring: water, sediments, and biota, and for the latter mostly for mussels. Issues related to the protection of human health for consumers of foodstuff of marine origin were addressed in previous chapters (e.g., Chapter 3) and will not be discussed here.

There are several alternative approaches for the derivation of scientifically sound EQC/S. The first and most common one, sometimes called **mechanistic EQC/S**, is based on laboratory toxicity testing with model species, which are assumed to be representative surrogates of the actual species to be protected in the field. From the parameters describing sensitivity of those model species to a given chemical obtained in dose:response experiments, a safe concentration is extrapolated by using statistical methods (**probabilistic EQC/S**). When the amount of data precludes this option the criterion is derived by applying to the **critical value** an assessment factor intended to account for the uncertainties of the extrapolation (**deterministic EQC/S**). The critical value is here the toxicity threshold obtained for the most sensitive species tested.

An alternative approach common in the development of sediment quality criteria is based on information gathered from matching measurements of chemical contents and biological effects (bioaccumulation, toxicity, benthic indices) in field samples. When large data sets covering the full range of pollution from pristine to highly impacted samples are available, multivariate analysis or other statistical techniques may be applied to relate contents of individual chemicals to toxicity or any other biological effect, and derive so-called **empirical EQC/S**. The well-known **Effects Range-Low (ERL)/Effects Range-Medium (ERM)**[45] and **Threshold Effects Level (TEL)/Probable Effects Level (PEL)**[46] sediment quality criteria (SQC), obtained applying rather simple algorithms to relatively large data sets of matching sediment chemistry and acute sediment toxicity in laboratory, belong to this type of criteria. Although this technique has been used mainly for sediments, in principle it is applicable to water and biota also.

Mechanistic approaches rely on the basic principle of toxicology that relates exposure to increasing amounts of a chemical with appearance of toxic effects, while in empirical approaches causality is not taken into account. In principle both approaches are complementary, and in fact field data are useful for validation of laboratory-based EQCs, considering the concerns raised by the extrapolation from laboratory to the natural environment. Unfortunately, toxicity testing with field samples also needs to rely on laboratory practices and concerns about laboratory to field extrapolation still hold.

The earliest attempts to derive probabilistic WQCs used the widely available LC_{50} parameter obtained from lethal toxicity tests. Assuming a log-logistic distribution of the LC_{50} values of a chemical for different species, more treatable than a log-normal that spans from $-\infty$ to $+\infty$, Kooijman[47] calculated the maximum allowable concentration of that chemical innocuous for the most sensitive species of an ecosystem with a given probability, customarily 95%. Although ideally those LC_{50} values should be obtained from chronic toxicity

Sidenotes:

Mechanistic *EQC/S* (either probabilistic or deterministic) are based on toxicity testing, while empirical *EQC/S* are based on field surveys information also

Derivation of probabilistic water quality criteria (WQC) assumes a log-logistic distribution of sensitivities among species…

tests, in practice for most chemicals of interest only acute toxicity data are available, and those were used for the calculations. Some modifications were later introduced[48]: (1) to increase protective value, sensitivity is best described by the toxicity threshold (TT) compared to LC_{50}, (2) given the statistical limitations of the NOEC/LOEC approach, such as the dependence on the replication, experimental design, and choice of statistical test, the use of effective concentrations for a low level of effect, such as EC_{10}, is advised to estimate TT,[49] and (3) given the functional redundancy present in most ecosystems, a more practical (less overprotective) parameter is obtained by seeking the protection of a given percentage of the species, e.g., 95% (termed hazardous concentration for 5% of the species, HC_5), rather than using the value obtained from the most sensitive of all species.

... and takes the lower end of the CI of the fifth percentile in the TT distribution as criterion

Under the premises, a scientifically sound WQC can be obtained from the fifth percentile of the distribution of EC_{10} values adjusted to a log-logistic model (the HC_5). Since this parameter is an estimate obtained from a limited number of laboratory tested species it has a 50% probability of being higher (and thus less protective) than the actual HC_5 value in the ecosystem. This is considered unacceptable and thus the lower end of the 95% confidence interval is a better choice for the criterion,[50] that would become,

$$WQC = HC_5 - 95\%CI \qquad\qquad (Eq.\ 19.2)$$

This criterion would protect 95% of the species with a 95% confidence. As illustrated in Fig. 19.4, this value ($HC_5 - 95\%CI$) becomes higher as the number of test species (n) increases. This provides an incentive for more complete testing that balances economic and ecological motivations, since the so derived WQC will become both higher and more reliable.

Uncertainties associated to environmental standards derive from interspecific variability and extrapolation from laboratory to the field

The assumptions underneath the procedures to derive EQCs raise both statistical and ecological concerns that should be taken into account. The latter involve all intrinsic (biological) and environmental factors modifying toxicity (discussed in Section 13.4) that must be fixed in standard laboratory testing but obviously vary in the field. The original statistical approaches previously discussed relied on the random choice of species for testing, necessary to apply standard statistical inference. In practice it is more appropriate that the **set of species** from which toxicity data are extracted was oriented toward enhancing representativeness of the main functional groups and life forms present in the ecosystem to be protected. This requires a comprehensive data set where all major phylogenetic groups should be represented. Even in that case, most biological models used in experimental aquatic ecotoxicology are **freshwater organisms** such as *Rhaphidocellis* (formerly *Selenastrum*), daphnia, or zebra fish. For marine species, toxicity data are scarcer, and therefore, they are frequently extrapolated from freshwater species.[51] This approach has been questioned because species from continental waters may not represent the sensitivity of saltwater organisms,[52] increasing in any case the uncertainty of the actual degree of protection for marine ecosystems. For example, the HC_5

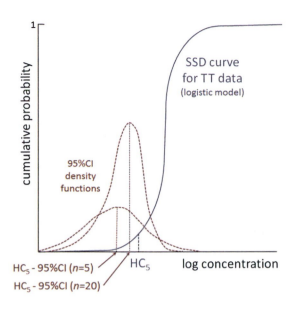

FIGURE 19.4

The continuous blue line shows a Species Sensitivity Distribution (SSD) curve, i.e., the cumulative probability of toxicity of a substance to a given proportion of the species of an ecosystem, fitted to a logistic function of the logarithm of the toxicity thresholds (TT). The fifth percentile (HC_5) of the distribution is also showed. On the left side of the SSD curve, the brown dotted lines show the density functions for the lower end of the HC_5 95% CI estimated for two sample sizes (n = 5 and n = 20). Notice that higher sample sizes, i.e., higher number of species tested, produce higher values for the WQC.

values obtained for pesticides according to acute toxicity to marine organisms using the United States Environmental Protection Agency (US-EPA) data base were on average fivefold lower than those obtained from freshwater species, although this is partially biased by the taxonomic differences for each set of test species.[53] A similar comparison made with a more limited dataset showed that higher sensitivity in terms of EC_{50} for saltwater species (both invertebrates and fish) versus freshwater species was twice more often than the opposite.[54] It seems clear than for organic chemicals marine species are generally more sensitive than their freshwater counterparts.

In designing the battery of toxicity tests, the following considerations are noteworthy:

- Primary producers, consumers, and saprotrophs should be included to represent the main trophic groups.
- Different phylogenetic groups with differing anatomical design and physiology should be represented, including bacteria, algae, invertebrates (at least bivalves, crustaceans, echinoderms), and fish.
- Different exposure routes (at least dissolved and particulate) derived from different feeding characteristics and habitats should be taken into account. Ecological significance is paramount here. It may not be

The species chosen for toxicity testing intended to derive EQC should represent the full range of systematic and functional variability

necessary to test waterborne toxicity of an extremely insoluble chemical, but exposure via food must then be tested.
- Subchronic or chronic effects should be also directly tested, using acute to chronic ratios only as last resort.

While the need to cover the full range of systematic variability is universally recognized, specific guidance on how to do it lacks consensus. Inspection of Table 19.2 comparing US, Canadian, and EU advice illustrates the highly arbitrary recommendations of standard protocols. US-EPA[55] required the use of eight different families for the derivation of saltwater criteria, with fish, crustacean, and algae representatives compulsory. Canadian guidelines[56] show solid scientific grounds, placing emphasis in EC_{10} values as preferred endpoint and SSD curves as preferred method, interestingly, including the possibilities of adapting the set of species and fitting bimodal curves for selective toxicants. Most guidelines stress the usefulness of crustaceans, but EC[57] overlooks the need of photosynthetic organisms. Seeking a balance between representativeness and data availability, I. Durán and R. Beiras[58] used for the derivation of acute marine WQC at least one representative of the following major groups of marine taxa: algae, bivalves, nondecapod crustaceans, decapod crustaceans, echinoderms, and chordates, and advocated the use of early life stages to improve sensitivity. Their choice of taxa was somewhat forced by the scarcity of data concerning macroalgae and polychaetes, and, to a minor extent, fish.

Using early life stages (ELS) and sublethal responses increases protective value of the criteria...

Regarding biological variability, we have to pay attention not only to the species but also to the life stage chosen for testing. To maximize sensitivity, and thus protective value of the WQC, toxicity data should be based on sublethal endpoints (algal population growth, reproduction, embryogenesis, and larval development) and early life stages (ELS) (embryos, larvae, neonates, juveniles), the weakest link in the life cycle of marine organisms. In fact ELS tests with aquatic organisms are extremely useful since they show similar sensitivity than much longer and labor consuming lifecycle tests.[59] Therefore, the use of ELS may contribute to derive ecologically sound EQC/S in a less data intensive way. Regarding sensitivity, the use of extremely tolerant species such as *Artemia* spp., though very practical for laboratory use because they can be easily stored and posted as dry cysts, is not advised.

...but laboratory testing is limited in terms of scale, number, and duration of the experiments

The use of indigenous species is in conflict with the methodological standardization needed to ensure robust and reproducible results. Standard acute toxicity testing procedures have been developed for the bacteria *Vibrio fischeri*, the microalgae *Phaeodactylum* and *Skeletonema*, several bivalve and sea-urchin embryos, copepods (e.g., *Acartia*, *Tigriopus*), and mysids, but other major groups such as marine phanerogams, macroalgae, polychaetes, and fish are still underrepresented in marine data bases, and practical procedures for testing chronic effects and sublethal endpoints such as **endocrine disruption** are

Table 19.2 Recommendations for the Design of a Representative Battery of Toxicity Tests Suitable for the Derivation of Water Quality Criteria

	US-EPA (1985)	CCME (2007)	EC (2011)	Durán & Beiras (2017)
Criteria Targeted				
	Acute and chronic, fresh and saltwater.	Short-term and long-term exposure, freshwater, and marine.	Maximum admissible concentration and annual average, continental and other waters.	Marine acute criteria.
General Requirements				
	Data should be available on the four major kinds of possible adverse effects: acute toxicity to animals, chronic to acute ratios, alga/plant toxicity, and bioaccumulation only if relevant.	EC_{10} data for chronic effects, while LC_{50}/EC_{50} data are acceptable for acute effects. Probabilistic criteria preferred. Deterministic criteria accepted if data scarce.	Chronic NOEC or EC_{10}, ideally statistically and ecologically representative of the community of interest. Marine and freshwater data can be combined unless evaluation shows otherwise.	Short term (up to 5 d) toxicity using when possible sublethal endpoints (algal population growth, reproduction, embryogenesis and larval development) and early life stages (embryos, larvae, neonates).
Specific Requirements				
	At least 10 species covering 8 taxa, the following would normally need to be represented: Fish, a second Chordata family, a crustacean, an insect, a family in a phylum other than Arthropoda or Chordata, a family in any order of insect or any phylum not already represented, an alga, a higher plant. When saltwater and freshwater datasets cannot be pooled insects and higher plants may be replaced by more typical marine taxa.	At least three fish species, two invertebrate species (three for freshwater), and one alga/plant. Two (acute) or three (chronic) alga/plant species for phytotoxic substances.	At least eight different families such that all of the following are included: 1. two families in the phylum Chordata; 2. a family in a phylum other than Arthropoda or Chordata; 3. either the Mysidae or Penaeidae family; 4. three other families not in the phylum Chordata (may include Mysidae or Penaeidae, whichever was not used above); 5. any other family.	At least one representative of: algae (microalgae), molluscs, nondecapod crustaceans, decapod crustaceans, echinoderms, chordates. Polychaetes and macroalgae are advisable but available data are scarce. Fish data on ELS are advisable but available data are scarce.

urgently needed. We must be aware that current EQC/S do not take into account this kind of effects revealed only after long exposure periods.

The impossibility in single species toxicity tests to take into account the complex species interactions that affect ecological fitness in the natural environment has also been criticized. Despite their costs, the use of **mesocosm experiments** has been advocated to partially overcome this limitation.

Environmental standards may be expressed as a function of conservative variables rather than as fixed values

The extrapolation from laboratory to field conditions raise concerns due to the effects of natural factors such as temperature, pH, salinity, or dissolved organic matter that modify the toxicity of chemicals in the field. This is especially relevant for metals, and some theoretical models have been developed to take this into account, particularly for freshwater environments but also for estuarine and coastal waters. US-EPA proposes to use the biotic ligand model (BLM, see Section 10.4) to correct the aquatic life estuarine and coastal WQC for copper depending on pH, DOC, and salinity.[60] In other cases the EQC can be expressed as a function of a single explicative natural variable rather than as fixed values. For example, background levels for nitrate and ammonium in the Minho estuary (Northwest Iberian Peninsula) could be modeled as a function of a conservative variable, salinity, according to the following equations,[61]

$$N = 1.6415 \, e^{-0.059 \cdot S} \qquad\qquad \text{(Eq. 19.3)}$$

$$A = 0.1052 \, e^{-0.052 \cdot S} \qquad\qquad \text{(Eq. 19.4)}$$

where N is nitrate concentration (mg/L), A is ammonium (mg/L), and S is salinity (psu). This allows the use of reference values across the whole estuary and with independence of the tidal cycle.

Advantages of sediments to assess marine pollution

Despite the well-known advantages of **sediments** for monitoring chemical pollution in aquatic ecosystems[62] (see Section 17.2), which dates back at least to the 1970s,[63] this basic ecotoxicological notion was ignored by the WFD and derived legislation up to date. The use of sediments rather than water is especially encouraged for organic chemicals with a sediment:water partition coefficient (K_P) such that log $K_P \geq 3$, since for those chemicals dealing with suspended particulate matter in water samples is a serious analytical issue.[64] Sediments have clear advantages for pollution assessment compared to water and even to biota. First, they may serve as records of **historical trends** in pollution. As early as 1956, H. Züllig[65] used sediment cores to track the trophic and geodynamic changes in alpine lakes many centuries backward. Secondly, for most chemical pollutants, and particularly for organics, the **enrichment factors** (concentrations in polluted sites divided by background values) in sediments are markedly higher than the same ratios in biota, even for well-known bioaccumulators such as mussels. Finally, concentrations in sediments are rather **stable in time** in a much higher degree than water concentrations, highly dependent on pluviometry, hydrodynamics and other natural events, and independent of spawning and other seasonal events that interfere in the levels recorded in biological samples.

P. Chapman[66] proposed the use of three sources of ecotoxicological information, namely chemical analyses, laboratory toxicity bioassays and field studies of community structure, to develop ecologically sound SQC, according to the **sediment quality triad** approach (see also Section 19.3). The two first components of the triad were used to derive the empirical criteria **ERL/ERM**[67] and **TEL/PEL**.[68] Based on the database assembled by E. Long et al.[69] SQC were calculated for trace metals, PAHs (aggregated and individual values) and chlorinated organic hydrocarbons. The procedure consisted of arranging concentrations for each substance in ascending order, and taking the 10th percentile, the **ERL**, and the 50th percentile, the **ERM** of the distribution for those sites showing adverse effects in concurrent sediment toxicity tests, normally conducted with amphipods. The US NOAA adopted these values as informal guidelines for the assessment of aquatic sediments at national scale.[70] Working on the same database but with an alternative approach that used also information from the no-effect cases, D. MacDonald and coworkers[71] derived the **TEL** and the **PEL**. For each substance, TEL was the geometric mean of the 15th percentile of the effects data set and the 50th percentile of the no-effects data set, whereas PEL was the geometric mean of the 50th and the 85th percentiles in the respective data sets. This resulted in similar (yet slightly lower) values for metals, but substantially lower—and more useful—values for PAHs.

Despite their simplicity (e.g., bioavailability issues or effects of mixtures are not considered), and geographical and ecotoxicological limitations of the data set, those empirical SQC have been embraced as useful assessment tools for sediment chemistry records worldwide, stressing the usefulness of this kind of tools for environmental managers. The need to develop local SQC values adapted to regional characteristics has been frequently advocated. J. Bellas and co-workers[72] tested the validity of ERL/ERM and TEL/PEL criteria with a completely different data set obtained from ecotoxicological studies conducted over a decade in the Galician Rias (Southwest Europe), using the sea-urchin embryo test rather than the amphipod bioassay as method for effects assessment. As shown in Table 19.3, despite broad differences in geographical areas, biological models, and mathematical methods, the low and medium levels of pollution calculated with the Galician Rias data set are quite consistent with the North America SQC values.

A more theoretical approach for developing EQC/S particularly intended for sediments tries to account for differences in bioavailability caused by factors such as organic matter content or sulfides. This approach is based on the **equilibrium partitioning** of the toxicant between bioavailable (pore waters) and generally not bioavailable (particulate phase) fractions. Since WQC are usually handy, SQC can be derived from WQC according to this approach simply by taking into account the organic fraction of the whole sediment (f_{OC}), and assuming that sorption to organic particles is directly dependent on the K_{OC}

ERL/ERM and TEL/PEL are empirical Sediment Quality Criteria (SQC) used world-wide

Bioavailability issues are considered in the theoretical criteria based on equilibrium partitioning

Table 19.3 Comparison Between Empirical Sediment Quality Criteria (SQC) Values Derived in N America (ERL/ERM and TEL/PEL) and SW Europe (Low/Medium), Obtained From Databases Matching Sediment Chemistry and Sediment Toxicity. Units Are mg/Kg DW for Metals and µg/Kg DW for PAHs

	ERL/ERM	TEL/PEL	Low/Medium
Cu	34–270	18.7–108	39.6–241
Zn	150–410	124–271	139–471
Pb	46.7–218	30–112	58.7–191
$\sum$PAHs	2643–15460	630–5873	299–3011

Data from: Long et al. (1995) op. cit; MacDonald et al. (1996) op. cit.; Bellas et al. (2011) op. cit.

of the substance, which quantifies the partition of the substance between organic carbon and water in equilibrium.[73] Thus,

$$SQC = WQC \cdot K_{OC} \cdot f_{OC} \qquad \text{(Eq. 19.5)}$$

It is easy to see that for a given substance the higher the organic content of the sediment (f_{OC}, ranging from 0 to 1) the higher the SQC value. The term f_{OC} can be eliminated from Eq. (19.5) by expressing SQC values on organic carbon weight basis (g/Kg OC). Then,

$$SQC = WQC \cdot K_{OC} \qquad \text{(Eq. 19.6)}$$

For hydrophobic chemicals log K_{OC} is frequently > 3, and this reflects in SQC values > 1000-fold higher than WQC.

The equilibrium partitioning method has also been applied for divalent trace **metals** taking into account that in the sediment they tend to form insoluble sulfides. The amount of reactive sulfide can be measured by cold acid extraction yielding the **acid-volatile sulfide (AVS)** content (see Section 10.4). The molar concentration of five metals (Cd, Cu, Ni, Pb, and Zn) can be summed up and the molar ratio of metals to AVS used to predict toxicity, which should take place when the ratio >1.[74] Whereas this works when individual metals are the only source of pollution,[75] the model cannot be extended to other contaminants. Equilibrium partitioning-based criteria in general show lower predictive ability—more false negatives—than empirical SQC, and remain to be properly validated.[76]

EQC/S for biota are emerging in the EU legislation, but values need further refinement

Only recently the European legislation for the protection of aquatic ecosystems paid attention to the concept of bioaccumulation and introduced biota EQS for some priority substances: Hg, BaP, Flu, BDEs, PFOS, and some organochlorines. According to Directive 2013/39/EU, these biota EQS are applicable to fish, with no indication of species. Surprisingly, Hg and BaP EQS are orders of magnitude more strict than European standards for foodstuff intended to the protection of consumer health. In the case of BDEs and some organochlorines, the EQS are in the order of parts per trillion (pg/g), which is analytically

Table 19.4 Marine Environmental Quality Criteria (EQC) Useful for the Assessment of Chemical Pollution in Coastal Ecosystems, and the Calculation of Chemical Pollution Indices

Contaminant	Seawater Criteria (μg/L)	Sediment Criteria (mg/Kg DW)	Mussel Criteria*** (mg/Kg DW)
Hg	0.05[#]	0.15[*]	0.2
Cd	0.2[#]	1.2[*]	2
Cu	1.39[##]	34[*]	10
Pb	7.2[#]	58.7[§]	3
Zn	8.24[##]	150[*]	200
BaP	–	$88.8 \cdot 10^{-3}$[**]	$5 \cdot 10^{-3}$
sum PAH	–	$1684 \cdot 10^{-3}$[**]	$250 \cdot 10^{-3}$
sum PCB	–	$22.7 \cdot 10^{-3}$[*]	$20 \cdot 10^{-3}$
sum DDTs	–	$1.58 \cdot 10^{-3}$[*]	$10 \cdot 10^{-3}$
lindane	–	$0.32 \cdot 10^{-3}$[**]	$5 \cdot 10^{-3}$
TBT	–	–	0.1

[#] *WFD Chronic;* [##] *Durán & Beiras (2013) op. cit.;* [*] *ERL;* [**] *TEL;* [***] *Norway, Level 1;* [§] *Bellas et al. (2011) op. cit.*

challenging and, for BDEs, well below levels shown by the most common bio-monitor, the marine mussel, in most European coasts. OSPAR has developed specific EAC values for mussels, intended to protect biota even from chronic exposures. However, at the light of the coastal monitoring results those values seem also in many instances too strict. For example, mussels from nonpolluted coastal water bodies in remote regions of Northern Norway far from any significant anthropogenic input fail to be compliant with the standards for BDEs, Hg, TBT, and PCBs. It is thus questionable whether these criteria will be operational in monitoring networks.[77] More useful criteria for mussels issued by the Climate and Pollution Agency are in use at national level in Norway.[78]

Table 19.4 summarizes some marine EQC/S values useful for the assessment of chemical pollution in seawater, sediments, and mussels, and for calculation of chemical pollution indices (CPI) according to Eq. 1.3 in Chapter 1.

Endnotes

1. Secretariat of the Convention on Biological Diversity. The ecosystem approach, (CBD guidelines) Montreal: secretariat of the convention on biological diversity, 2004; 50 p. https://www.cbd.int/doc/publications/ea-text-en.pdf.

2. Schei PJ. Chairman's Report. The Norway/UN Conference on the Ecosystem Approach for Sustainable Use of Biological Diversity. Trondheim. 1999.

3. Arkema KK, Abramson SC, Dewsbury BM. Marine ecosystem-based management: from characterization to implementation. Front Ecol Environ 2006;4(10):525–532. Morishita J. What is the ecosystem approach for fisheries management? Mar Policy 2008;32:19–26.

4. Christie DR. Implementing an Ecosystem-Approach to Ocean management: An Assessment of Current Regional Governance Models. 16 Duke Environmental Law & Policy Forum Spring 2006;117–142. Available online at: http://scholarship.law.duke.edu/delpf/vol16/iss2/1lastvisit: 12/01/2016.

5.   US-EPA 2016. https://www.epa.gov/wqc/national-recommended-water-quality-criteria.

6.   Canadian Council of Ministers of the Environment 2016. http://www.ccme.ca/en/resources/canadian_environmental_quality_guidelines/.

7.   ANZECC. Australian and New Zealand guidelines for fresh and marine water quality. Australian and New Zealand Environment and Conservation Council, Canberra; 2000. http://www.mincos.gov.au/publications/national_water_quality_management_strategy.

8.   EEA. Europe's environment. An assessment of assessments. Copenhagen: European Environment Agency; 2011. 197 pp.

9.   Aldenberg T, Slob W. Confidence Limits for Hazardous Concentrations Based on Logistically Distributed NOEC Toxicity Data. Ecotox Environ Safe 1993;25(1):48–63.

10. Smith EP, Cairns J. Extrapolation methods for setting ecological standards for water quality: Statistical and ecological concerns. Ecotoxicology 1993;2:203–219.

11. See Endnote. 8.

12. COMMISSION DECISION (EU) 2017/848.

13. Rosenberg R, Blomqvist M, Nilsson HC, et al. Marine quality assessment by use of benthic species-abundance distributions: a proposed new protocol within the European Union Water Framework Directive. Mar Poll Bull 2004;49:728–739.

14. European Environment Agency. European waters - assessment of status and pressures; 2012. EEA Report No 8/2012. ISSN 1725–9177.

15. Puente A, Juanes JA, García A, et al. Ecological assessment of soft bottom benthic communities in northern Spanish estuaries. Ecol Indicat 2008;8(4):373–388.

16. Weisberg SB, Ranasinghe JA, Dauer DM, et al. An estuarine benthic index of biotic integrity (B-IBI) for the Chesapeake Bay. Estuaries 1997;20(1):149–158.

17. Hunt JW, Anderson BS, Phillips BM, et al. A large-scale categorization of sites in San Francisco Bay, USA, based on the sediment quality triad, toxicity identification evaluations, and gradient studies. Environ Toxicol Chem 2001;20:1252–1265.

18. See Endnote. 14.

19. Muxika I, Borja A, Bald J. Using historical data, expert judgement and multivariate analysis in assessing reference conditions and benthic ecological status, according to the European Water Framework Directive. Mar Pollut Bull 2007;55:16–29.

20. Marín-Guirao L, Cesar A, Marín A, et al. Establishing the ecological quality status of soft-bottom mining-impacted coastal water bodies in the scope of the Water Framework Directive. Mar Pollut Bull 2005;50:374–387. Labrune C, Amouroux JM, Sarda R, et al. Characterization of the ecological quality of the coastalGulf of Lions (NWMediterranean). A comparative approach based on three biotic indices. Mar Pollut Bull 2006;52:34–47. Quintino V, Elliot M, Rodrigues AM. The derivation, performance and role of the univariate and multivariate indicators of benthic change: Case studies at differing spatial scales. J Exp Mar Biol Ecol 2006; 330:368–382. Puente et al. (2008) op. cit.

21. Blanchet H, Lavesque N, Ruellet T, et al. Use of biotic indices in semi-enclosed coastal ecosystems and transitional waters habitats—implications for the implementation of the European Water Framework Directive. Ecol Ind 2008;8:360–372.

22. Kilgour BW, Somers KM, Barton DR. A comparison of the sensitivity of stream benthic community indices to effects associated with mines, pulp and paper mills, and urbanization. Environ Toxicol Chem 2004;23(1):212–221.

23. Sigovini M, Keppel E, Tagliapietra D. M-AMBI revisited: looking inside a widely-used benthic index. Hydrobiologia 2013;717:41–50.

24. Ryan PG, Moore CJ, van Franeker JA, Moloney CL. Monitoring the abundance of plastic debris in the marine environment. Phil Trans R Soc B 2009;364:1999–2012.

25. Van Franeker JA, Blaize C, Danielsen J, et al. Monitoring plastic ingestion by the northern fulmar Fulmarus glacialis in the North Sea. Environ Pollut 2011;159:2609–2615.

26. Long ER, Chapman PM. A Sediment quality triad: measures of sediment contamination, toxicity and infaunal community composition in puget sound. Mar Pollut Bull 1985; 16(10):405—415.

27. Chapman PM, Dexter RN, Long ER. Synoptic measures of sediment contamination, toxicity and infaunal community composition (the Sediment Quality Triad) in San Francisco Bay. MEPS 1987;37:75—96.

28. E.g. Carr RS, Chapman DC, Howard CL, et al. Sediment quality triad assessment survey of the Galveston Bay, Texas system. Ecotoxicology 1996;5:341—364. Del Valls TA, Forja JM, Gómez-Parra A. Integrative assessment of sediment quality in two littoral ecosystems from the Gulf of Cádiz, Spain. Environ Toxicol Chem 1998;17(6):1073—1084. Anderson BS, Hunt JW, Phillips BM, et al. Sediment quality in Los Angeles harbor, USA: A Triad assessment. Environ Toxicol Chem 2001;20(2):359—370.

29. Piva F, Ciaprini F, Onorati F, et al. Assessing sediment hazard through a weight of evidence approach with bioindicator organisms: A practical model to elaborate data from sediment chemistry, bioavailability, biomarkers and ecotoxicological bioassays. Chemosphere 2011; 83:475—485.

30. Thomas JF, Hall TJ. A comparison of three methods of evaluating aquatic community impairment in streams. J Freshwater eco 2006;21(1):53—63. Reynoldson TB, Norris RH, Resh VH, et al. The reference condition: a comparison of multimetric and multivariate approaches to assess water-quality impairment using benthic macroinvertebrates. J N Am Benthol Soc 1997;16(4):833—852. Reynoldson TB, Thompson SP, Milani D. Integrating multiple toxicological endpoints in a decision-making framework for contaminated sediments. Human Ecolgic Risk Assessment 2002;8(7):1569—1584. Kilgour et al. (2004) op. cit.

31. Warwick RM, Clarke KR. A Comparison of some methods for analysing changes in benthic community structure. J Mar Biol Assoc UK 1991;71:225—244.

32. See Endnote. 23.

33. Reynoldson et al. (2002) op. cit.

34. Gotelli NJ, Ellison AM. A primer of ecological statistics. Sunderland MA: Sinauer Associates Inc. 2004. 510 pp.

35. Kruskal JB, Wish M. Multidimensional scaling. Beverly Hills: Sage Publications; 1978. 93 pp.

36. Landis WG, Matthews RA, Matthews GB. Design and analysis of multispecies toxicity test for pesticide registration. Ecol Applicat 1997;7(4):1111—1116.

37. Matthews G, Matthews R, Landis W. Nonmetric conceptual clustering in ecology and ecotoxicology. AI Applications 1995;9:41—48.

38. Dillon WR, Goldstein M. Multivariate analysis. Methods and applications. New York: John Wiley & Sons; 1984.

39. Salmaso N. Seasonal variation in the composition and rate of change of the phytoplankton community in a deep subalpine lake (Lake Garda, Northern Italy). An application on nonmetric multidimensional scaling and cluster analysis. Hydrobiologia 1996;337:49—68.

40. Shin PKS, Fong KYS. Multiple discriminant analysis of marine sediment data. Marine Pollution Bulletin 1999;39(1—2):285—294.

41. Bellas J, Ekelund R, Halldórsson HP, et al. Monitoring of organic compounds and trace metals during a dredging episode in the Göta Älv Estuary (SW Sweden) Using Caged Mussels. Water Air Soil Pollut 2007;181:265—279.

42. Beiras R, Durán I. Objective classification of ecological status in marine water bodies using ecotoxicological information and multivariate analysis. Environ Sci Pollut Res 2014;21: 13291—13301.

43. Legendre L, Legendre P. Numerical ecology, 2nd ed. Amsterdam:Elsevier, 1998.

44. Pardal MA, Cardoso PG, Sousa JP, et al. Assessing environmental quality: a novel approach. Mar Ecol Prog Ser 2004;267:1—8.

45. Long ER, MacDonald DD, Smith SL, et al. Incidence of Adverse Biological Effects Within Ranges of Chemical Concentrations in Marine and Estuarine Sediments. Environ Manage 1995;19(1):81–97.

46. MacDonald DD, Carr RS, Calder FD,et al. Development and evaluation of sediment quality guidelines for Florida coastal waters. Ecotoxicology 1996;5:253–278.

47. Kooijman SALM. A safety factor for LC_{50} values allowing for differences in sensitivity among species. Wat Res 1987;21(3):269–276.

48. Van Straalen NM, Denneman CAJ. Ecotoxicological evaluation of soil quality criteria. Ecotox Environm Saf 1989;18:241–251.

49. OECD. OECD series on testing and assessment. number 10: report of the OECD workshop on statistical analysis of aquatic toxicity data. Paris: Organisation for Economic Co-operation and Development; 1998. Reiley, MC, Stubblefield, WA, Adams, WJ, Di Toro DM, Hodson PV, Erickson RJ, Keating Jr., FJ, editor. Reevaluation of the state of the science for water-quality criteria development. Pensacola FL, USA: Society of Environmental Toxicology and Chemistry (SETAC); 2003. 224 pp. Vighi M, Altenburger R, Arrhenius A, et al. Water quality objectives for mixtures of toxic chemicals: problems and perspectives. Ecotoxicol Environ Saf 2003(54): 139–150.

50. Aldenberg & Slob (1993) op. cit.

51. EC. Technical guidance for deriving environmental quality standards. Common Implementation Strategy for the Water Framework Directive (2000/60/EC). Guidance document n 27. Technical report-2011-055. European Commission; 2011.

52. Leung KMY, Morritt D, Wheeler JR, et al. Can saltwater toxicity be predicted from freshwater data? Mar Pollut Bull 2001;42:1007–1013. But see also: Robinson PW. The toxicity of pesticides and organics to mysid shrimps can be predicted from *Daphnia* spp. Water Res 1999;33: 1545–1549.

53. Crane M, Sorokin N, Wheeler J, et al. European approaches to coastal and estuarine risk assessment. pp. 15–39. Chapter 2. In: Newman MC, Roberts Jr MH, Hale RC, editor. Coastal and estuarine risk assessment. Boca Raton: Lewis Publishers; 2002.

54. Hutchinson TH, Scholz N, Guhl W. Analysis of the ECETOC aquatic toxicity (EAT) database. IV- comparative toxicity of chemical substances to freshwater versus saltwater organisms. Chemosphere 1998;36(1):143–153.

55. US-EPA. Guidelines for deriving numerical national water quality criteria for the protection of aquatic organisms and their uses. PB85–227049. Office of Research and Development. Environmental Research Laboratories. Duluth, Minnesota: United States Environmental Protection Agency; 1985.

56. CCME. A protocol for derivation of water quality guidelines for the protection of aquatic life 2007. Canadian Council of Ministers of the Environment; 2007. Available online: https://www.ccme.ca/files/Resources/supporting_scientific_documents/protocol_aql_2007e.pdf.

57. See Endnote 52.

58. Durán I, Beiras, R. Acute water quality criteria for polycyclic aromatic hydrocarbons, pesticides, plastic additives, and 4-Nonylphenol in seawater. Environ Pollut 2017;224:384–391.

59. Hutchinson TH, Solbé J, Kloepper-Sams PJ. Analysis of the ECETOC aquatic toxicity database (EAC). III- comparative toxicity of chemical substances to different life stages of aquatic organisms. Chemosphere 1998;36(1):129–142.

60. US EPA. EPA-822-P-16–1001. Draft aquatic life ambient estuarine/marine water quality criteria for copper – 2016. Washington: US Environmental Protection Agency; 2016.

61. Beiras R. Assessing ecological status of transitional and coastal waters; current difficulties and alternative approaches. Front Mar Sci 2016. https://doi.org/10.3389/fmars.2016.00088.

62. USEPA, United States Environmental Protection Agency. Methods for collection, storage and manipulation of sediments for chemical and toxicological analyses: technical manual. EPA 823-B-01-002. Washington, DC: Office of Water; 2001. Förstner U, Salomons W. Mobilization of metals from sediment. In: Merian E, editor. Metals and their compounds in the environment. Weinheim:VCH; 1991. 379—398.

63. Webb JS. Regional biochemical reconnaissance in medical geography. Geol Soc Amer Man 1971;123:31—42.

64. Coquery M, Morin A, Bécue A, et al. Priority substances of the European Water Framework Directive: analytical challenges in monitoring water quality. Trends Anal Chem 2005;24(2): 117—127.

65. Zullig H., Sedimenteals Ausdruck des Zustandeseines Gewässers. Schweiz Z Hydrol 1956;18: 7—143.

66. Chapman PM. Sediment quality criteria from the sediment quality triad: an example. Environ Toxicol Chem 1986;5:957—964.

67. See Endnote. 46.

68. See Endnote. 47.

69. Long et al. (1995) op. cit.

70. NOAA. Sediment quality guidelines developed for the national status and trends program; 1999. Available online: http://ecoelectrica.com/wp-content/uploads/2015/12/Sediment-Quality-Guidlines.pdf.

71. MacDonald et al. (1996) op. cit.

72. Bellas J, Nieto O, Beiras R. Integrative assessment of coastal pollution: development and evaluation of sediment quality criteria from chemical contamination and ecotoxicological data. Continental Shelf Res 2011;31:448—456.

73. Ingersoll CG. Sediment tests. Chapter 8. In: Rand GM, editor. Fundamentals of aquatic toxicology. 2nd ed. Taylor & Francis; 1995. pp. 231—255.

74. Hansen DJ, Berry WJ, Mahony JD, et al. Predicting the toxicity of metal-contaminated field sediments using interstitial concentration of metals and acid-volatile sulfide normalizations. Environ Toxicol Chem 1996;15:2080—2094.

75. DiToro DM, Mahony JD, Hansen DJ, et al. Acid volatile sulfide predicts the acute toxicity of cadmium and nickel in sediments. Environ Sci Technol 1992;26:96—101.

76. Allen Burton G, Jr. Sediment quality criteria in use around the world. Limnology 2003;3: 65—75.

77. Beyer J, Green NW, Brooks S, et al. Blue mussels (*Mytilus edulis* spp.) as sentinal organisms in coastal pollution monitoring: A review. Mar Environ Res 2017;130:338—365.

78. Molvær J., Knutzen J., Magnusson J., et al. Klassifisering av miljøkvalitet I fjorder og kystfarvann. Veiledning. Classification of environmental quality in fjords and coastal waters. A guide. Norwegian Pollution Control Authority. TA no. TA-1467/1997; 1997. 36; pp. ISBN 82-7655-367—2.

Suggested Further Reading

- CCME. A protocol for derivation of water quality guidelines for the protection of aquatic life 2007. Canadian Council of Ministers of the Environment; 2007. Available online: https://www.ccme.ca/files/Resources/supporting_scientific_documents/protocol_aql_2007e.pdf.

- Nipper M. Chapter 43. The development and application of sediment toxicity tests for regulatory purposes. pp. 631—643. In: Wells PG, Lee K, Blaise C, editors. Microscale testing in aquatic toxicology. New York: CRC Press; 1998.

- Quintino V, Elliot M, Rodrigues AM. The derivation, performance and role of the univariate and multivariate indicators of benthic change: Case studies at differing spatial scales. J Exp Mar Biol Ecol 2006;330:368−82.

- Rosenberg R, Blomqvist M, Nilson HC, et al. Marine quality assessment by use of benthic species-abundance distributions: a proposed new protocol within the European Union Water Framework Directive. Mar Pollut Bull 2004;49:728−39.

- Ryan PG, Moore CJ, van Franeker JA, et al. Monitoring the abundance of plastic debris in the marine environment. Phil Trans R Soc B 2009;364:1999−2012.

- Smith EP, Cairns Jr J. Extrapolation methods for setting ecological standards for water quality: statistical and ecological concerns. Ecotoxicology 1993;2:203−19.

- Warwick RM, Clarke KR. A Comparison of some methods for analysing changes in benthic community structure. J Mar Biol Assoc UK 1991;71:225−44.

Glossary

Acceptability criteria Minimum performance of the control, maximum variability among replicates or any other criteria intended to control the quality of a bioassay, including the biological material and the experimental procedures.

Acetylcholinesterase (AChE) Enzyme involved in the correct transmission of the nervous signal across cholinergic synapses whose inhibition is considered a biomarker of general neurotoxicity.

Acid-volatile sulfides (AVS) Labile Mn^{2+} and Fe^{2+} sulfides that can be measured by cold acid extraction in the sediments. They remove trace metals from interstitial water forming non-bioavailable insoluble sulfides.

Action level Concentration of a contaminant that when exceeded triggers a given management measure or remedial action. For example, foodstuff exceeding the action level for any contaminant can be removed from the market by the competent authorities, or dredged sediments exceeding any action level are not allowed to be dumped in the sea.

Activated sludge Method of depuration of wastewaters based on the mineralization of the organic matter by aerobic heterotrophic microorganisms in suspension in the effluent.

Activation Metabolic process characterized by an increase in the toxicity of a substance as a consequence of a phase I detoxification reaction.

Acute to chronic ratio (ACR) The ratio between the concentrations of a given substance causing short- and long-term toxicity, respectively. ACR values taken from toxicity databases are used to replace actual chronic toxicity p, which increases uncertainty for so-derived risk estimations and environmental quality criteria.

Acute toxicity Toxicity caused after short-term exposures. In marine organisms this is normally 48—96 h.

Additivity Type of interaction between two or more substances where the overall toxicity of the mixture can be predicted from the sum of toxic units.

Adduct See DNA adduct.

Agonist Chemical that activates a molecular receptor by binding to it. The opposite is an antagonist.

AhR See Aryl-hydrocarbon receptor.

Aerosol Collection of liquid or solid particles suspended in the air, or in other gas.

δ-Aminolevulinic Acid Dehydratase (ALAD) Enzyme involved in the synthesis of hemoglobin, specifically inhibited by Pb and thus used as a biomarker of exposure to this substance.

Anammox Anaerobic ammonium oxidation, used as tertiary treatment for wastewaters: $NH_4^+ + NO_2 \rightarrow N_2 + 2H_2O$.

Androgenic chemical Chemical with similar effects on an organism to the natural male hormone testosterone. Antiandrogenic chemicals cause the opposite effect.

Anoxia Absence of dissolved oxygen.

Antagonism Interaction between two chemicals characterized by a lower combined effect compared to individual exposure. Se and Hg have been reported to show this interaction.

Apoptosis Programmed cell dead consisting of a sequence of biochemical steps that lead to dead and removal of damaged cells in multicellular organisms.

Aromatic compound Organic chemical with at least one six-carbon ring with alternating single and double bonds. The three double bonds can switch positions because the corresponding six electrons are delocalized between the six carbon atoms forming a circle. This feature, termed aromaticity, keep the ring structure flat and give stability to the molecule.

Aryl-hydrocarbon receptor Molecular receptor activated by certain coplanar aromatic hydrocarbons (e.g., TCDD, dioxin-like PCBs, BaP) that induce the expression of the cytochrome P450-1A1 gen.

Assessment criteria Cut-off values that allow the division of continuous variables measured in chemical analyses, biomarkers, and bioassays into discrete categories of pollution or toxicity.

Assessment Factor Numerical value—normally multiple of 10—used in quantitative estimation of ecological risk or derivation of environmental standards intended to obtain conservative estimates accounting for the sources of uncertainty (acute to chronic and laboratory to field extrapolations, etc.). Thus, the higher the uncertainty the higher the AF.

Atomic number Number of protons in the nucleus of an atom, determining the identity of the element.

Background assessment concentration (BAC) Statistical tools derived from background concentrations data that enable testing of whether observed concentrations exceed background concentrations.

Background concentration (BC) The concentration of a pollutant in a given environmental matrix not affected by human activities. This can be measured in pristine areas or in records from preindustrial times.

Ballast Load, normally water, taken on board a cargo ship below the water line to provide stability and allow navigation when traveling empty.

Benthos The community of organisms living in contact with the bottom or inside the sediments. For animals those are termed epifauna and infauna respectively.

Benthic index Index built with information on the composition and abundance of benthic taxa and used to assess environmental quality in aquatic ecosystems. A second generation of multivariate benthic indices include also in their calculation classical community indices such as Shannon diversity.

Benzo-a-pyrene hydroxylase (BaPH) Enzyme catalyzing the hydroxylation of BaP within the phase I of biotransformation, used as biomarker of exposure to PAHs.

Bioaccumulation Environmental process that results in concentrations of certain substances in the organisms markedly higher than in the surrounding environment. The substances showing this feature such as certain trace metals and organohalogenated compounds are termed bioaccumulative.

Bioaugmentation See Bioremediation.

Bioavailability Quantitative property of a given chemical form of a pollutant determined by the fraction of the total amount of chemical present in the environment that reaches the site of action in a given organism.

Biocide Chemical intended to kill living organisms.

Bioconcentration or Bioaccumulation Increase in the concentration of a pollutant achieved in the tissues of an organism compared to the physical environment where it leaves.

Bioconcentration (or Bioaccumulation) Factor (BCF or BAF) Concentration in the organism divided by the concentration in the environment. Bioconcentration should be used for waterborne pollutants taken up from the water only, while bioaccumulation takes into account additional uptake of pollutant from the digestive system via food and sediment.

Biogenic PAHs Those naturally synthetized by bacteria, fungi, and plants.

Biological Early Warning Systems (BEWS) See biomonitor.

Biological indicator, or bioindicator Any observation conducted at any level of biological organization (from biomolecules to communities) that provides useful information regarding the pollution status of a study area.

Biological (or Biochemical) Oxygen Demand (BOD) A laboratory method to estimate the organic matter in a water sample consisting of measuring in standard conditions the consumption of oxygen by the heterotrophic microorganisms naturally present in the water, normally after an incubation period of five days (BOD_5).

Biomagnification Phenomenon observed for certain chemicals in a food web characterized by an increase in the chemical body burden directly related to the trophic level occupied by the organisms. This results in elevated concentrations of the chemical in top predators.

Biomagnification factor (BMF) Antilog of the slope of the linear regression: concentration of pollutant (in log scale) versus trophic level (see Eqs. 11.15 and 11.16).

Biomarker A molecular, cellular, physiological, or anatomical quantitative response measured in an organism naturally exposed to the site under study that serves as indicator of the exposure to environmental pollutants.

Biomonitor In a broad sense any organism or biological response used for environmental monitoring. In a more restricted sense, organism used as bioaccumulator to provide a biotic matrix for chemical analysis of pollutants; most often the marine mussel.

Bioremediation Technique used to enhance the biodegradation of oil in the environment based on the creation of the ideal conditions for its microbial mineralization by adding adequate N and P amounts, facilitating oxygen diffusion and, in some cases, adding a microbial inoculum. The latter is termed bioaugmentation.

Biosensor In situ deployed device including a living organism that can record in continuum a biological response enabling the operation of an automatic electronic system for the on-line detection of impaired water quality. Heart beating, respiration rate, or valve closure in fish and bivalves are responses that have been used with this purpose.

Biota-Sediment Accumulation Factor (BSAF) Particular case of BAF for infaunal organisms taking the contaminants from the surrounding sediments. It is defined as the concentration in the organism normalized to the lipid content divided by the concentration in the sediment normalized by the organic carbon.

Biotic index See Benthic index.

Biotransformation In vivo metabolic pathways intended to prevent toxic effects of xenobiotic substances and eventually eliminate them from the organism.

Body burden The total mass (or concentration) of a toxic substance accumulated in an organism.

Bottom-up control Control in the abundance or biomass exerted by the scarcity of nutrients or food.

Builders Components of synthetic detergents that facilitate the action of the surfactant by sequestering Ca^{2+} and Mg^{2+}.

Carcinogen Any agent producing malignant tumors, i.e., cancer.

Chemical Oxygen demand (COD) A laboratory method to estimate the organic matter in a water sample consisting of measuring the amount of some strong oxidizing agent such as potassium dichromate consumed to mineralize the sample.

Chemotroph Organism that obtains the energy from the oxidation of reduced inorganic molecules such as H_2S or methane.

Chromosome aberrations Secondary DNA lesions visible at the microscope, including breakage, rearrangement, and loss of segments.

Chronic toxicity Toxicity caused after long-term exposures representing a substantial part of the total duration of the life stage; for marine organisms, typically weeks or months.

Civil Liability Convention (CLC) International convention hosted by IMO, adopted in 1969 following the Torrey Canyon accident and updated in 1992 to establish the liability of the owner of a ship causing pollution due to incidental oil spilling.

Clam reburial bioassay Sediment toxicity test based on recording the number of clams remaining on the sediment surface at short time intervals and calculating an ET_{50} value (time needed for complete burial of a half of the population).

Clastogen Chemical with the potential to damage the chromosome structure in a way visible at the microscope.

Clean Water Act (CWA) Federal Law in the US designed to control the point-source discharge of effluents into surface waters—originally defined as "navigable waters." It includes effluent standards issued by EPA that must be met to obtain permits of discharge.

Combined sewer system Urban sewer with a single system of pipes collecting both sanitary sewage and land runoff. After heavy rain this system may overflow or exceed the maximum flow accepted by the WWTPs.

Comet assay Method for DNA strand break detection in individual cells. The cells are embedded in agarose gel on a microscope slide to be lysed and their DNA unwound under alkaline conditions and subjected to gel electrophoresis. The broken DNA fragments or relaxed chromatin migrate away from the nucleus, and this results in a comet shape that can be visualized with a fluorescent dye.

Comprehensive environmental response, compensation, and liability act (CERCLA) 1980 US piece of legislation, also known as Superfund, aiming at the cleanup of inactive hazardous waste sites, and establishing the liability for cleanup costs on past and present agents related to the transport and disposal of hazardous substances.

Concentration addition Interaction between two chemicals characterized by additive effects when combined.

Concentration Response curves: classical experimental approach in aquatic toxicology that consists of recording a biological response by a predetermined time after exposure to different concentrations of a substance. The data so obtained is fitted to nonlinear mathematical models, such as probit or logistic, to obtain basic toxicity parameters, such as EC_{10} and EC_{50}.

Congener Each individual molecule member of a chemical family, frequently appearing together in environmental samples, e.g., PCB or PBDE congeners.

Conjugation Chemical reaction typical of phase II biotransformation of xenobiotics where an endogenous polar metabolite is added to the xenobiotic or a phase I metabolite of the xenobiotic.

Coordinated Environmental Monitoring Programme (CEMP) OSPAR monitoring program intended to coordinate the national initiatives and deliver comparable data across the OSPAR maritime area by defining a common set of measurements to be recorded, methods, and assessment criteria.

Criterion Continuous Concentration Term used in US regulations to refer to the chronic water quality criterion. It is defined by EPA as four-day average concentration, not to be exceeded more than once every three years. Notice that although the criterion is derived from chronic toxicity data the interval of measurement is short (four days), and from that standpoint results more protective than the annual average reflected in the European legislation.

Criterion Maximum Concentration Term used in US regulations to refer to the acute water quality criterion. Defined by EPA as 1-h average concentration.

Critical Body Burden (CBB) Total content or internal concentration in an organism (or target organ) at the time of mortality, or some other discrete sublethal endpoint.

Critical value In the derivation of environmental quality criteria from toxicity test data with a set of test species, the critical value is the lowest toxic concentration corresponding to the most sensitive species in the set.

Cultural eutrophication See Hypereutrophication.

Cycloalkane Hydrocarbon whose molecule show a ring of carbon atoms joint by single bonds.

Cytochrome P450 (CYP) Inducible hemoproteins that act, along with other proteins of the SER membrane, as electron transport proteins in the enzymatic oxidation of organic xenobiotics during phase I biotransformation processes. The ensemble of proteins is frequently termed CYP complex, or more properly CYP-monooxygenase complex. Different families of CYP act specifically in certain biotransformation routes. The CYP 1A1 complex for example catalyzes the phase I metabolism of benzo-*a*-pyrene and other polycyclic coplanar hydrocarbons.

$\delta^{15}N$ Nitrogen stable isotope ratio, $^{15}N:^{14}N$. This ratio increases as N passes through certain metabolic routes and thus it may characterize different pools of nitrogen. It also increases in upper trophic levels.

Denitrification Transformation of nitrate into nitrogen gas mediated by anaerobic bacteria.

Depuration Reduction of the levels of fecal microorganisms present in shell stock by maintenance in a clean controlled aquatic environment in land-based facilities.

Dioxin Common name of the TCDD and related congeners.

Dioxin-like PCB Nonortho or monoortho substituted PCB with a coplanar structure that confers them high toxicity similar to other coplanar organochlorines such as PCDD and PCDF.

Dissolved Organic Matter (DOM) Organic matter present in the water in the dissolved phase, operationally defined as that passing through 0.45 μm filters.

Dissolved oxygen (DO) Amount of oxygen (mg/L, ml/L or mol/L) dissolved in a water sample, an important water quality indicator.

Diversity See species diversity; ecological diversity.

DNA adduct The result of covalent binding of a xenobiotic chemical to DNA, which can trigger carcinogenesis.

Dose Response curves: the most classical experimental approach in toxicology that consists of recording a biological response by a predetermined time after giving different oral or injection doses of a substance. The data so obtained is fitted to nonlinear mathematical models to obtain the LD_{50}.

Dunnett's test Parametric post hoc statistical test that compares, for a factor identified as significant by ANOVA, each level of the factor with the control. It is commonly used to identify NOEC and LOEC.

Early life stages (ELS) The initial stages in the life cycle of marine organisms, particularly sensitive to pollution. Normally embryos and larvae, although gametes, neonates, and juveniles may also be included. Within the framework of risk assessment, ELS tests frequently provide critical values.

Ecological diversity Diversity of species plus diversity of habitats in an ecosystem.

Ecological Risk Assessment Technical framework to provide environmental managers with objective and quantitative information intended to help them make decisions on activities affecting the environment that do not pose a threat to ecosystem structure and functioning.

Ecosystem approach In legislation, strategy for the integrated management of environmental resources based on the application of appropriate scientific methodologies focused on levels of biological organization which encompass the essential processes, functions and interactions among organisms and their environment.

Ecosystem-based assessment Strategy of integrated environmental monitoring based on the ecosystem approach.

Ecosystem services Benefits provided by natural ecosystems to humankind, including regulation of conditions needed for life in the biosphere, provision of food, and aesthetic values. These services are not included in the conventional cost:benefit balances of economists.

Ecotoxicological bioassay Bioassay where the toxicity of environmental samples (water, sediment, elutriates, extracts) of in principle unknown composition is tested, frequently in serial dilutions, to assess their degree of pollution.

Effects Range-Low (ERL) Empirical sediment quality criteria corresponding to a low probability of toxicity developed on the basis of matching sediment chemistry and sediment toxicity information. The value corresponds to the 10th percentile in the distribution of concentrations with effects.

Effects Range-Median (ERM) Empirical sediment quality criteria corresponding to a medium probability of toxicity developed on the basis of matching sediment chemistry and sediment toxicity information. The value corresponds to the 50th percentile in the distribution of concentrations with effects.

Elutriates Aqueous extracts of the sediments obtained by vigorously mixing the sediment sample with clean seawater and taking the supernatant. Standard methods frequently use one part sediment with four parts water and 30 min shaking. They are intended to reproduce the potential remobilization of contaminants from the sediments to the water column.

Endocrine Disrupting Compound (EDC) Exogenous substance or mixture that alters function of the endocrine system and consequently causes adverse health effects in an organism or its progeny.

Endpoint Variable recorded in a toxicity test. It is useful to consider the difference between the assessed endpoint, or generic response (survival, growth, reproduction, etc.), and the recorded endpoint, or specific variable (motility after probing, length increase, number of eggs laid, etc.) chosen to standardize the test.

Enrichment factor (EF) Pollution assessment tool obtained for a given natural substance by dividing observed concentration by background concentration.

Environmental Assessment Criteria (EAC) Environmental Quality Criteria developed by OSPAR for abiotic (sediment) and biotic samples. They are based on ecotoxicological information and are intended to be the levels of a given pollutant or group of pollutants below which no chronic effects are expected to occur in marine organisms, including the most sensitive species.

Environmental quality criteria (EQC) Maximum concentrations of pollutants that should not be exceeded to avoid harmful effects on the organisms under a limited number of environmental scenarios (including different habitats, seasonal conditions, species to be protected, uses of water, etc.), derived by scientific consensus.

Environmental quality standards (EQS) Maximum concentrations of pollutants, enforced by law, which must not be exceeded in given environmental matrices, commonly water, to be given a certain use (drinking water, bathing water, fishing, aquaculture, etc.).

Environmental half-life ($T_{1/2}$) Time needed under specific environmental conditions for a half of the amount of a pollutant to disappear from an abiotic environmental compartment. Oxygen availability and temperature greatly affect this parameter.

Erika **packages I and II** Legislative initiatives adopted by the EU after the *Erika* and *Prestige* oils spills mainly intended to improve tanker navigation safety.

Estrogen receptors Molecular receptors present in cells of ovary, brain and other tissues of vertebrates, activated by estradiol and estrogenic chemicals. Activation of these receptors triggers the development of female sexual characteristics.

Estrogenic chemical Chemical with similar effects on an organism to the natural female hormone β-estradiol.

Ethinylestradiol Synthetic drug with similar properties to estradiol used in contraceptive pills.

Ethoxyresosufin-O-deethylase (EROD) Enzyme belonging to the family of cytochrome P450 1A1 monooxygenases, involved in the Phase I of the biotransformation of organic xenobiotics, induced by exposure to chemicals with a coplanar polycyclic structure such as PAHs, PCBs, and dioxins.

European Maritime Safety Agency (EMSA) European agency created in 2002 after the *Erika* oil spill with the aim to ensure prevention of pollution caused by ships, and response to marine pollution caused by ships and oil and gas installations.

Evenness Equitable share of individuals (or biomass) among species in a community.

Evenness index (J), When diversity is quantified using the Shannon index (H_S), then the evenness, ranging from 0 to 1, will be $J = H_S/H_{Smax}$, where $H_{Smax} = \log_2 S$.

Exposure Contact between an organism and an environmental pollutant. Time and route of exposure greatly affect the toxicity outcome. In a priori ERA, exposure assessment is a basic step where predicted concentrations to which an organism is going to be exposed are estimated from data on the environmental fate of a substance.

Fingerprinting Technique used to identify the origin of an oil stock based on the quantification of the proportions of minority components highly resistant to biodegradation and characteristic of each type of oil.

Flow-through Experimental setup in aquatic testing where the organisms are kept in a chamber with a constant flow of water of the required characteristics, including concentration of testing chemicals, in opposition to static conditions.

Fugacity The escaping tendency of a chemical from an environmental compartment. This is the basis of the fugacity models, thermodynamic models that predict the distribution of a chemical among compartments on the basis of their physicochemical and biological properties.

Genotoxicity Deleterious change in the DNA caused by a physical or chemical agent as a consequence of direct interaction with the DNA structure or by interference with the cellular apparatus involved in its replication.

Glucuronic acid Carbohydrate that can be conjugated to xenobiotics or their metabolites by the UDP-glucuronosyltransferase (UDPGT) in phase II biotransformation.

Glutathione (GSH) Tripeptide composed by glutamic acid, cysteine, and glycine with multiple roles in defense against chemicals. It may directly neutralize reactive oxygen species and other strong oxidants through its conversion to GSSG, serve as substrate for the antioxidant enzyme glutathione peroxidase (GPx), or take part in phase II biotransformation of xenobiotic metabolites as substrate of the glutathione transferase (GST).

Glutathione transferases (GST) Enzymes that take part in the Phase II of the biotransformation of organic xenobiotics by covalent binding of glutathione (GSH) to Phase I metabolites.

Half-life See environmental half-life and metabolic half-life.

Hardness Concentration of calcium and magnesium ions in continental waters. This concept is not used in seawaters, where it may be replaced by salinity. Notice that some regulations intended for saltwaters incorrectly reflect it.

Hazard identification Evaluation of the potential adverse effects that a substance has an inherent capacity to cause.

HELCOM Baltic Marine Environment Protection Commission, or Helsinki Commission; the governing body of the Convention on the Protection of the Marine Environment of the Baltic Sea Area, known as the Helsinki Convention. The contracting parties are Denmark, Estonia, the European Union, Finland, Germany, Latvia, Lithuania, Poland, Russia, and Sweden.

Heterotroph Organism that obtains the energy from the oxidation of organic matter.

Hormesis In a dose:response experiment, stimulatory effect of low doses of a toxicant resulting in a biphasic curve with increased values of the endpoint (growth, reproduction, etc.) at low doses compared to control.

Hypereutrophication Ecological imbalance caused by an excessive input of anthropogenic nutrients in aquatic ecosystems, resulting in primary production increased above the consumption capacity of herbivores, in decomposition of plant biomass, and in oxygen deficit.

Hypoxia Ecological condition of an aquatic ecosystem characterized by levels of DO < 2 mg/L.

Imposex Reproductive syndrome in marine gastropods characterized by the superimposition of male characters, such as penis and vas deferens, in female individuals, caused by organotin compounds.

Indicator species A species whose absence, presence, or abundance reflects a specific environmental condition, in particular, environmental pollution of a certain kind.

Inhabitant equivalents Units used to design wastewater treatment and project the dimension of WWTP facilities. They are based on average the amount of BOD_5 or suspended solids produced by one person in the sewage water. This allows accounting for other nonurban liquid wastes such as those from farms or food industry (See also Sect. 2.2).

Interaction Factor Within the context of the quantification of the joint toxicity of mixtures, parameter that asses the existence and type (synergism, antagonism) of interaction between toxic substances.

International Maritime Organization (IMO) International body (formerly Inter-Governmental Maritime Consultative Organization) created in 1948 to promote cooperation in technical matters affecting shipping engaged in international trade, including maritime safety, efficiency of navigation, and prevention and control of marine pollution from ships.

Interstitial water See Pore water.

Isoboles In a plot of proportions of individual EC_{50} values for two substances, lines joining combinations of equal effect in mixture.

Isotopes Forms of the same chemical element with the same number of protons but different number of neutrons. For example, nitrogen shows two stable (nonradioactive) isotopes, with 14 (>99%) and 15 neutrons.

K_{OW} See Octanol-water partition coefficient.

Life-cycle test Toxicity tests comprising the full life cycle of the test species, and thus taking into account effects on reproduction, development, growth, and survival of the organism.

Lipophilic Applied to a molecule with affinity for lipids that tend to move from water to biological membranes and fat deposits.

Lixiviates See Elutriates.

LOAEL See LOEC.

LOEC (lowest observed effect concentration) The lowest concentration tested whose mean response shows statistically significant differences with control.

LOEL See LOEC.

Lysosomes Cellular organelles that contain hydrolytic enzymes playing a central role in intracellular digestion, resorption, and excretion processes.

Margalef species richness index (d_M) Given a sample with a total of N individuals belonging to S species, the index takes the value: $d_M = (S-1)/\ln N$. This allows a quantification of species richness less sample-size dependent than simply taking S.

MARPOL International Convention for the Prevention of Pollution from Ships. The MARPOL 73/78 convention adopted, among others, regulations to limit the amount of oil that may be discharged into the sea and required larger tankers to have segregated ballast tanks.

Maximum Acceptable Toxicant Concentration (MATC) A mathematical estimation of the toxicity threshold, consisting of the geometric mean of NOEC and LOEC. See also: Toxicity threshold.

Maximum Admissible Concentration (MAC) Term used in the EU legislation to refer to the chronic water quality standard.

MED POL Monitoring program for the Mediterranean Sea within the umbrella of the UNEP Mediterranean Action Plan that stems from the Barcelona Convention for the protection of the Mediterranean against pollution, signed by all countries with Mediterranean shoreline and the EU in 1976.

Median effective concentration (EC_{50}), or half maximal effective concentration Theoretical concentration of a substance causing a 50% reduction in the biological response (survival, growth, reproduction, or any surrogate for any of them) in exposed organisms compared to control. The value is obtained by fitting the concentration—response data to mathematical models such as probit or logistic. When survival is recorded, it is termed median lethal concentration (LC_{50}).

Median lethal dose (LD_{50}) Theoretical dose causing a 50% mortality in a population of exposed organisms, calculated by fitting dose:response experimental data to mathematical models such as probit.

Median lethal time (LT_{50}) Theoretical time required for 50% of the individuals to die at a given exposure concentration.

Meiofauna Benthic animals in the size range of 0.1—0.5 mm.

Mesocosm Medium size multispecies experimental system, frequently outdoors.

Metabolic half-life Time required for the concentration of a substance in an organism to be reduced by a half due to biotransformation and/or excretion.

Metallothionein Low molecular weight protein rich in cysteine residues induced by exposure to metals capable to sequester divalent metal ions.

Micronuclei Intracytoplasmic genetic material originated from chromatin lagged in anaphase as a consequence of chromosome breakage or malfunction of the spindle apparatus. Their abundance is increased upon exposure to genotoxic agents.

Microsomal fraction Cellular compartment obtained after tissue homogenization and differential centrifugation at up to 100,000 g, consisting of fragments of membranes from the endoplasmic reticulum and other organelles.

Minamata disease Human poisoning epidemic that extended for several decades from 1950s in Japan caused by the consumption of fish and shellfish from a chronically polluted area affected by industrial organic mercury discharges.

Mixed Function Oxidase (MFO) See Monooxygenase.

Mixing zone Section of a water course directly affected by an effluent of residual waters where the discharges undergo initial dilution. Values of water quality parameters must be recorded downstream the mixing zone.

Monitoring program Scheme of sampling and measurements to assess environmental quality; specifically detect temporal trends, classify areas into groups of environmental status, identify sources of pollutants, and plan remediation measures accordingly.

Monotonic Applied to a mathematical function, a function whose values entirely increase or entirely decrease. For example, survival is a monotonic decreasing function of toxicant concentration except when hormesis enhances survivals at low concentrations and produces a biphasic response. Standard concentration—response models (probit, logistic, etc.) are only suitable for monotonic responses.

Monooxygenase Enzyme belonging to the cytochrome p450-monooxygenase complex located in the SER membranes and responsible for oxidation reactions in phase I biotransformation of xenobiotics in eukaryotes. The name refers to the fact that, unlike in prokaryote dioxygenases, only one atom of the oxygen molecule is transferred to the xenobiotic.

Most Probable Number (MPN) Statistical method to estimate the microbial load of a liquid sample based on incubating serial dilutions of the sample in appropriate media at a defined temperature. This method was traditionally used to estimate the levels of total and fecal coliforms but was gradually replaced by more specific methods that identify *Escherichia coli* after membrane filtration of the sample.

Multidimensional Scaling (MDS) Multivariate method of ordination that uses the rank ordering of the original distances or dissimilarities among observations. This allows application to any distance or dissimilarity measure with robust results.

Multivariate methods Methods of statistical analyses of data sets composed by more than one response variables (for example, the abundance of m species at n sites) in opposition to univariate methods (for example, the dissolved oxygen in n sites). Thus, multivariate methods work with $n \times m$ matrices.

Mussel Watch The first comprehensive effort to establish a biomonitoring program to assess coastal status in the United States, initiated by the United States Environmental Protection Agency (US-EPA) in the period 1976—78, and continued by NOAA starting in 1986. The program measures more than 140 chemical contaminants in mussels and oysters from 300 sites from the Atlantic, Pacific, and Great Lakes coasts. By extension, the term refers to similar biomonitoring initiatives in other areas.

Mutagen An agent with the potential to alter the sequence of nucleotide bases of DNA.

Narcosis General mechanism of nonreactive toxicity typical of hydrophobic chemicals based on their effect altering cell membranes, particularly those of neurons.

Neoplasia Uncontrolled cellular proliferation that causes tumors. Malignant neoplasia is known as cancer.

NOAEL See NOEC.

NOEC (no-observed effect concentration) The highest concentration tested showing no statistically significant differences with control.

NOEL See NOEC.

Nontarget species See Target species.

Octanol-water partition coefficient (K_{OW}) Ratio of concentrations of a substance reached at equilibrium between the organic solvent *n*-octanol and water. It is a measure of the lipophilicity of a substance and it is normally expressed in log scale.

Oil Pollution Act Act passed in the United States in 1990, in the aftermath of the *Exxon Valdez* oil spill, with the purpose to set practical tools for the prevention of spills and abatement of their consequences.

OSPAR Commission Commission for the Protection of the Marine Environment of the North-East Atlantic, created in 1992 to prevent pollution originated by dumping from ships, land-based sources, and the offshore industry, later extended to cover also nonpolluting human activities that can adversely affect the sea.

OSPAR Convention International agreement signed in 1992 by 13 countries for the protection of the marine environment of the Northeast Atlantic, as a result of the merging of the Oslo Convention (1972), for the prevention of marine pollution by dumping from ships, and the Paris Convention (1974), for the prevention of marine pollution by discharges from land-based sources.

Oxidative stress Biochemical condition characterized by potential damage to biomolecules due to the formation of reactive oxygen species.

Particulate Organic Matter (POM) Organic matter present in the water in the particulate phase, operationally defined as that retained in 0.45 μm filters.

Partition coefficient (K_{AB}) Ratio of concentrations of a substance reached at equilibrium between two adjacent phases or environmental compartments A and B. $K_{AB} = C_A/C_B$. Phases may be water, an organic solvent, organic carbon, etc. Environmental compartments may be the air, a water body, the bottom sediments, the biota body lipids, etc.

Persistent pollutants Those that remain unchanged under environmental conditions for periods of many months or years. Regulations for marine environments frequently consider a pollutant as persistent when $T_{1/2} > 60$ d in water and >180 d in sediments.

Petrogenic PAHs Those originating from fossil fuels.

Phase I reactions Biochemical reactions in the first step of the metabolic biotransformation of organic xenobiotics, frequently consisting of oxidations catalyzed by the CYT-monooxygenase complex.

Phase II reactions Biochemical reactions in the second step of the metabolic biotransformation of organic xenobiotics, consisting of conjugation to endogenous polar metabolites catalyzed by transferases.

Photosynthetic organism Organism that obtains the energy from light and can produce organic molecules from inorganic nutrients.

Pielou's evenness index See Evenness index.

Polymerase chain reaction (PCR) A molecular biology technique to produce millions of copies of DNA segments, i.e., to amplify DNA. The quantitative version of the technique is called real time PCR.

Pore water Water occupying space between sediment particles. It can be extracted for toxicity testing.

Positive control In a toxicity test experimental treatment to which a given amount of a toxicant was added or it is known to be present for the purpose of comparison among experiments, checking biological materials or procedures.

Potentiation See Synergism.

Precautionary principle Legal policy that prescribes taking the most environmentally conservative option to prevent potential harm when effects on the environment may occur, even in absence of scientific evidence, particularly regarding release of contaminants.

Predicted environmental concentrations (PEC) Concentration of a chemical predicted by environmental distribution models in a given compartment, on the basis of the amounts released and the properties of the chemical that determine its environmental fate. It is used in risk-assessment studies.

Predicted no-effect concentration (PNEC) Denomination of the toxicity threshold (TT) used in risk assessment studies.

Primary treatment First phase in the treatment of wastewaters in a WWTP based on the removal of floating oil and grease using skimmers, and removal of suspended solids by sedimentation.

Priority substances Substances identified as hazardous by national and international institutions, and issued in priority lists, on the basis of their properties of environmental persistence, bioaccumulation, and toxic effects. Environmental quality standards are available for these substances.

Probable Effects Level (PEL) Empirical sediment quality criteria corresponding to a medium probability of toxicity developed on the basis of matching sediment chemistry and sediment toxicity information. The value corresponds to the geometric mean of the 50th and the 85th percentiles in the effects and no effects data sets respectively.

Probits Abbreviation for "normal probability units." Transformed units adding 5 (to avoid negative values) to the normal deviates: $y = 5 + (x-\mu)/\sigma$. These units are used to fit dose–response curves under the assumption that the individual susceptibility to a chemical follows a normal distribution in a population, when doses are expressed in log scale. See Sect. 13.3.

Pyrogenic PAHs Those originating from combustion of organic matter.

QSAR (Quantitative Structure-Activity Relationship) model Models intended to predict the biological activity (bioaccumulation, toxicity, etc.) of chemicals as a function of chemical properties (partition coefficients, water solubility, molecular weight, etc.) dependent on the structure of the molecule.

REACH Regulation concerning the Registration, Evaluation, Authorization, and Restriction of Chemicals, passed in 2006 in the European Union, imposing an a priori risk assessment to place in the market a potentially hazardous chemical.

Reactive Oxygen Species (ROS) Strongly oxidative radicals such as superoxide anions ($^{\bullet}O_2^-$), hydroxyl radicals ($^{\bullet}OH$) and singlet oxygen ($^1O_2$) that may cause cellular damage if not counteracted by the cellular defenses against oxidative stress such as antioxidant enzymes and glutathione.

Reference conditions Values of environmental conditions corresponding to sites not impacted by human activities.

Reference sites In a sampling campaign, sites representing pristine areas not impacted by human activities but similar in natural characteristics to the remaining sampling sites.

Reference sediment In a sampling campaign, a sediment collected in a reference site within the same geographical area than the other samples. Control sediment is also a clean sediment but not necessarily representative of the same area.

Relaying Moving the shell stock to growing areas in a better microbiological quality status classification to achieve natural depuration and allow the stock to meet the microbiological standards required for commercialization.

Reproductive toxicity Within the context of identification of priority substances, a substance is considered to cause human reproductive toxicity if it can cause adverse effects on the reproductive ability of adult males and females (onset of puberty, gamete production and transport, reproductive cycle normality, sexual behavior, fertility, parturition, premature reproductive senescence, etc.) and developmental toxicity in the offspring.

Residence time The average amount of time spent in a given volume or compartment by the particles of a fluid, normally represented by the Greek letter *tau* (τ). For a given water volume (V) submitted to a natural or artificial flow (Q), residence time is $\tau = V/Q$.

Risk characterization Evaluation of the incidence and severity or the probability of adverse effects caused (a posteriori risk assessment) or likely to occur (a priori) under specific exposure conditions.

Risk quotient (RQ) Ratio between the actual (or predicted) environmental concentrations (PEC) and the toxicity thresholds for a given chemical under specific exposure conditions. $RQ = PEC/PNEC$. Risk is considered as acceptable when $RQ \ll 1$ (safety margins of several orders of magnitude), and unacceptable as RQ approaches 1.

Safe concentration See Toxicity threshold.

Salinity (S) Concentration of salts dissolved in the water. Despite this simple definition its exact measurement is extremely complicated. Traditionally S was expressed as parts per 1000 in weight, but gravimetrical measurements are inexact, and S is currently measured indirectly by using electrical conductivity. However, since each salt has a different conductivity, KCl was taken as reference and the ratio of conductivities between the seawater sample and KCl is used. Thus expressed, salinity is dimensionless.

Scope for Growth (SFG) Physiological parameter that represents the balance in terms of rates between the energy gains (feeding, assimilation) and loses (excretion, respiration).

Secchi disk A disk of standard size and color, frequently alternating black and white quadrants, used mostly in continental waters to measure turbidity as the inverse of Sechi depth, or depth at which it disappears from sight for an observer from the surface.

Secondary treatment Second phase in the treatment of wastewaters in a WWTP based on the mineralization of the organic matter by aerobic heterotrophic microorganisms either adhered to surfaces (trickling filters, biological contactors) or in suspension in the effluent (activated sludge). In the latter case the process includes recuperation of the microorganisms from the effluent.

Sediment quality criteria or guidelines (SQC) EQC for aquatic sediments.

Sediment Quality Triad Approach to marine pollution assessment based on the combined study of concentration of contaminants in the sediment, toxicity tests using ecotoxicological bioassays, and benthic fauna community structure.

Selective toxicity Toxicity due to mechanisms affecting only to certain taxa. For example, within marine invertebrates cadmium shows selective toxicity to crustaceans. This selectivity may be used to develop pesticides. The sensitive organisms intended to be affected are termed target species.

Self-purification capacity of natural waters The set of physical, chemical, and biological natural processes that allow a water body to recover its original ecological structure and functioning after the discharge of liquid wastes.

Sentinel species An autochthonous or transplanted (frequently caged) species used for the biological monitoring of pollution.

Shannon diversity index (H_S) An index of diversity used in ecology but taken from information theory, independently derived by C.E. Shannon and N. Wiener. If each species i has a given share p_i of the total number of individuals (or total biomass) in the community, then diversity in bits will be $H_S = -\sum p_i \log_2 p_i$.

Single-cell gel electrophoresis See Comet assay.

Sister chromatid exchange Secondary alteration of DNA consisting of exchange of DNA between the two chromatids of a chromosome as a consequence of DNA breakage followed by reunion and crossing over of DNA segments.

Sorption Association of a compound in solution to a particle. This loose term is used when the nature of the interaction is unknown. Adsorption implies adhesion to the particle's surface, whilst absorption implies permeation into the particle bulk.

Spiked sediment A sediment to which known amounts of contaminants were added for experimental purposes.

Species Abundance curves: graphical description of the structure of a community by plotting the distribution of the species in classes of abundance grouped in a logarithmic scale. Frequently these curves follow a normal distribution and thus may be linearized by plotting the cumulative percentage of species in each class of abundance in probability units (probits).

Species diversity An important characteristic of a community with two components: species richness, i.e., number of species in the community, and evenness, i.e., equitable share of individuals (or biomass) among species. The most frequent index to measure diversity is the Shannon index.

Species Sensitivity Distribution (SSD) Statistical distribution of the values of tolerance to given substance among the different species and life stages in an ecosystem. The distribution often has a sigmoidal shape and it is frequently fit to a log-logistic model.

SSD curves See Species Sensitivity Distribution.

Standard Legal limit (normally concentration) permitted for a substance in water or biota in relation with a specified use (swimming, discharge, foodstuff, etc.).

Stress proteins Proteins inducible upon physical or chemical cellular stress with a role in protein homeostasis, including protein transport, three-dimensional folding, defense against denaturation, and structural repair.

Superfund See CERCLA.

Surfactant Substance that lowers the surface tension between two phases because it contains a part of the molecule with affinity to each phase. In a detergent, this is achieved by molecules with polar heads and hydrophobic tails. These molecules form micelles that can emulsify hydrophobic organic substances in water.

Suspended Solids (SS or TSS) Particulate matter in an effluent or receiving waters gravimetrically quantified after filtration.

Synergism Interaction between two chemicals characterized by a higher combined effect compared to individual exposure. Oil and dispersants have been reported to show this interaction.

Target species In a pesticide, species intended to be killed by the product. For example, an algicide is intended to kill algae and only algae. All other taxa are nontarget species.

Teratogen An agent causing teratogenicity.

Teratogenicity Interference with normal embryogenesis, including malformations of the embryo, normally caused by chemical agents.

Tertiary treatment Any of the additional phases of treatment of wastewaters in a WWTP targeting N removal, P removal, elimination of particles by filtration, disinfection, elimination of electrolytes by osmosis or dialysis, etc.

Threshold Effects Level (TEL) Empirical sediment quality criteria corresponding to a low probability of toxicity developed on the basis of matching sediment chemistry and sediment toxicity information. The value corresponds to the geometric mean of the 15th and the 50th percentiles in the effects and no effects data sets respectively.

Top-down control Control in the abundance or biomass exerted by the consumers or predators. This includes cascade effects where the control on a trophic level may be exerted by changing the composition of the trophic chain several levels above. For example, hypereutrophication may be abated by introducing carnivorous fish that reduce the population of planktivorous fish consuming zooplankton grazers.

Total Organic Carbon (TOC) Amount of organic carbon in a water sample used to characterize both natural and residual waters. The TOC analyzers resort to combustion or catalytic oxidation of the organic matter and take advantage of the infrared absorption by the CO_2 produced.

Toxic Equivalency Factor (TEF) Weighing factor applied to the concentrations of PCDD, PCDF, and dioxin-like PCB in foodstuff to check whether the sum meets the maximum allowable concentration of 8 pg/g WW recommended by WHO. The TEF takes a value of 1 for the two most potent dioxins, and values ranging from 0.1 to 10^{-5} for the remaining PCDD, PCDF, and dioxin-like PCB.

Toxic Substances Control Act (TSCA) US piece of legislation regulating the placing in the market of chemicals and imposing an a priori risk assessment before commercialization.

Toxic Units (TU) Common currency used to predict combined effects of toxic mixtures of chemicals showing additive effects (no antagonism or synergism). Toxic units are calculated by dividing the concentration of each chemical in a mixture by its individual EC_{50}.

Toxicity test, biological assay, or bioassay Experiment conducted in laboratory where populations of individuals or cell cultures are exposed under standardized conditions to known levels of single or combined pollutants or environmental samples, and lethal or sublethal (growth, reproduction, or any surrogate for any of them) biological responses are recorded.

Toxicity threshold (TT) Theoretical minimum level of exposure to a substance above which harmful effects on exposed organisms begin to manifest. It lies between NOEC and LOEC. A similar concept is the safe concentration, the theoretical concentration in the environment that must not be exceeded to avoid any harmful effect on exposed organisms. While TT is derived from laboratory toxicity testing, the safe concentration is intended to be applied to environmental samples.

Toxicodynamics Set of processes that determine the relationship between the internal dose received by the target organ and the toxic effect. This includes interactions with molecular receptors, potential antagonisms, or synergisms, etc.

Toxicokinetics Set of processes that determine the relationship between the dose at which the organism is exposed and the dose reaching the target organ. This includes uptake, distribution, metabolism, and excretion.

Triad See Sediment Quality Triad.

Trophic magnification factor (TMF) See Biomagnification factor (BMF).

Trophic transfer factor (TTF) Parameter assessing the accumulation in a consumer (or predator) of a given pollutant present in the food (or prey). This is quantified by the expression: $TTF = C_{pred}/C_{prey}$.

Trophic web biomagnification factor (BMF_{TW}) See Biomagnification factor (BMF).

TSCA See Toxic Substances Control Act.

TSS See Suspended solids.

Uncertainty factor See Assessment factor.

Uptake The movement of a contaminant into the organism.

Vitellogenin (VTG) Egg yolk precursor protein synthetized in the liver and carried through the blood into the ovaries. The induction of VTG synthesis in males is a biomarker of xenoestrogens.

Water Framework Directive (WFD) European piece of legislation (Directive 2000/60/EC) that embraces the ecosystem approach and aims at the environmental protection of all underground and surface waters, including sustainable use and prevention of deterioration of ecological status.

Waterborne diseases A nonscientific term commonly used to refer to human diseases caused by pathogenic microorganisms and transmitted by the ingestion of contaminated water or consumption of contaminated seafood.

Water quality criteria (WQC) EQC for natural waters.

Weathering Set of physicochemical and microbiological processes that affect the composition and fate of oil once it is spilled in surface waters.

Whole sediment Sediment and associated pore water that have had minimal manipulation. Normally intended for toxicity testing.

Xenobiotic Atom or molecule with no known biological function in the organism.

Xenoestrogen Synthetic or natural but exogenous chemical with properties similar to the natural hormone estradiol. This can be identified through in vitro test quantifying the affinity of chemicals for the estrogen receptor, or in vivo through the induction of VTG synthesis.

Index

9780128137369